Y0-AAA-975

Analysis of Carbohydrates by GLC and MS

Editors

Christopher J. Biermann, Ph.D.
Assistant Professor, Pulp and Paper Science
Department of Forest Products
Oregon State University
Corvallis, Oregon

Gary D. McGinnis, Ph.D.
Professor of Wood Science and Technology
Forest Products Utilization Laboratory
Mississippi State University
Mississippi State, Mississippi

CRC Press, Inc.
Boca Raton, Florida

Library of Congress Cataloging-in-Publication Data

Analysis of carbohydrates by GLC and MS.
Bibliography: p.
Includes index.
1. Carbohydrates—Analysis. 2. Gas chromatography.
3. Mass spectrometry. I. Biermann, Christopher J.
II. McGinnis, Gary D., 1940-
QP701.A57 1988 547.7′8046 88-10580
ISBN 0-8493-6851-0

This book represents information obtained from authentic and highly regarded sources. Reprinted material is quoted with permission, and sources are indicated. A wide variety of references are listed. Every reasonable effort has been made to give reliable data and information, but the author and the publisher cannot assume responsibility for the validity of all materials or for the consequences of their use.

Direct all inquiries to CRC Press, Inc., 2000 Corporate Blvd., N.W., Boca Raton, Florida, 33431.

International Standard Book Number 0-8493-6851-0

Library of Congress Card Number 88-10580
Printed in the United States

PREFACE

Carbohydrates are prevalent in all aspects of our environment. They are major constituents of the tiny genes which determine our heredity and of the huge wooden beams spanning hundreds of feet to support a building. Research continues to show the importance of carbohydrates as structural materials, as a source of energy for biological processes, and as informational molecules of crucial importance in the regulation of numerous biological processes. The analysis of carbohydrates is fundamental to many areas of research. Studies involving dietary fiber, chemical processing of wood, structural analysis of blood group substances, and so forth, involve sophisticated analysis of carbohydrates. Gas chromatography and high performance liquid chromatography are two powerful techniques available for the separation and analysis of carbohydrates.

One is first overwhelmed by all of the literature available for separation of carbohydrates by these techniques. Although Churms (see the citation in Chapter 1) compiled an excellent review of the literature current to 1979 on all of these aspects, space limited the amount of detailed information on any one method and there have been some important advances since then. It was with this concern in mind that the idea for this volume was conceived. It is centered around analysis of carbohydrates by gas chromatography; however, since HPLC may have certain advantages for some types of analysis, it is judicious to include an introductory chapter on this technique as well. Mass spectrometry is a powerful tool, usually used in conjunction with gas-liquid chromatography, for analysis and structural determination of carbohydrates, so it too deserves consideration in this volume.

The editors sincerely hope that this volume is useful in clarifying what otherwise might seem to be a sea of confusion and anticipate that this volume will be particularly useful to those scientists who require analysis of carbohydrates but are not yet experts in gas chromatographic analysis of carbohydrates. We further believe that this volume will be useful to the more seasoned carbohydrate analyst as a reference book.

The authors of these chapters are all experts in their fields with many significant contributions to the carbohydrate chemistry literature. Furthermore, all of the authors had the opportunity to review each chapter and make suggestions. Therefore, each chapter represents the combined experiences of numerous, competent investigators. We are grateful for the time they have taken from their important research to contribute to this volume.

Christopher J. Biermann
Gary D. McGinnis
Editors

THE EDITORS

Christopher J. Biermann, Ph.D., is Assistant Professor of Pulp and Paper Science in the Department of Forest Products at Oregon State University, Corvallis, Oregon.

Dr. Biermann was educated at the University of Maine at Orono, where he obtained the B.S. in Forest Resources in 1980, and Mississippi State University, where he obtained the Ph.D. in Forest Resources in 1983. He obtained further training as a postdoctorate at the University of Wisconsin-Madison, working on carbohydrate metabolism in gnotobiotic and conventional flora rats in the Department of Nutritional Sciences, and at Purdue University working on grafting of polymers onto cellulose and cellulose derivatives in the Laboratory of Renewable Resources Engineering.

Dr. Biermann has published more than 30 papers on carbohydrate chemistry, biomass conversion, polymer chemistry, and interfacing of computers to laboratory equipment and has presented his work at numerous scientific meetings. He is a member of the American Chemical Society, Forest Products Research Society, the Society of Wood Science and Technology, and the Technical Association of the Pulp and Paper Industry.

Gary D. McGinnis, Ph.D., is Professor of Wood Science and Technology and Chemistry at Mississippi State University, Mississippi State, Mississippi.

Dr. McGinnis received the B.S. degree in Chemistry in 1962 from Pacific Lutheran University at Tacoma, Washington, the M.S. in Organic Chemistry at the University of Washington, and the Ph.D. at the University of Montana under Dr. Fred Shafizadeh in 1970.

Dr. McGinnis has published more than 60 papers on carbohydrate chemistry, biomass conversion, and analysis of plant constituents using a variety of chromatographic techniques. He has organized several national symposiums on the analysis of carbohydrates using gas chromatography and high performance liquid chromatography.

CONTRIBUTORS

Nicholas C. Carpita, Ph.D.
Associate Professor
Department of Botany and Plant Pathology
Purdue University
West Lafayette, Indiana

Anne Dell, Ph.D.
Reader of Carbohydrate Biochemistry
Department of Biochemistry
Imperial College of Science and Technology
London, England

Peter Englmaier, Ph.D.
University Assistant
Institute of Plant Physiology
University of Vienna
Vienna, Austria

Alvin Fox, Ph.D.
Associate Professor
Department of Microbiology and Immunology
School of Medicine
University of South Carolina
Columbia, South Carolina

James Gilbart, Ph.D.
Research Assistant Professor
Department of Microbiology and Immunology
School of Medicine
University of South Carolina
Columbia, South Carolina

Susumu Honda, Ph.D.
Professor
Faculty of Pharmaceutical Sciences
Kinki University
Kowakae, Higashi-osaka, Japan

Kazuaki Kakehi, Ph.D.
Associate Professor
Faculty of Pharmaceutical Sciences
Kinki University
Kowakae, Higashi-osaka, Japan

Murray L. Laver, Ph.D.
Associate Professor
Department of Forest Products
Oregon State University
Corvallis, Oregon

Stephen L. Morgan, Ph.D.
Associate Professor
Department of Chemistry
University of South Carolina
Columbia, South Carolina

Jean-Richard Neeser, Ph.D.
Glycoconjugate Specialist
Nestlé Research Centre
Nestec Ltd.
Lausanne, Switzerland

Jane E. Thomas-Oates, Ph.D.
Postdoctoral Research Assistant
Department of Biochemistry
Imperial College of Science and Technology
London, England

Karl-Gunnar Rosell, Ph.D.
Research Director
British Columbia Research
Industrial Chemistry Division
Vancouver, British Columbia, Canada

Thomas F. Schweizer, Ph.D.
Food Carbohydrate Expert
Nestlé Research Centre
Nestec Ltd.
Lausanne, Switzerland

Elaine M. Shea, Ph.D.
Research Associate
Department of Botany and Plant Pathology
Purdue University
West Lafayette, Indiana

DEDICATION

To those who hunger and thirst for what is righteous.

LIST OF ABBREVIATIONS

Ac	Acetyl
AO-TFA	*O*-Alkyloxime trifluoroacetate
Ara	Arabinose
Asn	Asparagine
BO-TMS	per-Trimethylsilylated *O*-benzyloxime
BSA	bis-(Trimethylsilyl)acetamide
BSTFA	bis-(Trimethylsilyl)trifluoroacetamide
BTFA	bis(Trifluoroacetamide)
CI	Chemical ionization
DMAP	4-(Dimethyl)aminopyridine
DMF	Dimethyl formamide
DMSO	Dimethylsulfoxide
DP	Degree of polymerization
DS	Degree of substitution
EI	Electron impact
FAB-MS	Fast atom bombardment mass spectrometry
FID	Flame ionization detector
Fuc	Fucose
Gal	Galactose
GC	Gas chromatography
GLC	Gas-liquid chromatography
Glc	Glucose
GlcA	Glucuronic acid
GlcN	2-amino-2-deoxyglucose (glucosamine)
GlcNAc	2-Acetamido-2-deoxyglucose (N-acetylglucosamine)
GlcU	Glucuronic acid
Hex	Hexose
HMDS	Hexamethyldisilazane
HMF	5-(Hydroxymethyl)furfural
HPLC	High-performance liquid chromatography
ID	Inside diameter
KDO	2-Keto-3-deoxyoctanoic acid
LPS	Lipopolysaccharide
Man	Mannose
MBTFA	*N*-Methyl-bis(trifluoroacetamide)
MOA	*O*-Methyloxime acetate
MO-TMS	per-Trimethylsilylated O-methyloxime
MS	Mass spectrometry
NANA	*N*-Acetylneuraminic acid
NeuNAc	*N*-Acetyl neuraminic acid
NeuNG	*N*-Glycolyl neuraminic acid
NMR	Nuclear magnetic resonance
PAAN	per-*O*-Acetylated aldononitrile
Pent	Pentose
PGC	Pyrolysis gas chromatography
Rha	Rhamnose
TFA	Trifluoroacetic acid
TFAA	Trifluoroacetic anhydride
TFAI	Trifluoroacetylimidazole
TMCS	Trimethylchlorosilane
TMS	Trimethylsilyl
TMSI	Trimethylsilylimidazole
Xyl	Xylose

TABLE OF CONTENTS

Chapter 1

INTRODUCTION TO ANALYSIS OF CARBOHYDRATES BY GAS-LIQUID CHROMATOGRAPHY (GLC)

Christopher J. Biermann

TABLE OF CONTENTS

I. APPLICATIONS OF CARBOHYDRATE CHEMISTRY

The field of carbohydrate chemistry has seen many advances in the last few decades. Sharon[1] points out that in 1953 only 4 amino sugars were known, in 1960 only 20 were known, but by 1970 50 amino sugars were known. The interest in carbohydrate chemistry is certainly not surprising considering the importance of carbohydrates in most aspects of our environment. It is only in the last few decades that the significance of carbohydrates, in their many forms, has truly been recognized. Carbohydrates have long been recognized for their roles as structural materials and sources of energy in the biological world, but their role as informational molecules has only relatively recently been understood, and we have much more to learn.

Some interesting facts regarding carbohydrates only hint at their remarkable properties.

1. Cellulose, a polymer of glucose, is the most abundant organic chemical on the face of the earth.
2. Some 400-billion lb of carbohydrates are produced annually by photosynthesis.
3. The typical diet consists of more than 60% (dry weight) of carbohydrates.
4. The major blood group types are determined by sequences and branching of carbohydrates.
5. Clearance of erythrocytes from the blood stream by the spleen is determined by the structure of oligosaccharides on the erythrocyte membrane.

Recent work by Albersheim and co-workers[2] shows another role of carbohydrates as informational molecules. These investigators found that oligosaccharides are capable of regulating physiological processes in plants and act as chemical messengers.

Carbohydrates have several unique roles in the world of bacteria. Brondz and Olsen[3] were able to differentiate between two closely related facultative, Gram-negative rods by analysis of the carbohydrate composition using gas-liquid chromatography (GLC) of their phenol-extracted lipopolysaccharides. Galactose is known to be an attractant for *Escherichia coli*; this is one example of bacterial chemotaxis which may be thought of as a primitive sensory system.[4] Wetwood, a bacterial infection of trees which attacks the heartwood giving it a distinctively bad aroma from volatile carboxylic acids produced by anaerobic fermentation, generally does little harm to the tree; it may first enter the plant in the germinating seed, perhaps attracted by the carbohydrates emanating from the seeds.

The carbohydrate moieties of glycosphingolipids have also been studied by GLC.[5] This class of compounds is a constituent of cell membranes and is implicated as being communication molecules involved in cell growth, blood group antigens, and cell surface receptors.[5] Specific oligosaccharide sequences are capable of being bound, and therefore recognized, by a class of compounds known as lectins. The binding properties of one such lectin were investigated by Wu, et al.[6] It is this property of lectins which make them useful in identification and purification of specific carbohydrate sequences, and molecules they may be attached to such as glycoproteins, in a form of affinity chromatography.

Many, if not most, proteins have carbohydrate moieties as an integral part of the protein. Such proteins are now known as glycoproteins. Indeed many polysaccharides are also glycoproteins. The functions of the carbohydrate moieties include control of the lifetime of glycoproteins in the blood stream, control of the uptake of certain glycoproteins by cells, protein conformation stabilization, and other functions. For a review of the structure and biological role of glycoproteins, the work of Sharon and Lis[7] can be consulted. For more information on complex carbohydrates in biochemistry, such as those mentioned in this section, the highly readable work by Sharon[1] is recommended.

It is quite apparent that the importance of carbohydrates in biochemistry and material

D-glucose	D-fructose	D-glucuronic acid
CHO	CH_2OH	CHO
HCOH	C=O	COH
HOCH	HOCH	HOCH
HCOH	HCOH	HCOH
HCOH	HCOH	HCOH
CH_2OH	CH_2OH	COOH

N-acetyl-glucosamine	2-amino-2-deoxyglucose	N-acetyl neuraminic acid
		COOH
		HC=O
		CH_2
CHO	CHO	HCOH
HCNHAc	$HCNH_2$	AcNHCH
HOCH	HOCH	HOCH
HCOH	HCOH	HCOH
HCOH	HCOH	HCOH
CH_2OH	CH_2OH	CH_2OH

FIGURE 1. Some examples of monosaccharides with various functional groups.

science cannot be overestimated. Further elucidation of the role of carbohydrates in nature will be in part dependent on analysis of oligo- and polysaccharides by high-performance liquid chromatography (HPLC) and GLC in combination with other powerful techniques such as nuclear magnetic resonance (NMR) and mass spectrometry (MS). These methods allow us to understand such things as the Maillard browning reaction,[8] the structure and function of gangliosides,[9] and the structure and mechanism of action of dietary fiber.[10] The uses of carbohydrate analyses are as limitless as the information obtained by them.

II. INTRODUCTORY CARBOHYDRATE CHEMISTRY

Generally speaking carbohydrates may be defined as polyhydroxy aldehydes or polyhydroxy ketones, occurring in their open chain forms or in their heterocyclic ring forms (the acetal or ketal forms). These are called neutral sugars as opposed to their derivatives (reduced to alditols, oxidized to sugar acids, with a phosphate group as sugar phosphates, etc.). The chemistry of carbohydrates is presented here in so far as it is of importance to analysis of these compounds by GC. For further information introductory biochemistry or organic chemistry texts are available. Many of these texts contain one or more chapters devoted to introductory carbohydrate chemistry.

A. Nomenclature

The monosaccharides, simple sugars that cannot be easily hydrolyzed into smaller units, are classified according to the number of carbon atoms in the molecule. This classification is usually used for carbohydrates with three to seven carbon atoms; that is with trioses, tetroses, pentoses, hexoses, and heptoses. Aldoses are monosaccharides which have an aldehyde when in the acyclic form (that is in the absence of the hemiacetal form to be discussed); ketoses are monosaccharides with a ketone when in the acyclic form (absence of the hemiketal bond). Glucose is an example of an aldohexose, and fructose is an example of a ketohexose or hexulose, a six-carbon ketose, as shown in Figure 1.

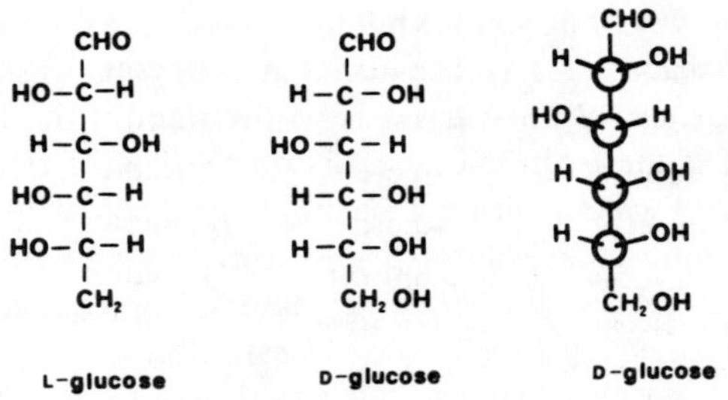

FIGURE 2. Fischer projections of L- and D-glucose and the equivalent ball-and-stick model of D-glucose.

Aldoses represent the simplest carbohydrates commonly analyzed by GC (Figure 1). If the terminal-CH_2OH of an aldose is oxidized to a carboxylic acid, then the monosaccharide is known as a uronic acid; if the aldehyde is oxidized to a carboxylic acid, the compound is referred to as an aldonic acid; and if both terminal carbon atoms are oxidized to carboxylic acids, the compound is referred to as an aldaric acid. Often a hydroxyl group, particularly the hydroxyl group of the carbon atom adjacent to the aldehyde, is replaced by an amine to give a 2-amino-2-deoxy sugar which may also occur in an acetylated form (2-acetamido-2-deoxy sugar, or *N*-acetyl amino sugar). Amino sugars in nature generally occur in the acetylated form. Chitin, which is the 2-acetamido-2-deoxy glucose (*N*-acetyl glucosamine, often abbreviated as GlcNAc) analogue of cellulose, makes up the exoskeleton of shell fish, insects, and other animals and is a linear polymer of *N*-acetyl glucosamine. Carbon atoms contained in derivatives of the hydroxyl groups are not counted in the classification of the monosaccharide; thus, *N*-acetyl-glucosamine is considered a hexose.

The sialic acids is the last group of monosaccharides that will be considered here, though there are other types of monosaccharides. Sialic acids are nine-carbon sugars which are *N*- and *O*- acetylated derivatives of α-ketopolyhydroxyamino acid. *N*-Acetylneuraminic acid is perhaps the most prevalent of these, and may be viewed as a condensation product of *N*-acetylmannosamine and pyruvic acid. Sialic acids are constituents of the glycoconjugates glycoproteins, mucins, and gangliosides.

Monosaccharide constituents of particular importance in woody plant cell wall polysaccharides are the pentoses arabinose and xylose, the hexoses glucose, mannose, and galactose, and the uronic acid (4-*O*-methyl) glucuronic acid. Galactose, mannose, glucose, fucose, *N*-acetylglucosamine, *N*-acetylgalactosamine, acetylneuraminic acids, xylose, and, in the case of plants, arabinose are important constituents in glycoproteins and other glycoconjugates. The uronic acids occur in pectin (galacturonic acid), mucopolysaccharides (glucuronic acid and iduronic acid), hemicellulose ([4-*O*-methyl] glucuronic acid), and other polysaccharides. Other monosaccharides of general importance are ribose and 2-deoxyribose (which comprise the backbone of RNA and DNA, respectively) and rhamnose.

B. Forms of Monosaccharides

Monosaccharides may be represented, as in Figure 1, by the Fischer projection, introduced by Emil Fischer in the late 19th century. The Fischer projection and the ball-and-stick model of D-glucose are presented in Figure 2, as well as the Fischer projection of L-glucose. Horizontal lines of the Fischer projection represent H and OH groups coming towards the viewer. Vertical lines represent bonds going into the plane of the paper. Although these models force the carbon backbone to be curved, it is projected onto the plane of the paper as flat. Carbon atoms are numbered consecutively from the top down. When n is the number of carbon atoms, aldoses have $n - 2$ and ketoses have $n - 3$ chiral substituted carbon atoms. The stereochemistry of the chiral carbon farthest from the anomeric carbon determines whether or not a monosaccharide is designated as D or L; in the case of glucose this is the C-5. If the hydroxyl group on this carbon is on the right of the Fischer projection, then it represents a D-series; if the hydroxyl group is on the left, it represents an L-series. Most naturally occurring monosaccharides are members of the D-series. Notable exceptions are arabinose, fucose, rhamnose, and iduronic acid. Since aldohexoses have four chiral carbons there are 2^4 (or 16) possible neutral monosaccharides, eight of the D-series and eight of the L-series. With ketohexoses there are 2^3 (or 8) possibilities.

In solution, monosaccharides do not remain as aldehydes and ketones, but form hemiacetal and hemiketal bonds, respectively. In the case of aldoses, one of the hydroxyl groups will attack the carbon atom of the carbonyl group to give a C–O bond; the carbon oxygen double bond of the aldehyde becomes a C–OH group. Usually a five-member ring (a furanose) or a six-member ring (a pyranose) is formed, which means that the hydroxyl group of the C-4 or the C-5, respectively, will react with the aldehyde group. The pyranose form is more common than the furanose form, especially in polysaccharides. In some cases both pyranose and furanose forms exist in equilibrium. Since stereochemistry is introduced at the aldehyde, there are two possible products, which are called anomers, for the pyranose or furanose. In all there may be five species (or more if other ring sizes are considered) present in solution: an acyclic form, two pyranose anomers, and two furanose anomers. The equilibrium between these possible forms depends on the stereochemistry of the monosaccharide (i.e., glucose is different from mannose, etc.), the composition of the solution and solvent, and the temperature of the solution. The ring forms of various monosaccharides are discussed in some detail by Angyal.[11] This concept is of great importance in anaylsis of carbohydrates by GC, since the method of derivatization will determine the number of derivatized products for each mono- or oligosaccharide.

The cyclic forms of carbohydrates are represented by Haworth projections. In fact, pyranoses exist in one of two possible chair conformations. The chair form with the most bulky groups in the equatorial position is greatly favored. The Haworth projection is a convenient approximation of the chair form to use. Figure 3 shows the Fisher projections, the corresponding Haworth projections, and the favored chair conformations of the two anomers of glucopyranose. α-Glucopyranoside is represented with the anomeric OH group down in the Haworth projection (trans to the terminal CH_2OH), and β-glucopyranoside is represented with the anomeric OH group in the down position. Any group that is on the right in the Fischer projection is projected down in the Haworth projection. The acid- or based-catalyzed conversion of one anomer into its equilibrium mixture of anomers is called mutarotation and is accompanied by change in the optical rotation. The equilibrium mixture of glucose in dilute aqueous solution at 20°C is about one third in the α-glucopyranose, two thirds in the β-glucopyranose form, and much less than 1% in the acyclic and furanose forms.

More complex carbohydrates occur in nature when two or more simple sugars are linked together. In nature, the linkages are glycosidic linkages, that is acetal or ketal bonds involving the carbonyl group of at least one of the monosaccharides involved. The carbon atom of the carbonyl group is referred to as the anomeric carbon atom. Glycosidic linkages, which are subject to hydrolysis in the presence of acid at reflux temperatures of water (and often at room temperature for more labile linkages), allow the formation of dimers, trimers, tetramers, and so forth. Carbohydrates with a degree of polymerization (DP, number of monosaccharide units) of two to eight are called oligosaccharides, those with more than eight units are considered polysaccharides. Sucrose (table sugar) and maltose are examples of disaccharides; cellulose and starch are examples of polysaccharides.

A glycosidic linkage involving the anomeric OH of an aldose is called an acetal linkage; if the anomeric OH of a ketose is involved, the bond is a ketal. The anomeric OH group may also react with an ROH group in general. The reaction with methanol produces a methyl

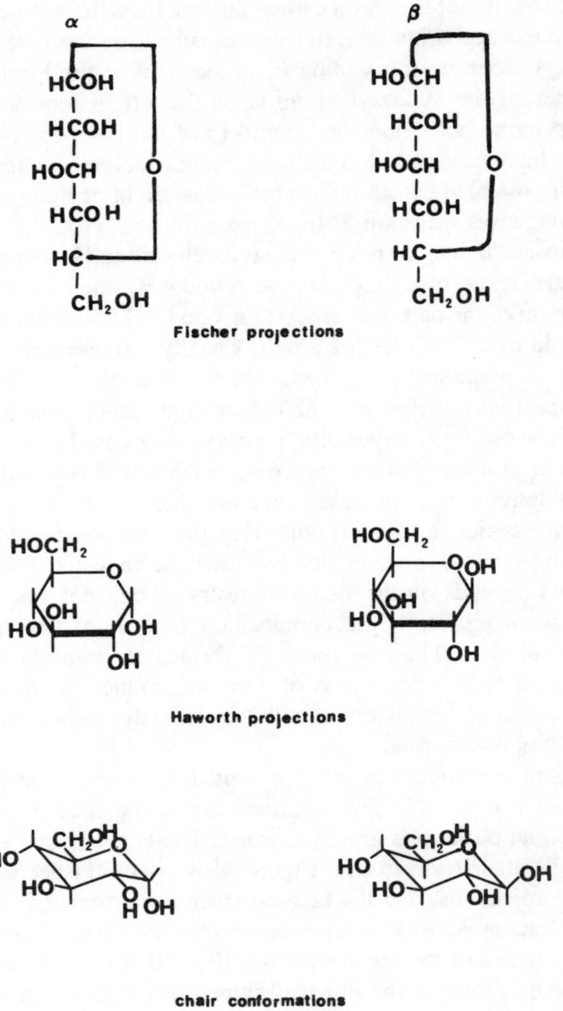

FIGURE 3. The Fischer and Haworth projections of the two glucopyranoses.

glycoside, the reaction with phenol produces a phenyl glycoside, and so forth. Once a glycosidic linkage is formed it is usually fairly stable; although, glycosidic linkages may be cleaved by acid hydrolysis or acid-catalyzed methanolysis at reflux temperatures or by the use of specific glycosidases. Figure 4 shows some examples of glycosidic linkages.

Uronic acids (and aldonic acids, though less common) have an additional complicating factor because they may form lactones. A lactone is an intramolecular esterification reaction occurring between a carboxylic acid and a γ or δ hydroxyl group, that is a hydroxy on the carbon atom three or four carbons from the carboxylic acid. In glucuronic acid, the lactone forms between the carboxylic acid and the hydroxyl group of the C-3, as shown in Figure 5; the lactone ring forms in addition to the normal pyranose ring anomers, thus there are four cyclic forms of D-glucuronic acid in solution (Figure 5). Galacturonic acid does not form a lactone since both the C-2 and C-3 hydroxyl groups are trans to the carboxylic acid at C-6 when in the ring form. In order to simplify analysis of uronic acids by GC, the sodium salt of the carboxylic acid is often formed prior to derivatization; this prohibits the formation of the lactone.

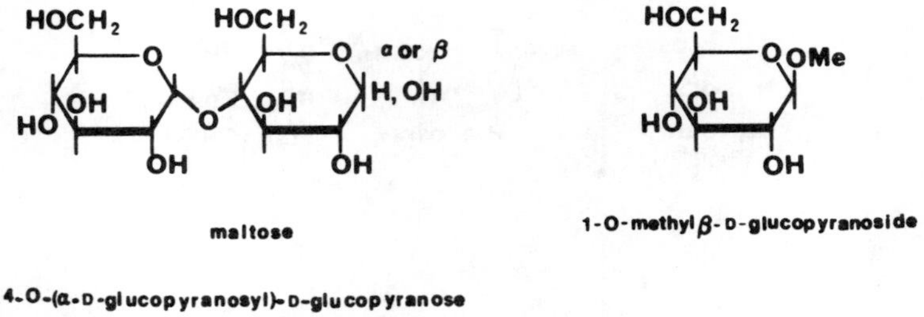

maltose

1-O-methyl β-D-glucopyranoside

4-O-(α-D-glucopyranosyl)-D-glucopyranose

FIGURE 4. Examples of glycosidic linkages.

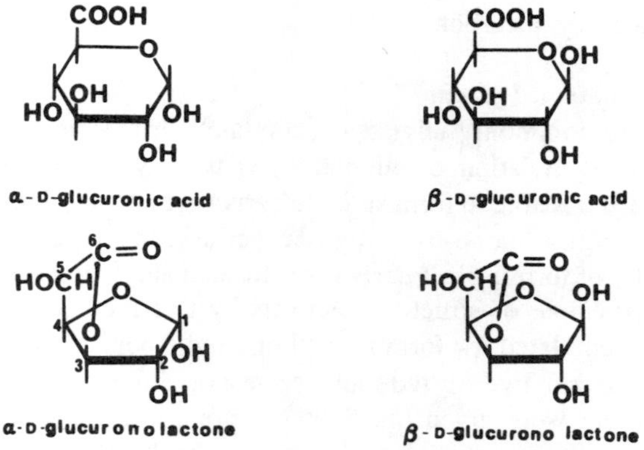

α-D-glucuronic acid

β-D-glucuronic acid

α-D-glucuronolactone

β-D-glucuronolactone

FIGURE 5. The four prominent forms of D-glucuronic acid in solution.

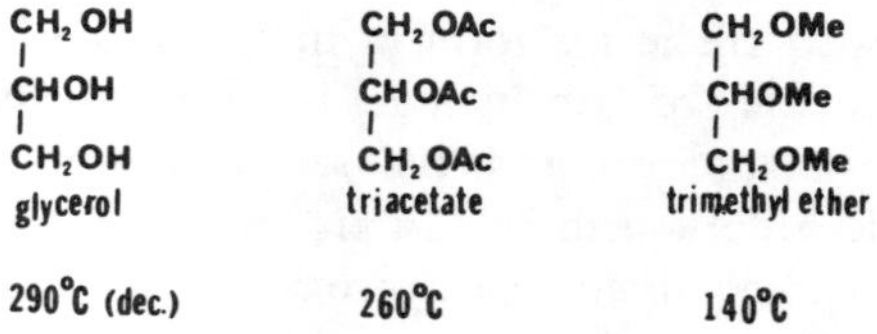

CH_2OH	CH_2OAc	CH_2OMe
$CHOH$	$CHOAc$	$CHOMe$
CH_2OH	CH_2OAc	CH_2OMe
glycerol	triacetate	trimethyl ether
290°C (dec.)	260°C	140°C

FIGURE 6. The boiling temperature of glycerol and two of its derivatives.

C. Purpose of Derivatization for GLC Analysis

The polar groups of carbohydrates (the hydroxyl, amino, and carboxylic acid groups) make carbohydrates nonvolatile. By making derivatives of these groups, the volatility is greatly increased. This concept is demonstrated by considering glycerol and two of its derivatives (Figure 6). Glycerol does not boil at atmospheric pressure, rather it decomposes at 290°C, but the triacetate boils at 260°C and the trimethyl ether boils at 140°C. The more volatile the derivative, the higher the oligosaccharide which can be analyzed by GLC. It is possible to analyze oligosaccharides up to the hexasaccharides by GLC using the more volatile types of derivatives (see Chapter 7).

D. Types of Derivatives Commonly Used

Types of derivatives usually employed in analysis of carbohydrates by GLC will be considered by functional groups. Derivatization of mono- and oligosaccharides involves consideration of all of the functional groups present in their various forms, as will become apparent in this discussion. There are numerous ways of combining the following derivatives

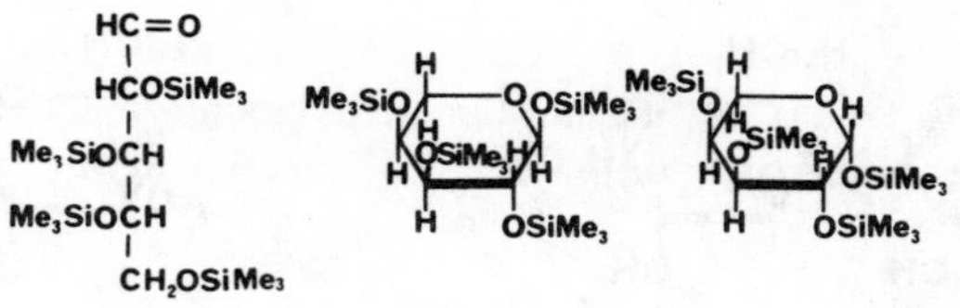

FIGURE 7. Products of direct silylation of arabinose from concentrated syrup.

when analyzing carbohydrates. Many chapters of this book deal with many of the common combinations used to analyze carbohydrates.

1. Derivatives of the Hydroxyl Group

Hydroxyl groups are commonly silylated, acetylated, trifluoracetylated, methylated, or ethylated. Alkylation (methylation or ethylation) is usually carried out by the Hakomori method.[12] Although it is possible to form suitable derivatives by alkylating all of the available OH groups (peralkylating) of mono- and oligosaccharides, this is seldom done anymore due to the relative difficulty of forming this derivative. Instead alkylation is reserved for structural analysis. The basic principle of structural analysis by this method is that polysaccharides are first methylated or ethylated for form methyl or ethyl ethers at the free hydroxyl groups. The polysaccharides are then hydrolyzed to the corresponding monosaccharides and free OH groups form where there were glycosidic linkages previously. These free OH groups are then derivatized by acetylation or silylation. Analysis by GLC, combined with MS of the products, is used to ascertain the original linkages. These topics are covered in detail in the appropriate chapters.

Sometimes polysaccharides are hydrolyzed in anhydrous methanolic hydrochloric acid. The aldoses that are liberated are in the form of methyl glycosides, that is the anomeric carbon is methylated. The mixture of carbohydrates is often derivatized in the form of methyl glycosides. An analogous situation occurs if polysaccharides are hydrolyzed in anhydrous acetic acid/acetic anhydride/sulfuric acid acid, a process called acetolysis.

For routine analysis of carbohydrates the hydroxyl groups are silylated, acetylated, or trifluoroacetylated. Silylation is usually carried out in pyridine with hexamethyldisilazane and trimethylchlorosilane at room temperature for 5 min or longer[13] or other procedures as outlined in the chapter on silylation. The sample must be dry or a very concentrated syrup since water will hydrolyze the reactants and products. The particular silylated derivative usually employed for GLC is the trimethylsilylated derivative, although others are occasionally used.

The acetate or trifluoroacetate derivatives are prepared with either pyridine or sodium acetate as a base catalyst and either acetic anhydride or trifluoroacetic anhydride. Derivatization is carried out at elevated temperatures for 15 min or longer. In the last 5 years, new acylation catalysts like *N*-methylimidazole[14] or 4-dimethylaminopyridine[15] have been incorporated in acetylation methods for analysis of carbohydrates. Aqueous solutions of carbohydrates may be used if excess acetic or trifluoroacetic anhydride is used.

If one attempts to silylate neutral aldoses without prior derivatization of the aldehyde function, then the aldoses are "locked" into their anomeric or acyclic forms. For this reason direct silylation of arabinose yields three major products,[13] as shown in Figure 7. The three major forms correspond to the α-pyranose, β-pyranose, and acyclic forms. If one analyzes a mixture of six monosaccharides, there may be 12 or more peaks to resolve. For silylated derivatives, which are separated solely by volatility since they are nonpolar, this is a difficult task. This problem may be circumvented, as will be shown, by derivatization of the aldehyde group prior to silylation. On the other hand, direct silylation offers a convenient method for quantitating the anomeric forms of carbohydrates in solution.

2. Derivatives of the Amino Group

Amino sugars usually occur in oligo- or polysaccharides in their *N*-acetylated forms. For analysis of the monosaccharide building blocks, these carbohydrates must be subject to acid hydrolysis prior to gas chromatographic analysis of the liberated monosaccharides. The amino group makes acid hydrolysis of the glycosidic linkages without undue degradation difficult. If successful, acid hydrolysis usually leads to deacetylation of the amine. The amine may be reacetylated prior to analysis, though this has been somewhat difficult to do quantitatively using older methods of acetylation.

To get around these problems, nitrous acid deamination has been used. 2-Amino hexoses are converted to their 2,5-anhydrohexose forms.[16,17] On the other hand, analysis of hexoses and hexosamines does not necessarily require deamination.[5,18] Recent work with a new method of hydrolysis (4 *M* trifluoracetic acid) and a new method of acetylation allows analysis of neutral and amino sugars at the same time, without deamination of the amino hexoses.[19,20] More information is available in the chapters on acetylated derivatives of carbohydrates.

3. Derivatives of the Carboxyl Group

Carboxylic acids may be derivatized by silylation, as described above. The ester linkage that is formed is particularly sensitive to hydrolysis, however. Acetates, of course, cannot be formed, since these involve formation of anhydride linkages. Carboxylic acids may be methylated, without methylation of the other hydroxyl groups with the exception of the anomeric hydroxyl group.

A new method for derivatization of carboxylic acids of aldonic acids has been developed by Lehrfeld.[21] Mixtures of aldoses and aldonic acids are subjected to sodium borohydride reduction to give alditols and unreacted aldonic acids. After the lactonization of the aldonic acids, treatment with *n*-propylamine converts the lactones to the corresponding *N*-(1-propyl) aldonamides. The aldonamides and alditols are then acetylated with acetic anhydride in pyridine.

Hydrolysis of glycosidic linkages involving uronic acids is also more difficult than hydrolysis of glycosidic linkages involving only neutral carbohydrates. One way to get around the problems of acid hydrolysis and subsequent derivatization is to reduce the carbodiimide-activated carboxyl group with $NaBH_4$.[22] Another method is to use methanolysis to hydrolyze the glycosidic linkages and protect the carboxyl group. Analysis of uronic acids by GLC and a colorimetric assay have been discussed.[23]

4. Derivatives of the Aldehyde Group

The derivatization of the aldehyde group is of particular importance. If it is not derivatized, then multiple peaks will be obtained for each carbohydrate. Multiple peaks are sometimes desired since the occurrence of two or more peaks with the proper ratio indicates a carbohydrate is actually present instead of a possible impurity. Yet the ratios of a mixture of anomers is not easy to reproduce in a consistent manner. As will be demonstrated, the use of oximes allows a better way to produce two compounds in a reproducible manner, which may or may not be separated by GLC.

Usually it is desirable to modify the anomeric carbon in such a fashion that no stereochemistry is introduced (other than the stereochemistry of oximes). There are four common derivatives (Figure 8):

1. Reduce the aldehyde with sodium borohydride to the corresponding alditol

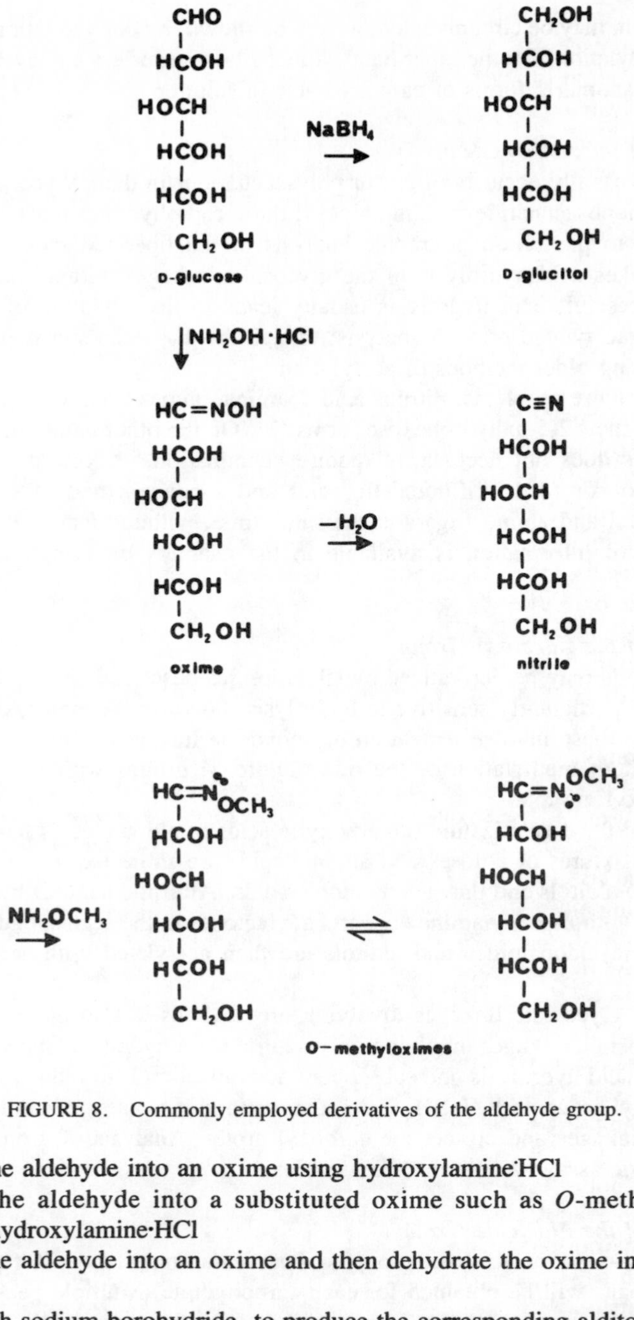

FIGURE 8. Commonly employed derivatives of the aldehyde group.

2. Convert the aldehyde into an oxime using hydroxylamine·HCl
3. Convert the aldehyde into a substituted oxime such as *O*-methyloxime using *O*-methylhydroxylamine·HCl
4. Convert the aldehyde into an oxime and then dehydrate the oxime into a nitrile

Reduction with sodium borohydride, to produce the corresponding alditol, is done under alkaline conditions for a period of about 1 hr.[18,24] Stereochemistry is not introduced at the anomeric carbon; however, two hexoses, two pentoses, etc. may lead to the same alditol. D-Arabinose and D-lyxose both yield D-arabitol; D-glucose and L-gulose yield D-glucitol (also called sorbitol), and so forth. In practice, this is not often a problem since the important hexoses (D-mannose, D-glucose, and D-galactose) each give unique alditols (from each other), as do the important pentoses (D-xylose, L-arabinose, and D-ribose). Also D-alditols cannot be separated from L-alditols without use of a chiral stationary phase or chiral derivative. If pyridine is used as an acetylation catalyst, then all of the boron must be removed with four additions of methanol followed by evaporation[24] or by use of ion exchange resin.[50] This tedious task is not necessary if *N*-methylimidazole or other powerful acylation catalysts are used as the acetylation catalyst.[25,26] Alditols may also be analyzed as silylated derivatives, though these derivatives are only poorly separated.

The formation of oximes has been catalyzed by pyridine in the past, though *N*-methylimidazole has been successfully applied to this as well.[26] A mixture of methanol, pyridine, 1-dimethylamino-2-propanol, and *O*-methylhydroxylamine hydrochloride was used to form the *O*-methyloximes of aldehydes and for subsequent acetylation[27] in what has proved to be a very useful derivative, the *O*-methyloxime acetates.[19,20] Oximes in general give two products corresponding to the *syn* and *anti* forms. In the case of *O*-methyloximes, the two forms are produced in a reproducible ratio.[19] Since the *O*-methyloxime gives two products with a reproducible ratio, this derivative is a useful tool for proving that one indeed has the particular carbohydrate present, as opposed to a possible impurity.

When the oxime of an aldose is acetylated, the oxime is converted to a nitrile. This probably occurs by acetylation of the free hydroxyl group of the oxime, which is then a leaving group for the dehydration reaction. *O*-Methyl oximes, on the other hand, do not form nitriles under the conditions of acetylation since the methoxyl group is a poor leaving group. The nitrile imparts no stereochemistry and each aldose yields a unique aldononitrile. Because the nitrile is formed from the oxime during acetylation, one will not encounter acetylated, simple oxime derivatives. Furthermore, during silylation, the oxime is silylated, but not dehydrated, to form the nitrile, so that trimethylsilyl aldononitriles are rarely studied.

III. SAMPLE PREPARATION FOR CARBOHYDRATE ANALYSIS

Sample preparation involves sample purification, preliminary determination of carbohydrates (often by colorimetric assays), and other steps to prepare the sample for gas chromatographic analysis. Not all of these steps may be required depending on the sample and the purpose of the analysis. One could write an entire volume just on these subjects. Here only some general concepts are presented as a starting point. There are many good literature references covering certain aspects of these subjects in detail for more information.

A. Purification of the Sample

Sample purification may take on many forms, depending on the purpose of the analysis. Glycoproteins may be purified using methodology for purification of proteins. Some useful review articles are listed.[28-32] If the glycoprotein contains a high percentage of carbohydrate, it may behave differently from other proteins; e.g., glycoproteins may have a higher solubility in water, depending on the amount of sugar moiety and, therefore, may not easily be salted out of solution.

In the case of polysaccharide analysis of wood subject to pretreatment prior to biomass conversion, sample purification is basically nil.[33] The reason is that wood is about 70% carbohydrates; the noncarbohydrate portions remain insoluble after acid hydrolysis and do not interfere with subsequent analysis. The composition of the monosaccharides indicates which components of the wood, cellulose or hemicellulose, are in the fraction of interest.[33]

The analysis of soluble oligosaccharides in foods is relatively straightforward as well. The food may be defatted, if necessary, with a nonpolar solvent such as hexane. Mono- and oligosaccharides, such as glucose, fructose, sucrose, maltose, and raffinose, from potatoes, breakfast cereals, etc. are usually extracted with 80% ethanol,[34] though 40% ethanol has been used for confectioneries,[35] and 80% methanol has been used for several foods.[36] Oligosaccharides up to a degree of polymerization (DP) of about 6 are easily soluble in these solvents, though interfering substances such as starch and proteins are not. An internal standard may be added and the solvent removed by evaporation. The isolation of gums from food products is more involved.[37]

Ion exchange techniques are useful for sample purification, although mono- and oligosaccharides should not be purified using strongly basic anion exchange resins since mono- and oligosaccharides are bound to such resins.[38] Weakly basic anion exchange resins may be used. Sample cleanup of sheep plasma for carbohydrate analysis has been described,[39] and diethyl amino ethyl cellulose has been used for fractionation of polysaccharides.[40] Sep-Pak C_{18} cartridges are useful for sample cleanup as well.[39,41] Aspinall has recently discussed isolation and purification of polysaccharides.[42]

B. Preliminary Determination of Carbohydrates

When one has isolated a material for analysis of carbohydrates, it is often useful to get a preliminary idea of the quantity and type of carbohydrates present. This is particularly true for proteins which are potentially glycoproteins and may be in very short supply. Colorimetric assays are sensitive, useful tests for this requirement.[43]

One of the most useful and widely used methods for this is the phenol/sulfuric acid assay for neutral pentoses and hexoses and uronic acids.[44] It is useful in the range of 10 to 70 μg of sugar. Many enzymatic hydrolysis reactions, such as the hydrolysis of cellulose by cellulase, are routinely followed by colorimetric methods such as the 3,5-dinitrosalicyclic acid or the Nelson-Somogyi methods. These two methods were recently compared,[45] though are only applicable to reducing sugars and would not be sensitive assays for oligo- and polysaccharides without acid hydrolysis. Some colorimetric assays are relatively specific. A colorimetric procedure for 3,6-anhydrogalactose in polysaccharides from red seaweeds has been developed.[46] Anthrone reagent was used to measure carbohydrates plus glycerol in deproteinated samples such as egg-yolk lipids.[47] Tetrazolium blue has been used to determine neutral and amino sugars with a sensitivity of 1 nmol.[48] Optimal conditions for reaction of aromatic hydrazides with specific carbohydrates including ketosamines has been done.[49]

There are several methods relatively specific to the amino sugars[50-52] or the uronic acids.[53,54] Often the neutral sugars will be quantitated individually by GLC, but the uronic acids will be determined as a group colorimetrically[23] in order to facilitate analysis. This is a useful approach if the uronic acids do not need to be quantitated individually. Amino sugars are also quantitated as a group colorimetrically in some cases. The approach for the overall analysis of carbohydrates depends on the samples being analyzed and the requirements of the experiment.

C. Preparation for Gas Chromatographic Analysis

Sample preparation has been covered to some extent up to this point. As mentioned, sample preparation for analysis of mono- and oligosaccharides in many foods is simply a matter of extracting and purifying, as necessary, the carbohydrates of interest. Evaporation of solvent is required for some methods of derivatization. Analysis of larger oligo- and polysaccharides involves some form of hydrolysis to cleave the glycosidic linkages to form manageable-sized mono- or oligosaccharides. For many purposes, this involves total acid hydrolysis with analysis of the component monosaccharides, as is covered in Chapter 3. Structural determination requries more sophisticated techniques such as partial hydrolysis, selective hydrolysis, and methylation and is covered in Chapter 9. Oligosaccharides moieties of glycoproteins may be isolated from the protein portion by β-elimination[1], which is discussed in Chapter 3.

IV. OVERVIEW OF APPLICATIONS OF GAS CHROMATOGRAPHY TO CARBOHYDRATE ANALYSIS

GLC of carbohydrates, after total acid hydrolyis, allows determination of the monosaccharides comprising an oligo- or polysaccharide. GLC may also be used to analyze oligosaccharides after selective or partial hydrolysis, giving information about the structure of the original polysaccharide. Combined with MS and alkylation of polysaccharides, GLC is also used for structural determination (Chapter 9). The absolute configuration of monosaccharides (i.e., whether the monosaccharide is a member of the D-series or L-series) may be determined by GLC using chiral stationary phases or by reacting the monosaccharide with a reagent which derivatizes the carbohydrate with a group containing a chiral center (Chapter 12).

Other uses of GLC (Chapter 12) include determination of the degree of polymerization of polysaccharides based on derivatization of the reducing end group (the end of the polysaccharide where the anomeric carbon is not part of a glycosidic linkage) prior to acid hydrolysis, quantitation of the anomeric forms of sugars in solution, direct determination of pentoses based on their conversion to furfural in the gas chromatograph, and analysis of miscellaneous carbohydrate derivatives such as phosphates, glycosides, and nucleosides.

V. CAPILLARY COLUMNS, SENSITIVITY, AND COLUMN POLARITY

An important improvement in the analysis of carbohydrates by GLC is the development of capillary columns. Although many separations do not require capillary columns and may be carried out on packed columns, packed columns can almost always be replaced by a capillary column with increased resolution and decreased analysis time. (Preferably packed columns should be made of deactivated glass, and on-column injection should be used, since metallic surfaces may decompose the derivatives at elevated temperatures.) Capillary columns with an I.D. of 0.25, 0.32, 0.53 or 0.75 mm are available. Those columns with I.D. of 0.75 mm are drawn from regular borosilicate glass, while the others are made from fused (vitreous) silica. Fused silica columns offer the advantage of being fairly flexible; they are easy to use without frequent breakage. Suppliers have recently begun selling the borosilicate glass columns in metal cages with fused silica leads for connection to the gas chromatograph. This feature has made the borosilicate glass columns easy to use.

The liquid phase is coated on the walls of the capillary tube in a film 0.2 μm to >1 μm, depending on the diameter of the capillary and the desired column sample capacity. The liquid phase typically covers about 75% of the capillary wall. In some cases the liquid phase is simply coated on the surface and is not chemically bonded; these are called nonbonded phases. Most phases are now chemically bonded to the inside surface of the capillary tube (bonded phases). Bonded phases allow larger sample sizes to be injected without concern for the integrity of the liquid phase layer; furthermore, when column performance decreases below usable levels, the columns may be washed by passing a suitable solvent through them. Though some of the liquid phase will be removed, a bonded-phase column can usually withstand several washings if necessary.

The choice of column diameter depends on several factors. The smaller the I.D., the greater the resolution; resolution, in terms of theoretical plates, is roughly inversely proportional to capillary diameter. On the other hand, sample capacity (in units of μg per component injected) is on the order of 50 for 0.25 mm I.D. columns, 500 for 0.32 mm I.D. columns, 1,500 for 0.53 mm I.D. columns, and 10,000 for 0.75 mm I.D. columns.

Because of the low sample capacity of the 0.25 mm I.D. capillary column, a sample splitter is usually required. With a splitter, only a portion of the sample, on the order of 1/100, is injected onto the column and most of the sample is vented. With a 0.32 mm I.D. column, up to 1 μℓ may be injected directly onto the column; with the two larger columns, up to 2 to 5 μℓ may be injected (if the phases are bonded).

It is well known that the resolution of an individual column depends on the carrier gas and its linear velocity. This fact, described by the Van Deemter equation, becomes important with capillary columns. Heavier gases, such as nitrogen, work best at low gas velocities.

Nitrogen works best at 20 to 25 cm/sec which corresponds to a flow rate of 0.6 to 0.7 mℓ/min in a 0.25 mm column and 3 to 4 mℓ/min in a 0.57 mm I.D. column.

Helium and hydrogen are most often used with capillary columns because they have optimal gas velocities of 50 to 60 cm/sec which correspond to flow rates of 1.5 to 1.7 and 7.5 to 9 mℓ/min for 0.25 and 0.57 mm I.D. columns, respectively, resulting in drastically shortened run times. With packed columns, nitrogen is commonly used at flow rates of about 20 mℓ/min in a 2 mm I.D. column.

The 0.53 and 0.75 mm I.D. columns are amenable to use in gas chromatographs normally operated with packed columns. The modification is quick and reversible. The injector side of the column is connected to a 1/16- to 1/8-in. or 1/4-in. stainless steel reducing union. A 1/4-in. or 1/8-in. short piece of glass (depending on the instrument) is used to connect the reducing union to the gas chromatograph. In most cases it is possible to use a 2 mm I.D., 1/4-in. O.D. glass tube which extends inside the injector directly. The glass tube should be cleaned frequently if it forms a char to protect the column. Appropriate ferrules are used depending on the operation temperature of the column.

The detector side of the capillary column is also connected as above except that there must be a separate tube welded to the stainless steel reducing union to supply a source of make up gas. The make up gas supplies the 20 to 30 mℓ/min of nitrogen or inert gas required for efficient and sensitive operation of the flame ionization detector (FID). Using a 0.75 mm I.D. capillary column, the column can be connected to the "A" side of the gas chromatograph using hydrogen as the "A" carrier gas, and the "B" side can be used as a source of nitrogen for the make up gas. When using hydrogen as the carrier gas, one should check for leaks frequently for safety reasons, though there is probably no more danger than using hydrogen in the detector. Hydrogen as a carrier gas saves having a separate helium tank and allows higher flow rates than nitrogen for quick analysis with good resolution.

An FID suits most purposes and has a typical detection limit of 5 ng, which is enough for most applications. An electron capture detector offers an increased sensitivity of 100- to 1000-fold, but can only be used with halogenated derivatives such as the pertrifluoroacetates, *N*-trifluoroacetylated amino sugars and bromomethyldimethylsilyl ethers. The use of N-FID is possible with amino sugars,[55] oximes, and aldononitriles, but such applications are not common. The sensitivity is typically increased by 10-fold over conventional FID detectors.

Separation of silylated sugars involves nonpolar stationary phases and is based on volatility. Separation of acetylated and trifluoroacetylated derivatives involves moderately polar stationary phases and is based on polar interactions and volatility.

VI. AVAILABLE LITERATURE OF INTEREST

There is much literature of interest regarding analysis of carbohydrates. The volume of carbohydrates in the CRC Handbook of Chromatography series is a very useful, comprehensive reference for separation of carbohydrates under various gas chromatographic conditions.[56] Other aspects of the chromatography of carbohydrates, such as HPLC, sample isolation, and derivatization, are covered (through 1978) as well. Every year, since the first volume was published in 1967, the carbohydrate literature published during the year is referenced according to subject material in the series *Carbohydrate Chemistry*.[57] The series *Methods in Carbohydrate Chemistry* is also a useful reference. Volumes 5, 7, and 8[58-60] are particularly useful for studying isolation, chemical analyses, derivatization, and structural determination of polysaccharides. The occurrence and properties of polysaccharides are reviewed in other volumes as well.[61-63] A recent general review of mono-, oligo- and polysaccharide, and glycoconjugate analysis is also available.[64] Other methods of analysis should be considered; i.e., NMR spectroscopy is a powerful tool in carbohydrate structural determination which should be considered if the equipment is available,[65-67] but is not necessary for more routine analysis.

REFERENCES

1. **Sharon, N.,** *Complex Carbohydrates: Their Chemistry, Biosynthesis, and Functions,* Addison-Wesley, Reading, Mass., 1975.
2. **Van, K. T. T., Toubart, P., Cousson, A., Darvill, A. G., Gollin, D. J., Chelf, P., and Albersheim, P.,** Manipulation of the morphogenetic pathways of tobacco explants by oligosaccharins, *Nature (London),* 314(18), 615, 1985.
3. **Brondz, I., and Olsen, I.,** Differentiation between *Actinobacillus actinomycetemcomitans* and *Haemophilus aphrophilus* based on carbohydrates in lipopolysaccharide, *J. Chromatogr.,* 310, 261, 1984.
4. **Ordal, G. W.,** Bacterial chemotaxis: a primitive sensory system, *BioScience,* 30(6), 408, 1980.
5. **Torello, L. A., Yates, A. J., and Thompson, D. K.,** Critical study of the alditol acetate method for quantitating small quantities of hexoses and hexosamines in gangliosides, *J. Chromatogr.,* 202, 195, 1980.
6. **Wu, A. M., Kabat, E. A., Gruezo, F. G., and Allen, H. J.,** Immunochemical studies on the combining site of the D-galactopyranose and 2-acetamido-2-deoxy-D-galactopyranose specific lectin isolated from *Bauhinia purpurea alba* seeds, *Arch. Biochem. Biophys.,* 204(2), 622, 1980.
7. **Sharon, N. and Lis, H.,** Glycoproteins: research booming on long-ignored, ubiquitous compounds, *Chem. Eng. News,* 59, 21, 1981.
8. **Wolfrom, M. L. and Kashimura, N.,** Gas-liquid chromatography in the study of the Maillard browning reaction, *Carbohydr. Res.,* 11, 151, 1969.
9. **Taki, T., Hirabayashi, Y., Ishikawa, H., Ando, S., Kon, K., Tanaka, Y., and Matsumoto, M.,** A ganglioside of rat ascites hepatoma AH 7974F cells, *J. Biol. Chem,* 261(7), 3075, 1986.
10. **Slavin, J. L. and Marlett, J. A.,** Evaluation of high performance liquid chromatography for measurement of the neutral saccharides in neutral detergent fiber, *J. Agric. Food Chem.,* 31(3), 1, 1983.
11. **Angyal, S. J.,** The composition of reducing sugars in solution, *Adv. Carbohydr. Chem. Biochem.,* 42, 15, 1984.
12. **Hakomori, S.,** A rapid permethylation of glycolipids, and polysaccharides catalyzed by methylsulfinyl carbanion in dimethyl sulfoxide, *J. Biochem. (Tokyo),* 55(2), 205, 1964.
13. **Sweeley, C. C., Bentley, R., Makita, M., and Wells, W. W.,** Gas liquid chromatography of trimethylsilyl derivatives of sugars and related substances, *J. Am. Chem. Soc.,* 85, 2497, 1963.
14. **Wachowiak, R. and Connors, K. A.,** *N*-methylimidazole-catalyzed acetylation of hydroxy compounds prior to gas chromatographic separation and determination, *Anal. Chem.,* 51(1), 27, 1979.
15. **Höfle, G., Steglich, W., and Vorbrüggen, H.,** 4-Dialkylaminopyridines as highly active acylation catalysts, *Angew Chem. Int. Ed. Engl.,* 17, 569, 1978.
16. **Porter, W. H.,** Application of nitrous acid deamination of hexosamines to the simultaneous GLC determination of neutral and amino sugars in glycoproteins, *Anal. Biochem.,* 63, 27, 1975.
17. **Varma, R. and Varma, R. S.,** A simple procedure for combined gas chromatographic analysis of neutral sugars, hexosamines and alditols: determination of degree of polymerization of oligo- and polysaccharides and chain weights of glycosaminoglycans, *J. Chromatogr.,* 139, 303, 1977.
18. **Whiton, R. S., Lau, P., Morgan, S. L., Gilbart, J., and Fox, A.,** Modifications in the alditol acetate method for analysis of muramic acid and other neutral and amino sugars by capillary gas chromatography-mass spectrometry with selected ion monitoring, *J. Chromatogr.,* 347, 109, 1985.
19. **Neeser, J.-R. and Schweizer, T. F.,** A quantitative determination by capillary gas-liquid chromatography of neutral and amino sugars (as *O*-methyl oxime acetates), and a study of hydrolytic conditions for glycoproteins and polysaccharides in order to increase sugar recoveries, *Anal. Biochem.,* 142, 58, 1984.
20. **Neeser, J. -R.,** G.L.C. of *O*-methyloxime and alditol acetate derivatives of neutral sugars, hexosamines, and sialic acids: "one pot" quantitative determination of the carbohydrate constituents of glycoproteins and a study of the selectivity of alkaline borohydride reductions, *Carbohydr. Res.,* 138, 189, 1985.
21. **Lehrfeld, J.,** Simultaneous gas-liquid chromatographic determination of aldonic acids and aldoses, *Anal. Chem.,* 57, 346, 1985.
22. **Taylor, R. L. and Conrad, H. E.,** Stoichiometric depolymerization of polyuronides and glycosaminoglycans to monosaccharides following reduction of their carbodiimideactivated carboxyl groups, *Biochemistry,* 11, 1383, 1972.
23. **Selvendran, R. R., March, J. F., and Ring, S. G.,** Determination of aldoses and uronic acid content of vegetable fiber, *Anal. Biochem.,* 96, 282, 1979.

24. **Albersheim, P., Nevins, D. J., English, P. D., and Karr, A.,** A method for the analysis of sugars in plant cell-wall polysaccharides by gas-liquid chromatography, *Carbohydr. Res.*, 5, 340, 1967.
25. **Henry, R. J., Blakeney, A. B., Harris, P. J., and Stone, B. A.,** Detection of neutral and aminosugars from glycoproteins and polysaccharides as their alditol acetates, *J. Chromatogr.*, 256, 419, 1983.
26. **McGinnis, G. D.,** Preparation of aldononitrile acetates using *N*-methylimidazole as catalyst and solvent, *Carbohydr. Res.*, 108, 284, 1982.
27. **Mawhinney, T. P., Feather, M. S., Barbero, G. J., and Martinez, J. R.,** The rapid, quantitative determination of neutral sugars (as aldononitrile acetates) and amino sugars (as *O*-methyloxime acetates) in glycoproteins by gas-liquid chromatography, *Anal. Biochem.*, 101, 112, 1980.
28. **Regnier, F. E. and Gooding, K. M.,** High-performance liquid chromatography of proteins, *Anal. Biochem.*, 103, 1, 1980.
29. **Regnier, F. E.,** High-performance ion-exchange chromatography of proteins: the current status, *Anal. Biochem.*, 126, 1, 1982.
30. **Furth, A. J., Bolton, H., Potter, J., and Priddle, J. D.,** Separating detergent from proteins, *Methods Enzymol.*, 104, 318, 1984.
31. **Hunkapiller, M. W., Strickler, J. E., and Wilson, K. E.,** Contemporary methodology for protein structure determination, *Science*, 226, 304, 1984.
32. **Varki, A. and Diaz, S.,** The release and purification of sialic acids from glycoconjugates: methods to minimize the loss and migration of *O*-acetyl groups, *Anal. Biochem.*, 137, 236, 1984.
33. **Biermann, C. J., Schultz, T. P., and McGinnis, G. D.,** Rapid steam hydrolysis/extraction of mixed hardwoods as a biomass pretreatment, *J. Wood Chem. Technol.*, 4(1), 111, 1984.
34. **Saura-Calixto, F., Canellas, J., and Garcia-Raso, A.,** Gas-chromatographic analysis of sugars and sugar alcohols in the mesocarp, endocarp, and kernel of almond fruit, *J. Agric. Food Chem.*, 32(5), 1018, 1984.
35. **Johncock, S. I. M. and Wagstaffe, P. J.,** Improvements in the determination of sugars in confectionery by high performance liquid chromatography, *Analyst*, 105, 581, 1980.
36. **Li, B. W., Schuhmann, P. J. and Wolf, W. R.,** Chromatographic determinations of sugars and starch in a diet composite reference material, *J. Agric. Food Chem.*, 33, 531, 1985.
37. **Lawrence, J. F. and Iyengar, J. R.,** Gas chromatographic determination of polysaccharide gums in foods after hydrolysis and derivatization, *J. Chromatogr.*, 350, 237, 1985.
38. **Baust, J. G., Lee, R. E., Jr., and James, H.,** Differential binding of sugars and polyhydric alcohols to ion exchange resins: inappropriateness for quantitative h.p.l.c., *J. Liq. Chromatogr.*, 5(4), 767, 1982.
39. **Owens, J. A. and Robinson, J. S.,** Isolation and quantitation of carbohydrates in sheep plasma by high-performance liquid chromatography, *J. Chromatogr.*, 338, 303, 1985.
40. **Siddiqui, I. R. and Wood, P. J.,** DEAE-cellulose carbonate form: a useful medium for fractionating polysaccharides, *Carbohydr. Res.*, 16, 452, 1971.
41. **Mort, A. J., Parker, S., and Kuo, M.-S.,** Recovery of methylated saccharides from methylation reaction mixtures using Sep-Pak C_{18} cartridges, *Anal. Biochem.*, 133, 380, 1983.
42. **Aspinall, G. O.,** Isolation and fractionation of polysaccharides, in *The Polysacchardies*, Vol. 1, Aspinall, G. O., Ed., Academic Press, New York, 1982.
43. **Ashwell, G.,** New colorimetric methods of sugar analysis, *Methods Enzymol.*, 8, 3, 1966.
44. **Dubois, M., Gilles, K. A., Hamilton, J. K., Rebers, P. A., and Smith, F.,** Colorimetric method for determination of sugars and related substances, *Anal. Chem.*, 28(3), 350, 1956.
45. **Breuil, C. and Saddler, J. N.,** Comparison of the 3,5-dinitrosalicyclic acid and the Nelson-Somogyi methods of assaying for reducing sugars and determining cellulase activity, *Enzyme Microb. Technol.*, 7(7), 327, 1985.
46. **Matsuhiro, B. and Zanlungo, A. B.,** Colorimetric determination of 3,6-anhydrogalactose in polysaccharides from red seaweeds, *Carbohydr. Res.*, 118, 276, 1983.
47. **Pons, A., Roca, P., Aguilo, C., Garcia, F. J., Alemany, M., and Palou, A.,** Method for simultaneous determination of total carbohydrate and glycerol in biological samples with anthrone reagent, *J. Biochem. Biophys. Methods*, 4, 227, 1981.
48. **Jue, C. K. and Lipke, P. N.,** Determination of reducing sugars in the nanomole range with tetrazolium blue, *J. Biochem. Biophys. Methods.*, 11, 109, 1985.
49. **Lever, M., Walmsley, T. A., Visser, R. S., and Ryde, S. J.,** Optimal conditions for 4-hydroxybenzoyl- and 2-furoylhydrazine as reagents for the determination of carbohydrates, including ketosamines, *Anal. Biochem.*, 139(1), 205, 1984.
50. **Varma, R. S., Varma, R., Allen, W. S., and Wardi, A. H.,** A gas chromatographic method for determination of hexosamines in glycoproteins and acid mucopolysaccharides, *J. Chromatogr.*, 93, 221, 1974.
51. **Swann, D. A. and Balazs, E. A.,** Determination of the hexoseamine content of macromolecules with manual and automated techniques using the *p*-dimethylaminobenzaldehyde reaction, *Biochim. Biophys. Acta*, 130, 112, 1966.

52. **Boas, N. F.,** Method for the determination of hexosamines in tissues, *J. Biol. Chem.*, 204, 553, 1953,
53. **Scott, R. W.,** Colorimetric determination of hexuronic acids in plant materials, *Anal. Chem.*, 51(7), 936, 1979.
54. **Blumenkrantz, N. and Asboe-Hansen, G.,** New method for quantitative analysis of uronic acids, *Anal. Biochem.*, 54, 484, 1973.
55. **Whenham, R. J.,** Sensitive assay for amino sugars using capillary gas chromatography with nitrogen-selective detection, *J. Chromatogr.*, 303, 380, 1984.
56. **Churms, S. C.,** *CRC Handbook of Chromatography, Carbohydrates,* Vol. 1, CRC Press, Boca Raton, Fla., 1982.
57. *Carbohydrate Chemistry,* The Chemical Society, Burlington House, London,
58. **Whistler, R. L., Ed.,** *Methods in Carbohydrate Chemistry,* Vol. 5, *General Polysaccharides,* Academic Press, New York, 1965.
59. **Whistler, R. L. and BeMiller, J. N.,** Eds., *Methods in Carbohydrate Chemistry,* Vol. 7, *General Methods, Glycosaminoglycans and Glycoproteins,* Academic Press, New York, 1976.
60. **Whistler, R. L. and BeMiller, J. N., Eds.,** *Methods in Carbohydrate Chemistry,* Vol. 8, *General Methods,* Academic Press, New York, 1980.
61. **Pigman, W., Horton, D., and Herp, A., Eds.,** *The Carbohydrates, Chemistry and Biochemistry,* Vol. 2B, Academic Press, New York, 1970.
62. **Pirson, A. and Zimmermann, M. H., Eds.,** *Encyclopedia of Plant Physiology,* Vol. 13B, *Plant Carbohydrates,* Springer-Verlag, New York, 1982.
63. **Ginsburg, V., Ed.,** *Methods in Enzymology,* Vol. 83, *Complex Carbohydrates, Part D.*, Academic Press, New York, 1982.
64. **Chaplin, M. F. and Kennedy, J. F., Eds.,** *Carbohydrate Analysis: a Practical Approach,* IRL Press, Washington, D. C., 1986.
65. **Gorin, P. A. J.,** Carbon-13 nuclear magnetic resonance spectroscopy of polysaccharides, *Adv. Carbohydr. Chem. Biochem.*, 38, 13, 1981.
66. **Vliegenthart, J. F. G., Dorland, L., and van Halbeek, H.,** High resolution, ^{1}H-nuclear magnetic resonance spectroscopy as a tool in the structural analysis of carbohydrates-related glycoproteins, *Adv. Carbohydr. Chem. Biochem.*, 41, 201, 1983.
67. **Bock, K., Pedersen, C., and Pedersen, H.,** Carbon-13 nuclear magnetic resonance data for oilgosaccharides, *Adv. Carbohydr. Chem. Biochem.*, 42, 193, 1984.

Chapter 2

HIGH-PERFORMANCE LIQUID CHROMATOGRAPHY (HPLC) OF CARBOHYDRATES

Gary D. McGinnis, Murray L. Laver, and Christopher J. Biermann

TABLE OF CONTENTS

I. INTRODUCTION

There are many excellent reviews on high-performance liquid chromatography (HPLC) and column chromatography as methods for analyses of carbohydrates;[1-11] therefore, only a brief introduction will be presented in this chapter. Separation of mono- and lower oligosaccharides will be especially considered, because it is with these compounds that an evaluation should be made between HPLC and gas-liquid chromatography (GLC). HPLC does not necessarily require derivatization, while GLC does. HPLC is a versatile method and sample preparation prior to injection is minimal in some cases. In addition, if derivatization is not required and if a nondestructive detector is used, the separated carbohydrates may be easily recovered. HPLC has been used to prepare pure oligo- or polysaccharides for subsequent GLC analysis, or HPLC methods may be used in conjunction with GLC for characterization of complex carbohydrates.[12-14]

It is helpful to view the various techniques used to separate carbohydrates by the general types of columns that are used (which is an indication of the separation mechanism). The important types of columns are silica gel, amine bonded to silica gel and amine modified silica gel, reverse-phase columns with various nonpolar molecules bonded to silica gel, ion-exchange resin-based columns, methods based on complexing with borates, and gel permeation chromatography. Finally detection of carbohydrates is considered.

The particle size of the packing material is given in some cases. It is important because the smaller and more uniform in size the packing material, the better the separation. The availability of packing materials of uniform size from 5 to 10 μm or even smaller is perhaps the single most important development in HPLC.

II. SILICA GEL

Separation of compounds on silica gel columns is based on polar interactions between the components to be separated and the column packing material; consequently, increasing the polarity of the solvent reduces the retention times, and the more polar a substance, the longer its retention time. For example, reduced (but otherwise underivatized [and polar in nature]) disaccharides from sulfated polysaccharides are separated with a fairly polar eluant of acetonitrile-methanol-0.5*M* ammonium formate (69:14:17 v/v),[15] whereas the relatively nonpolar alditol benzolates were separated on silica gel using *n*-hexane (85%)-dichloromethane-dioxane as the eluant.[16] Silica gel is available as 3-μm-particles, which allows good separations with short run times (typically <30 min).

III. AMINO-BONDED AND AMINE-MODIFIED PHASES

Microparticulate silica with amines bonded to the silica (amino-bonded phases) or, more recently, with the use of small quantities of amines in the mobile phase (amine modifiers) are useful methods for separation of purified, neutral mono, and lower oligosaccharides. In the case of the bonded amino phase, acetonitrile/water is used as the eluant with a ratio of 60:40 to 90:10 (v/v) depending on the required separation and the manufacturer of the column. A higher proportion of water elutes the carbohydrates more quickly and is useful for separation of larger oligosaccharides. These columns are particularly useful for analyses of lower saccharides extracted from dried food (as described in Chapter 1), where relatively pure oligosaccharide extracts are obtained from the foods, using either the bonded amine,[17] or the amine modifiers.[18] Some commercial products used in various studies are BioSil Amino 5Sj® Bio-Rad),[17] μBondapak® (Waters),[19] Lichrosorb-NH_2® (E. Merck),[20-22] Zorbax-NH_2® (DuPont),[23] Supelcosil LC-NH_2® (Supelco),[23] and MicroPak NH_2-10® (Varian).[24] The use of gradient elution in some of these studies was made possible by UV

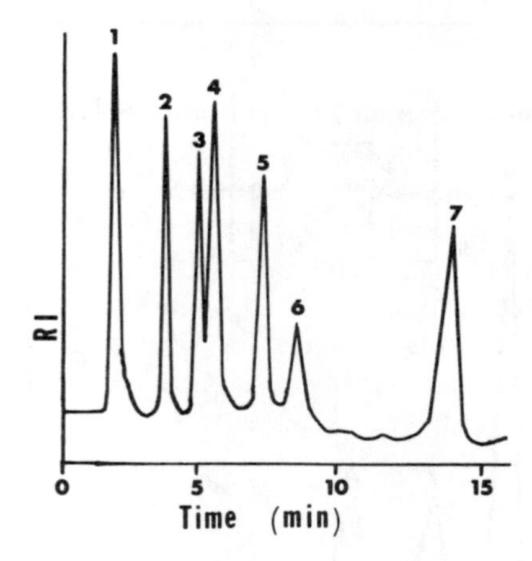

FIGURE 1. Separation of sugars on Bio-sil amino 5S column (Bio-Rad Laboratories, Richmond, Calif.), 250 × 4 mm acetonitrile/water 72:28 at 1.00 mℓ/min: 1, water (1.98 min); 2, erythritol (3.87 min); 3, fucose (4.99 min); 4, glucose (5.57 min); 5, sucrose (7.25 min); 6, maltose, cellobiose (8.55 min); and 7, maltotriose, raffinose (13.78 min).

detection.[20,22] A typical separation is shown in Figure 1. If acid was used to hydrolyze samples, it must be removed prior to injection of the sample onto the column.

IV. REVERSE PHASE

Microparticulate silica columns are modified with relatively nonpolar side chains to produce reverse phase (RP) columns. Besides silica, other materials such as cross-linked polystyrene may also serve as the base. These columns get their name because the separation is based on nonpolar interactions; hence, high polarity compounds are eluted ahead of low polarity compounds, which is the reverse of earlier forms of chromatography. This technique is so effective, however, that reverse phase is the most widely used of all HPLC methods, in general.

Since separation is based on nonpolar interactions, one might expect that unsubstituted carbohydrates are not effectively separated by this means, which is partially true. For example, a series of neutral hexoses have similar retention times, but series of maltodextrins, cyclodextrins, and isomaltosedextrins were separated on a RP column,[25] as were oligosaccharides from human milk.[26] The separation of maltodextrins is shown in Figure 2. Some of the advantages of using RP columns for oligosaccharide separations are their wide availability, high capacity, and use of water as an eluant.

Derivatized carbohydrates have been separated by RP chromatography. The formation of sugar-dansyl hydrazones[27] or pyridylamino derivatives[28] allows fluorescent detectors to be used with a sensitivity of <0.1 nmol per component. These separations required water/acetonitrile eluants. Other RP separations along these lines have been reported.[29-32]

V. USE OF ION-EXCHANGE RESINS

The usefulness of ion-exchange resins for the separation of carbohydrates has been known for about 40 years. In most cases the principal mechanism for separation of the carbohydrates is actually not ion-exchange, since the eluant is near neutral in pH, so ionization of neutral carbohydrates does not occur. The amino-bonded phases can be considered as weak anion-exchange resins, but were already considered in a separate section. The resin backbones are most often polystyrene cross-linked with 4 to 10% of divinylbenzene.

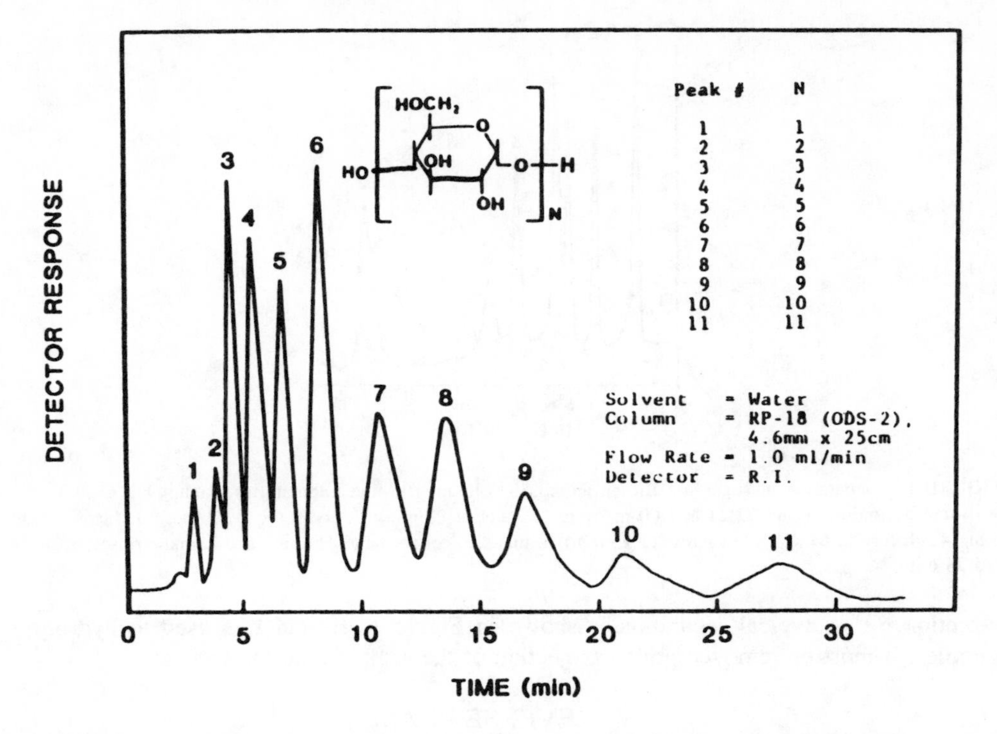

FIGURE 2. Separation of a series of maltodextrins on a RP column. (From McGinnis, G. D., Prince, S., and Lowrimore, J., *J. Carbohydr. Chem.*, 5, 83, 1986. With permission.).

The HPX series of columns (Bio-Rad) demonstrate important considerations in the use of ion-exchange columns for carbohydrate separations. Small particle size is very important for resolution so that the HPX-85 series (13 μm) has been replaced by the HPX-87 series (9 μm). Various cation forms are available including hydrogen ion (HPX-87H), calcium ion form (HPX-87C), and lead ion form (HPX-87P). These columns are all used with water eluants for the separation of mono- and disaccharides. Because large molecules are eluted prior to smaller molecules, the principal mechanism is size exclusion (gel permeation). The pore size is controlled by the amount of crosslinking; only a limited amount of crosslinking is allowable in order for the proper pore size to be realized. Hence, the physical strength of these resins is also somewhat limited, traditionally making their use somewhat difficult, although recently available products are fairly easy to use. The preparation of similar columns has been described.[33]

The hydrogen ion from is used with about 0.005 *M* aqueous sulfuric acid as the eluant. It is the most convenient form to use since carbohydrate mixtures may be analyzed directly after acid hydrolysis, but the separation is poor compared to other cation forms[34] so that this column is not often used for analyses of monosaccharides. It is a very useful column for analyses of organic acids, alcohols, and similar compounds.[35] Because the principal mechanism is size exclusion, interference from contaminants in complex samples is a problem (personal experience and also see chromatograms in Reference 37, for example).

The lead form of this column has been used for the analysis of carbohydrates following acid-hydrolysis and after removal of the acid.[36-38] Although good results are obtained (Figure 3), removal of the acid is a very laborious, time consuming operation. All cations must be removed or they will displace lead and even form precipitates within the column. Frequent regeneration of these columns, by pumping lead nitrate solution through the column, is required. Furthermore, these columns require elevated temperatures (typically 85°C). The

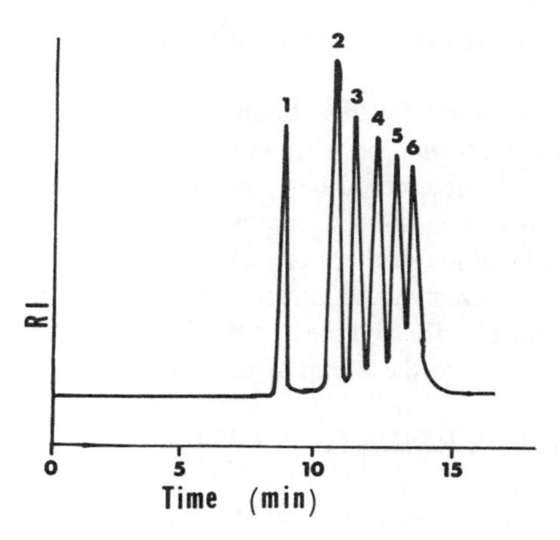

FIGURE 3. Separation of neutral mono- and disaccharides on a HPX-87P column, 0.6 mℓ/min, 85°C: 1, cellobiose; 2, glucose; 3, xylose; 4, galactose; 5, arabinose; and 6, mannose.

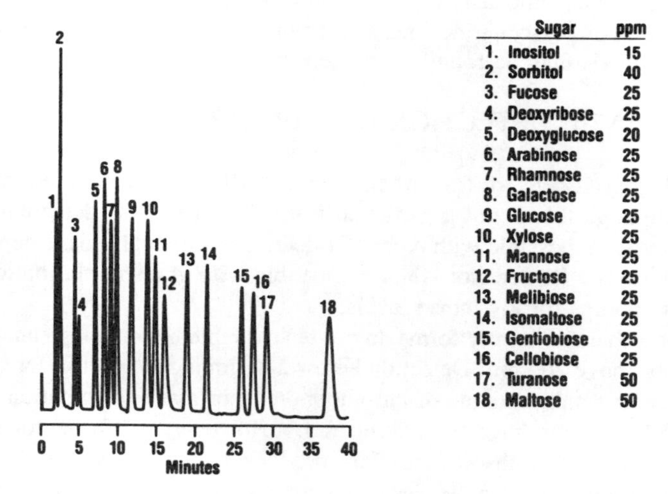

Sugar ppm
1. Inositol 15
2. Sorbitol 40
3. Fucose 25
4. Deoxyribose 25
5. Deoxyglucose 20
6. Arabinose 25
7. Rhamnose 25
8. Galactose 25
9. Glucose 25
10. Xylose 25
11. Mannose 25
12. Fructose 25
13. Melibiose 25
14. Isomaltose 25
15. Gentiobiose 25
16. Cellobiose 25
17. Turanose 50
18. Maltose 50
0 5 10 15 20 25 30 35 40
Minutes

FIGURE 4. Separation of carbohydrates on a Dionex ion-exchange column. Gradient elution, pulsed amperometric detection. (See text.) Courtesy of Dionex, Inc., Sunnyvale, Calif.

use of the calcium form is similar to that of the lead form.[39,40] Other manufacturers have similar columns.[41-43] The use of ion-exchange resins with organic solvent eluants changes the mechanism to that of normal phase partition chromatography, reversing the order of elution of compounds on ion-exchange resins.[1,44]

A paper recently described the use of a true anion-exchange mechanism for the separation of carbohydrates.[45] The eluant was 0.15 *M* sodium hydroxide with up to 0.2 *M* sodium acetate in distilled, deionized water. The column was a prototype HPIC-AS6 (Dionex). Good separation was achieved by this setup, and remarkable sensitivity (100 ppb) was achieved by use of a pulsed amperometric detector. A system is now commercially available, although it is somewhat expensive. The method looks very promising, but there are few publications at present on the topic. A recent one describes the separation of wood hydrolysates.[46] Pulsed amperometric detection is used with this method.[47,48] One separation is shown in Figure 4.

VI. BORATE BASED METHODS

The separation of carbohydrates by their borate complexes is possible. The method has the drawback that the complex formation is slow; therefore, low flow rates and long analyses times are required. However, the method has good selectivity. Control over the pH, buffer molarity, and column temperature, allows separation of many combinations of sugars. This method is not widely used, although there are some fairly recent publications on the subject.[49-51] Borates tend to be corrosive and abrasive on HPLC equipment. Constant attention is required; e.g., pumps must be flushed clean before they are turned off to preclude borate precipitation which might damage the pump seals and other parts.

VII. GEL-PERMEATION CHROMATOGRAPHY

Gel-permeation chromatography involves separation of compounds based on size, so that large compounds are eluted first. Some of the ion-exchange resin methods described previously fall into this category, though they are not generally considered as such. Besides the ion-exchange resins mentioned above, the HPX-42A (silver form) is useful for separating oligosaccharides such as maltodextrins.[52] The use of many types of gel-filtration materials has been described, with emphasis on separation of large dextrans.[4] The use of BioGel P-4 for separation of oligosaccharides has also been described,[9] and further information of gel-permeation methods may be found in the general references.[5-8]

VIII. DETECTION OF CARBOHYDRATES

The most widely used detector for carbohydrates in HPLC eluants is the refractive index (RI) detector. Although this is not a particularly sensitive detector, detection limits of <1 μg per carbohydrate are possible with recent RI detectors, although this is dependent on the eluant composition. The RI detector is not compatible with gradient elution techniques and requires pulseless pumps for low noise levels.

Neutral monosaccharides in ring forms do not have double bonds; they cannot be detected with UV detectors above 200 nm. Detection below 200 nm is not specific for carbohydrates; solvent impurities or small amounts of chromophores can make quantitation difficult. The use of UV and RI detection have been compared.[53] However, the formation of derivatives prior to separation allows UV absorbence of the products with a high sensitivity. Post column derivatization techniques can also be used in conjunction with UV or fluorescent detectors.[5] Another detector that has received attention is the mass detector, although it has not been used widely.[54] The pulsed amperometric detector[47,48] was considered in Section V.

REFERENCES

1. **Samuelson, O.,** Chromatography of oligosaccharides and related compounds on ion-exchange resins, *Adv. Chromatogr.*, 16, 113, 1978.
2. **Whistler, R. L. and BeMiller, J. N., Eds.,** *Methods in Carbohydrate Chemistry*, Vol. 8, *General Methods*, Academic Press, New York, 1980.
3. **McGinnis, G. D. and Fang, P.,** High-performance liquid chromatography, *Methods Carbohydr. Chem.*, 8, 33, 1980.
4. **Whistler, R. L. and Anisuzzaman, A. K. M.,** Gel permeation chromatography, *Methods Carbohydr. Chem.*, 8, 45, 1980.
5. **Churms, S. C.,** *Handbook of Chromatography, Carbohydrates*, CRC Press, Boca Raton, Fla., 1982.
6. **Chaplin, M. F. and Kennedy, J. F., Eds.,** *Carbohydrate Analysis: a Practical Approach*, IRL Press, Washington, D.C., 1986.

7. **Honda, S.**, High-performance liquid chromatography of mono- and oligosaccharides, *Anal. Biochem.*, 140, 1, 1984.
8. **Kakehi, K. and Honda, S.**, Profiling of carbohydrates, glycoproteins and glycolipids, *J. Chromatogr.*, 379, 27, 1986.
9. **Yamashita, K., Mizuochi, T., and Kobata, A.**, Analysis of oligosaccharides by gel filtration, *Methods Enzymol.* 83, 105, 1982.
10. **Pirisino, J. F.**, High-pressure liquid chromatography of carbohydrates in foods: fixed-ion resin and amino-bonded silica columns, *Food Sci. Tech. Ser. Monogr.*, 11, 159, 1984; *Chem. Abstr.*, 101:149859y.
11. **Hanai, T.**, Liquid chromatography of carbohydrates, *Adv. Chromatogr.*, 25, 279, 1986; *Anal. Abstr.*, 11D170.
12. **Valent, B. S., Darvill, A. G., McNeil, M., Robertsen, B. K., and Albersheim, P.**, A general and sensitive chemical method for sequencing the glycosyl residues of complex carbohydrates, *Carbohydr. Res.*, 79, 165, 1980.
13. **McNeil, M., Darvill, A. G., Åman, P., Franzén, L.- E., and Albersheim, P.**, Structural analysis of complex carbohydrates using high performance liquid chromatography, gas chromatography, and mass spectrometry, *Methods Enzymol.* 83, 3, 1982.
14. **Reinhold, V. N., Coles, E., and Carr, S. A.**, New techniques for oligosaccharide sequencing, *J. Carbohydr. Chem.*, 2, 1, 1983,
15. **Lee, G. J.-L., Liu, D.-W., Pav, J. W., and Tieckelmann, H.**, Separation of reduced disaccharides derived from glycosaminoglycans by high-performance liquid chromatography, *J. Chromatogr.*, 212, 65, 1981.
16. **Oshima, R. and Kumanotani, J.**, Determination of neutral, amino, and *N*-acetyl amino sugars as alditol benzolates by liquid-solid chromatography, *J. Chromatogr.*, 265, 335, 1983.
17. **Van Den, T., Biermann, C. J., and Marlett, J. A.**, Simple sugars, oligosaccharides, and starch concentrations in raw and cooked sweet potato, *J. Agric. Food Chem.*, 34, 421, 1986.
18. **Wade, N. L. and Morris, S. C.**, Rapid determination of sugars in cantaloupe melon juice by high-performance liquid chromatography, *J. Chromatogr.*, 240, 257, 1984.
19. **Zsadon, B., Otta, K. H., Tüdös, F., and Szejtli, J.**, Separation of cyclodextrins by high-performance liquid chromatography, *J. Chromatogr.*, 172, 490, 1979.
20. **Boersma, A., Lamblin, G., Degand, P., and Roussel, P.**, Separation of complex mixture of oligosaccharides by HPLC on a bonded-primary amine packing using a linear-gradient solvent system, *Carbohydr. Res.*, 94, C7, 1981.
21. **Blanken, W. M., Bergh, M. L. E., Koppen, P. L., and van den Eijnden, D. H.**, High-pressure liquid chromatography of neutral oligosaccharides: effects of structural parameters, *Anal. Biochem.*, 145, 322, 1985.
22. **Bergh, M. L. E., Koppen, P. L., van den Eijnden, D. H., Arnarp, J., and Lõnngren, J.**, High pressure liquid chromatography of isomeric oligosaccharides that form part of the complex-type carbohydrate chains of glycoproteins, *Carbohydr. Res.*, 117, 275, 1983.
23. **Nikolov, Z. L., Meagher, M. M., and Reilly, P. J.**, High-performance liquid chromatography of disaccharides on amine-bonded silica columns, *J. Chromatogr.*, 319, 51, 1985.
24. **Yang, M. T., Milligan, L. P., and Mathison, G. W.**, Improved sugar separations by high-performance liquid chromatography using porous microparticle carbohydrate columns, *J. Chromatogr.*, 209, 316, 1981.
25. **McGinnis, G. D., Prince, S., and Lowrimore, J.**, The use of reverse-phase columns for separation of unsubstituted carbohydrates, *J. Carbohydr. Chem.*, 5, 83, 1986.
26. **Dua, V. K. and Bush, C. A.**, Identification and fractionation of human milk oligosaccharides by proton-nuclear magnetic resonance spectroscopy and reverse-phase high-performance liquid chromatography, *Anal. Biochem.*, 133, 1, 1983.
27. **Alpenfels, W. F., Mathews, R. A., Madden, D. E., and Newsom, A. E.**, The rapid determination of neutral sugars in biological samples by high-performance liquid chromatography, *J. Liq. Chromatogr.*, 5, 1711, 1982.
28. **Takemoto, H., Hase, S., and Ikenaka, T.**, Microquantitative analysis of neutral and amino sugars as fluorescent pyridylamino derivatives by high-performance liquid chromatography, *Anal. Biochem.*, 145, 245, 1985.
29. **Blumberg, K. Liniere, F., Pustilnik, L., and Bush, C. A.**, Fractionation of oligosaccharides containing *N*-acetyl amino sugars by reverse-phase high-pressure liquid chromatography, *Anal. Biochem.*, 119, 407, 1982.
30. **Kennedy, I. R., Mwandemele, O. D., and McWhirter, K. S.**, Estimation of sucrose, raffinose, and stachyose in soybean seeds, *Food Chem.*, 17, 85, 1985.
31. **Eggert, F. M. and Jones, M.**, Measurement of neutral sugars in glycoproteins as dansyl derivatives by automated high-performance liquid chromatography, *J. Chromatogr.*, 333, 123, 1985.
32. **Batley, M., Redmond, J. W., and Tseng, A.**, Sensitive analysis of aldose sugars by reversed-phase high-performance liquid chromatography, *J. Chromatogr.*, 253, 124, 1982.

33. **Anderson, A. W. and Tsao, G. T.,** Analysis of 2-keto-gulonic acid and fermentation substrates by HPLC, *Biotechnol. Bioeng.*, 26, 374, 1984.
34. **Bonn, G.,** High-performance liquid chromatography of carbohydrates, alcohols and diethylene glycol on ion-exchange resins, *J. Chromatogr.*, 350, 381, 1985.
35. **Pecina, R., Bonn, G., Burtscher, E., and Bobleter, O.,** High-performance liquid chromatographic elution behaviour of alcohols, aldehydes, ketones, organic acids and carbohydrates on a strong cation-exchange resin, *J. Chromatogr.*, 287, 245, 1984.
36. **Slavin, J. L. and Marlett, J. A.,** Evaluation of high-performance liquid chromatography for measurement of the neutral saccharides in neutral detergent fiber, *J. Agric. Food Chem.*, 31, 467, 1983.
37. **Paice, M. G., Jurasek, L., and Desrochers, M.,** Simplified analysis of wood sugars, *TAPPI*, 65, 103, 1982.
38. **Pettersen, R. C., Schwandt, V. H., and Effland, M. J.,** An analysis of the wood sugar assay using HPLC: a comparison with paper chromatography, *J. Chromatogr. Sci.*, 22, 478, 1984.
39. **McBee, G. C. and Maness, N. O.,** Determination of sucrose, glucose, and fructose in plant tissue by high-performance liquid chromatography, *J. Chromatogr.*, 264, 474, 1983.
40. **Makkee, M., Kieboom, A. P. G., and Van Bekkum, H.,** H.p.l.c. analysis of reaction mixtures containing monosaccharides and alditols, *Int. Sugar J.*, 77, 55, 1985.
41. **DiCesare, J. L.,** The analysis of carbohydrates on high efficiency columns using an aqueous mobile phase, *Chromatogr. Newsl.*, 8, 52, 1980.
42. **Murata, K. and Yokoyama, Y.,** A high-performance liquid chromatography for constituent disaccharides of chondroitin sulfate and dermatan sulfate isomers, *Anal. Biochem.*, 146, 327, 1985.
43. **Josić, Dj., Hofermaas, R., Bauer, C., and Reutter, W.,** Automatic amino acid and sugar analysis of glycoproteins, *J. Chromatogr.*, 317, 35, 1984.
44. **Kawamoto, T. and Okada, E.,** Separation of mono- and disaccharides by high-performance liquid chromatography with a strong cation-exchange resin and an acetonitrile-rich eluent, *J. Chromatogr.*, 258, 284, 1983.
45. **Rocklin, R. D. and Pohl, C. A.,** Determination of carbohydrates by anion exchange chromatography with pulsed amperometric detection, *J. Liq. Chromatogr.*, 6, 1577, 1983.
46. **Edwards, W. T., Pohl, C. A., and Rubin, R.,** Determination of carbohydrates using pulsed amperometric detection combined with anion exchange separations, *TAPPI*, 70(6), 138, 1987.
47. **Johnson, D. C. and Polta, T. Z.,** Amperometric detection in liquid chromatography with pulsed cleaning and reaction of noble metal electrodes, *Chromatogr. Forum*, Nov.-Dec., 37, 1986.
48. **Neuburger, G. G. and Johnson, D. C.,** Pulsed amperometric detection of carbohydrates at gold electrodes with a two-step potential waveform, *Anal. Chem.*, 59, 150, 1987.
49. **Mopper, K., Dawson, R., Liebezeit, G., and Hansen, H.-P.,** Borate complex ion exchange chromatography with fluorimetric detection for determination of saccharides, *Anal. Chem.*, 52, 2018, 1980.
50. **Honda, S., Matsuda, Y., Takahashi, M., and Kakehi, K.,** Fluorometric determination of reducing carbohydrates with 2-cyanoacetamide and applications to automated analysis of carbohydrates as borate complexes, *Anal. Chem.*, 52, 1079, 1980.
51. **Hagemeier, E., Boos, K.-S., and Schlimme, E.,** Synthesis and applications of a boronic acid-substituted silica for high-performance liquid affinity chromatography, *J. Chromatogr.*, 268, 291, 1983.
52. **Warthesen, J. J.,** Analysis of saccharides in low-dextrose equivalent starch hydrolysates using high-performance liquid chromatography, *Cereal Chem.*, 61, 194, 1984.
53. **Binder, H.,** Separation of monosaccharides by high-performance liquid chromatography: comparison of ultraviolet and refractive index detection, *J. Chromatogr.*, 189, 414, 1980.
54. **Macrae, R., Trugo, L. C., and Dick, J.,** The mass detector: a new detection system for carbohydrate and lipid analysis, *Chromatographia*, 15, 476, 1982.

Chapter 3

HYDROLYSIS AND OTHER CLEAVAGE OF GLYCOSIDIC LINKAGES

Christopher J. Biermann

TABLE OF CONTENTS

I. INTRODUCTION

Cleavage of glycosidic linkages of larger oligo- and polysaccharides is necessary to determine their monosaccharide constituents. Hydrolysis — cleavage by the addition of a water molecule across a bond — is the most common method for cleavage of glycosidic linkages. Hydrolysis is carried out in aqueous solutions with an acid catalyst, although some special purpose hydrolyses, such as the liberation of carbohydrate chains from glycoconjugates, require alkaline catalysts. Common acid catalysts, in decreasing acidity, are hydrochloric, sulfuric, and trifluoroacetic acid.

Glycosidic linkages may also be cleaved in other solvents such as methanol. In this case, water is rigorously excluded so methanol is added across the glycosidic linkage to form the methyl glycosides. Methanolysis is usually catalyzed by dry hydrogen chloride. Other solvents such as acetic anhydride-acetic acid-sulfuric acid (acetolysis) and formic acid (formolysis) are occasionally used for special purposes.

With any of these methods, there is always the tradeoff between incomplete cleavage of the glycosidic linkage under relatively mild conditions and decomposition of the liberated monosaccharides under more severe conditions. Figure 1 shows the liberation of neutral monosaccharides during 0.5 *M* sulfuric acid hydrolysis (100°C, refluxing) of an algal exopolysaccharide. Even after 6 hr of hydrolysis, additional galactose is being liberated, though some xylose and mannose are lost due to decomposition. Aldopentoses and deoxy sugars are particularly susceptible to acid decomposition compared to aldohexoses. Uronic and aldonic acids are subject to decomposition by reactions such as decarboxylation, while amino sugars are relatively stable, although if they are acetylated, there is loss of acetyl groups (de-*N*-acetylation) during hydrolysis.

To correct for the relatively small amount of decomposition of liberated monosaccharides under the conditions presented in Figure 1, it is possible to subject a mixture of monosaccharides to identical hydrolysis conditions and measure the decomposition compared to a mixture not subjected to conditions of acid hydrolysis; this will obtain a correction factor for hydrolysis losses. Alternately, the hydrolysis may be continued for an extended period of time to measure decomposition, and the concentration of monosaccharide extrapolated back to time zero. Once correction factors have been determined for a set of hydrolysis conditions, it is only necessary to recheck the correction factors occasionally. However, correction factors are only approximate because the decomposition of free monosaccharides is usually higher than that of monosaccharides in polymers; consequently it is always important to minimize decomposition.

To determine the correction factors, an internal standard is usually used, and the detector response of each component (relative to the internal standard) is measured. If the ratio of a monosaccharide to the internal standard decreases after hydrolysis of a test mixture of monosaccharides, then decomposition is occurring and a recovery factor should be determined. It is preferable to add the internal standard after hydrolysis in case it is subject to decomposition. The internal standard must not appear in the samples and must be resolved from other components in the sample, as with any internal standard.

In practice it turns out that the polysaccharides containing only neutral monosaccharides are fairly easily hydrolyzed with $<10\%$ decomposition of some of the more labile monosaccharides. The presence of monosaccharides with carboxyl or amino groups in polysaccharides makes hydrolysis of their glycosidic linkages more difficult, and the liberated uronic acids themselves are much more susceptible to degradation, particularly by decarboxylation. Additionally, the amino groups in many naturally occurring polysaccharides are acetylated, so that the fate of these acetyl groups is also of consequence. Fortunately, many of the hydrolysis procedures developed recently are able to circumvent these problems for many types of samples.

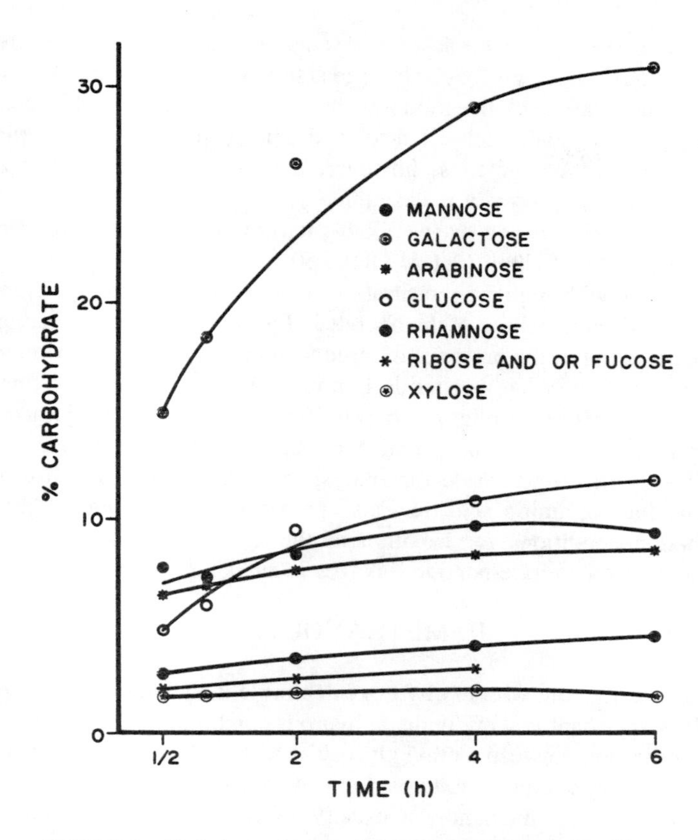

FIGURE 1. Hydrolysis of extracellular polysaccharide produced by *Anabaena flos-aquae* A-37 with 0.5 *M* sulfuric acid under reflux.

The neuraminic acids (also called sialic acids), which occur in the terminal positions of carbohydrate moieties of many glycoconjugates, must be hydrolyzed under very mild conditions. For example 0.01 *M* hydrochloric acid hydrolysis for 30 min at 100°C causes 20% destruction of *N*-acetylneuraminic acid.[1] Sialic acids are more tolerant to methanolysis, or sialic acids may be removed enzymatically. (See Chapter 4, Section III.F)

It is useful to arrange a review of the literature concerning acid hydrolysis by the type of substrate and then by the type of acid within a substrate. Many of the references for cleavage of glycosidic linkages of a particular substrate contain methods for the separation (and derivatization, if employed) of the liberated monosaccharides. In most cases, literature where hydrolysis or methanolysis recoveries are specified will be used. Studies that do not specify recoveries will be cited if there is not much other literature available for hydrolytic conditions for those types of samples. This review is not comprehensive, but does discuss many of the more important studies.

II. LIBERATION OF *N*- AND *O*-LINKED CARBOHYDRATE CHAINS

In glycoconjugates the carbohydrate moieties are linked to the protein portion by *N*-glycosylic or glycosidic (*O*-glycosylic) linkages. Generally, *N*-acetylglucosamine (GlcNAc) is linked to asparagine by an *N*-glycosylic linkage (Asn-GlcNAc). This linkage involves the anomeric carbon of GlcNAc in the β-pyranose form and the amide of asparagine. Common glycosylic linkages include D-xylose and D-galactose to L-serine, D-mannose and *N*-acetyl galactosamine to L-serine of L-threonine, D-galactose to hydroxy-L-lysine, and L-arabinose to hydroxy-L-proline. Except for the last two examples, the above linkages are cleaved by dilute alkali,[1a,2] although the *N*-glycosylic linkage is somewhat more stable than the glycosidic linkage. For characterization of the monosaccharide constituents of glycoconjugates, the entire glycoconjugate is usually subjected to total acid hydrolysis. For detailed structural analysis of the carbohydrate moieties, however, the isolation of the intact carbohydrate moieties from the remaining protein is desirable.

Glycosidic linkages are cleaved by the alkali-catalyzed β-elimination reaction with relatively low concentrations of alkali (0.1 *M* OH^-, 50°C), whereas *N*-glycosylic linkages are cleaved by hydrolysis with higher concentrations of base at higher temperatures (*M* OH^-, 100°C).[2] To prevent decomposition of the liberated oligosaccharides by "peeling" reactions, the reaction is carried out in the presence of a reducing agent.[2] Since the formerly reducing-end-monosaccharide of the oligosaccharide is now reduced, it may be distinguished from the remaining constituents of the oligosaccharide.[3] Alkali cleavage of oligosaccharide chains from glycoconjugates in the presence of tritium-labeled or deuterated sodium borohydride identifies the linking monosaccharide for analysis by radioactivity or mass spectrometry, respectively. The linking amino acid may also be ascertained.[1a] Two studies looked at a variety of reduction conditions for isolation of the oligosaccharide portions.[2,4] A more thorough review of carbohydrate-peptide linkages is available.[1a]

III. METHANOLYSIS

Methanolysis is carried out with dry hydrogen chloride in anhydrous methanol at elevated temperatures. It is important to work under anhydrous conditions since the presence of water will set up an equilibrium between methyl glycosides and the free forms of sugars in aqueous solutions, leading to very complex mixtures. Some workers prefer methanolysis to hydrolysis of glycosidic linkages, since methanolysis usually results in little degradation of the methyl glycosides liberated.

However, a complex mixture of methyl glycosides is the result. Also, *N*-acetylated carbohydrates are partially deacetylated; little work has been done on peracetylation of carbohydrates after methanolysis as a method of analysis by gas-liquid chromatography (GLC), thus, a separate step for re-*N*-acetylation is necessary. Re-*N*-acetylation with acetic anhydride in the presence of pyridine results in partial acetylation of the hyroxyl groups, making a second, though milder, methanolysis necessary to remove them.[5,6] Typically, after methanolysis, samples are neutralized with silver carbonate, re-*N*-acetylated, subjected to mild methanolysis by some workers[7] (0.1 *M* HCl in methanol, 65°C for 0.5 hr), dried, and trimethylsilylated. HCl may also be neutralized with Amberlite IRA-400 (HCO_3^-) resin[8] or simply evaporated with cosolvent.[7,9] Deaminataion prior to methanolysis is also possible,[9] in which case *N*-acetylation is no longer a concern. Often, only one or two of the major peaks of a particular monosaccharide are used for quantitation (for simplification) with the assumption that, for a particular set of methanolysis conditions, the ratio of products from a monosaccharide will be constant.[9] High-performance liquid chromatography of the methyl glycosides without derivatization (in a study of guar and other plant gums)[8] or after perbenzoylation (in a study of glycoproteins) has also been accomplished.[7]

A. Monosaccharides

The study of monosaccharides subjected to conditions of methanolysis is considered for two reasons,. First, the decomposition of monosaccharides is indicative of the decomposition of monosaccharides liberated during methanolysis; second, the ratio of methylglycosides of a particular standard monosaccharide is the same as for the same monosaccharide released during methanolysis (for a particular set of methanolysis conditions). Up to four methyl glycosides (corresponding to the α- and β-anomers of the pyranose and furanose rings) of a particular monosaccharide may be formed. The ratios of methyl glycosides of ten monosaccharides subjected to methanolysis (1 *M* HCl, 80°C, 24 hr) has been reported.[8] Similar information is also available in another study.[7]

Mega and Ikenaka[5] used methanolysis (0.8 *M* HCl, 90°C) to determine the number of *N*-glycosylically linked carbohydrate chains and the amount of glycosidic *N*-acetylglucosamine (GlcNAc) in glycoproteins. They employed *N*-acetylglucosamine with specific linkages as model compounds. *N*-Linked GlcNAc of Asn-GlcNAc gave 44% GlcN after methanolysis for 8 hr or longer. The recovery of GlcN itself after methanolysis was 70%, the remainder probably being in the form of methylglycosides. This work shows that methanolysis is capable of cleaving *N*-glycosylic linkages. Analysis of ovalbumin, flavoprotein, and taka-amylase A glycoproteins by methanolysis (0.8 *M* HCl, 90°C) up to 24 hr shows that 24 hr is a reasonable length of time for complete cleavage of the glycosidic linkages.

Chambers and Clamp[6] studied methanolysis of monosaccharides and glycoproteins with various concentrations of hydrogen chloride and various times at 85°C and 100°C. None of the monosaccharides showed any decomposition when subjected to methanolysis employing *M* HCl at 85°C for 24 hr. In methanolic 2 *M* hydrochloric acid at 100°C, the following decompositions were realized: hexuronic acids, 15%; xylose, 11%; and *N*-acetylneuraminic acid, 5%; while fucose, hexoses, and aminohexoses remained stable. This is an important reference for the decomposition of monosaccharides subjected to methanolysis with up to 6 *M* HCl and for the release of monosaccharides from glycopeptides under various methanolysis conditions. Analysis of glycopeptides and oligosaccharides of known composition indicated that methanolysis in *M* HCl was complete after 3 hr at 85°C.

Methanolysis of standard uronic acids has been studied by Inoue and Miyawaki[9] in regards to the depolymerization of chondroitin sulfate and dermatan sulfate. It was found that the glucuronidic linkage in galactosamine are rather resistant to methanolysis, but more efficiently cleaved after deamination of galactosaminide with its conversion of 2,5-anhydrotalose. For iduronic, glucuronic, and mannuronic acids released from a polymer, it was found that the peaks monitored for these acids, relative to an internal standard, increased during the first 8 hr and remained constant up to 20 hr of methanolysis (1 *M* HCl, 100°C). This indicates that 8 hr are required for methanolysis, and the liberated monosaccharides are stable to conditions of methanolysis.

B. Glycoconjugates

Reference has been made to methanolysis of glycosidic linkages involving peptides in the preceding section. The detailed work of Chambers and Clamp shows that 1 *M* methanolic hydrogen chloride at 85°C for 3 hr is sufficient to release the monosaccharides from glycopeptides and oligosaccharides.[6] Quantitation of uronic acids in acid mucopolysaccharides has been investigated by Inoue and Miyawaki using 1 *M* methanolic HCl at 100°C for 8 to 20 hr;[9] whale intestine, beef lung, and umbilical samples were investigated. Jentoft[7] studied the standard glycoconjugates sialyllactose, pig submaxillary mucin, and fetuin, and found methanolysis with 1 *M* HCl containing 1 *M* methyl acetate to be equally effective at 80°C for 4 hr or 65°C for 16 hr.

Chaplin has used methanolysis for the analysis of carbohydrates of glycoproteins.[10] His method is a variation of the above procedures with an improvement of using *tert*-butyl alcohol to remove HCl by co-evaporation instead of the prolonged trituration with silver carbonate. His method is useful for samples containing uronic acids and lipids. Mononen studied methanolysis followed by deamination and borohydride reduction for the determination of monosaccharide constituents of glycoconjugates.[11] This method was applied to a lipid-free, protein fraction of rat brain.

The composition of lipopolysaccharides (LPS) of bacterial origin has been investigated in two studies.[12,13,13a] Both studies used 2 *M* methanolic HCl at 85°C for 16 to 24 hr. The method allows analysis of carbohydrates, fatty acids, and 2-keto-3-deoxyoctanoic acid (KDO), which normally occur in LPS samples, as their trifluoroacetylated methyl glycosides or methyl esters.[13,13a] This method has been applied to differentiate taxonomically related bacteria.[12] Pyruvated monosaccharides of bacterial origin have been studied using methanolysis with 1 *M* HCl at 82°C for 16 hr.[14] Silylated methyl glycosides were prepared after re-*N*-acetylation of the amino sugars.

IV. TOTAL ACID HYDROLYSIS

In the section on monosaccharides, losses of monosaccharides subjected to conditions of acid hydrolysis are considered. In the remaining sections, methods of hydrolysis for various types of samples are considered. It should be kept in mind that monosaccharides of polysaccharides exist in an "anhydro" form since water is added across the glycosidic linkage during hydrolysis. For example, the molecular weight of glucose is 180, and the formula weight for anhydroglucose units of cellulose is 162, so that 162 g of cellulose ideally yields 180 g of glucose upon hydrolysis.

A. Monosaccharides

Subjecting monosaccharides to condition of acid hydrolysis is of importance in measuring the expected decomposition during hydrolysis of oligo- and polysaccharides. Hydrolysis losses may be predicted based on either the absolute or relative decomposition of monosaccharides. Absolute decompositions are based on decomposition of monosaccharides. Relative decompositions are based on studies where several methods of hydrolysis were applied to the same samples for various lengths of time; in this section, these are classified under the type of acid which resulted in the least decomposition (that is the largest yield of liberated monosaccharides), since this is usually the acid of principal concern in the particular study.

1. Sulfuric Acid

Hydrolysis with sulfuric acid has a drawback. Its removal prior to derivatization of the monosaccharides requires the use of barium carbonate (or other material which precipitates sulfate and neutralizes the acid, which may absorb some carbohydrates) and/or ion-exchange resin followed by filtering and so forth. Often the removal of the sulfuric acid is much more laborious than the derivatization itself. However, it is possible to form aldononitrile acetates of neutral sugars in the hydrolyzate without removal of the sulfuric acid (Chapter 6).

Figure 1 shows decomposition of monosaccharides liberated by 0.5 *M* H_2SO_4 hydrolysis of an algal exopolysaccharide. There appears to be some decomposition of xylose and mannose; however, to get accurate decomposition rates, the hydrolysis should have been continued much longer, since monosaccharides are still being liberated from the polysaccharide, and there are not enough data points to accurately extrapolate back to the origin in those cases where there is apparent decomposition.

Selvendran et al. studied 1 *M* H_2SO_4 hydrolysis of some neutral hexoses and pentoses for 2, 5, 8, 12, and 18 hr of hydrolysis at 100°C.[15] After 5 hr of hydrolysis, recoveries were galactose, 95%; arabinose, 96.3%; xylose, mannose, and glucose, 98% each; and 100% of the starting rhamnose. After 18 hr of hydrolysis, recoveries were xylose, 67%; arabinose, 75.3%; rhamnose, mannose, and galactose, 81% each; and 82% of the starting glucose. Most of the decomposition occurred during 12 to 18 hr of hydrolysis. Since hydrolysis of neutral polysaccharides with 1 *M* H_2SO_4 rarely requires more than 4 hr, sulfuric acid hydrolysis of neutral polysaccharides is an excellent method.

2. Hydrochloric Acid

Hydrolysis with hydrochloric acid is usually performed in tubes purged with nitrogen and then sealed. The presence of oxygen increases the decomposition of sugars. Niedermeier[16] added monosaccharides prior to hydrolysis of the human immunoglobulin IgG to study monosaccharide recovery. He found that in the case of 1 *M* HCl hydrolysis at 100°C, 98% of the fucose was recovered after 1 hr, 91% of the mannose and 95% of the galactose was recovered after 4 hr, and over 100% of glucosamine was recovered after 16 hr, the same hydrolysis times used to quantitate each of these monosaccharides of IgG. Torello et al.[17] used 1 *M* HCl for hydrolysis of monosaccharides at 100°C and obtained the following decompositions: galactose, 26%; glucose, 20%; GlcNAc, 14%; and GalNAc, 4% during the first 4 hr of hydrolysis and little additional decomposition up to 16 hr. Since the samples were analyzed as alditol acetates, de-*N*-acetylation was not considered. The authors do not mention purging the sample tubes with nitrogen prior to acid hydrolysis; perhaps the presence of oxygen or impurities account for much of the decomposition during the first 4 hr.

Griggs et al.[18] studied 3 *M* HCl hydrolysis for 3 hr at 100°C of canine submaxillary mucin (CSM) and CSM spiked with a standard sugar mixture. Sugar recoveries were frucose, 76.0%; mannose, 76.6%; galactose, 84.3%; *N*-acetylglucosamine, 82.5%; and *N*-acetylgalactosamine, 87.7%.

3. *Trifluoroacetic Acid*

Hydrolysis with trifluoroacetic acid (TFA) has the advantage that it may easily be removed after hydrolysis by evaporation under reduced pressure or with a stream of dry nitrogen. This allows a method for concentrating the sample when high sensitivity is required. TFA hydrolysis, like HCl hydrolysis, is usually carried out in sealed tubes purged with nitrogen to reduce any decomposition which might occur in the presence of oxygen.

Analysis of the decomposition rates of monosaccharides by 2 *M* TFA has been accomplished in two studies[19,20] by continuing hydrolysis of polysaccharides after most of the monosaccharides have been liberated. Albersheim et al.[19] studied the hydrolysis of pinto bean hypocotyl cell walls (10 mg/mℓ) at 121°C for up to 6 hr, although the monosaccharides, except glucose and mannose, and been liberated after 1 hr. After 6 hr, more than 50% of the xylose and arabinose and more than 25% of the galactose, rhamnose, and fucose had been destroyed; however, since most of the monosaccharides were liberated within 1 hr, these decomposition rates are tolerable.

Honda et al.[20] studied the hydrolysis of urinary glycoconjugates at 100°C for up to 15 hr. The investigators found that 6 hr of hydrolysis gave the maximum yield of aldoses and uronic acids. Recovery of monosaccharides added prior to 6 hr hydrolysis were GlcN, 93%; xylose, 94%; Glc, 95%; GlcA, 96%; GalN, 97%; fucose, 98%; mannose, 102%; and galactose, 105%. The recovery for glucuronic acid is very high considering how labile this monosaccharide is. This method may be useful for any sample containing uronic acids such as pectin and some plant gums. Recoveries for monosaccharides added prior to 4 *M* HCl hydrolysis at 100°C for 6 hr gave similar yields, though not quite as high. In either case, the yields are very high for accurate determination of monosaccharide constituents of glycoconjugates. The investigators favored the use of TFA for aldoses and uronic acids, but HCl for hexosamines since hexosamines are hydrolyzed with more difficulty.

Neeser and Schweizer studied 4 *M* TFA hydrolysis at 125°C for 1 hr of neutral sugars.[21] They studied two concentrations of sugars, about 50 and 2.5 m*M*. It was found that at the higher concentrations recoveries were generally much lower. This may be explained on the basis of reversion products, i.e., the formation of new glycosidic linkages. Because 4 *M* TFA is only about 50% water, an equilibrium shift towards the formation of glycosidic linkages becomes favorable since this reaction produces water. Of course, acid which catalyzes the hydrolysis of glycosidic linkages necessarily catalyzes the formation of glycosidic linkages (principle of microscopic reversibility). The formation of reversion products is potentially important under any set of hydrolysis conditions when the concentration of sugars is high (For example, when concentrating sugar solutions with traces of nonvolatile acid present). By keeping the concentration of monosaccharides relatively low, this problem is averted. Under the conditions of dilute concentration, yields of monosaccharides were rhamnose, 82%; fucose, 89%; arabinose, 86%; xylose, 72%; xylitol, 88%; 3-*O*-methyl-glucose, 97%; mannose, 86%; Glc, 97%; galactose, 81%; myoinositol and GlcN, 100%; GalN, 92%; and ManN, 95%.

B. Wood and Other Plant Polysaccharides

Hydrolysis of woody materials usually involves the liberation of only neutral monosaccharides, except for a small amount of 4-*O*-methyl glucuronic acid. In some gums and certainly in polyuronic acids, uronic acids are also liberated. Amino sugars are not often studied in these substrates. In addition to the work presented below, Neeser and Schweizer[21] studied two soluble vegetable fibers by 4 *M* TFA hydrolysis at 121°C for 1 hr and found this to be superior to 0.5 *M* H_2SO_4 hydrolysis at 100°C for 3 hr. There is an excellent review available on the isolation and characterization of plant cell wall components, including uronic acids.[62]

1. Woody Materials

Sulfuric acid hydrolysis of woody materials is usually accomplished by a two-step hydrolysis procedure. In the first step the sample is swelled with 72% H_2SO_4 (w/w) for about 1 hr at 30°C, 1.5 hr at 25°C, or 2 hr at 20°C. In the second step the sulfuric acid is diluted to about 1 *M* and refluxed for 2 to 5 hr. This method is often referred to as Saeman hydrolysis,[22] although 72% sulfuric acid has been used since 1910.[23] There are many variations to this method.[15,22-29] In some cases 77% sulfuric acid hydrolysis has been used by researchers to obtain better dissolution of wood pulps.[23,24] In the case of wood and fiber samples, the insoluble residue remaining is often weighed, being attributable to lignin, and is referred to as Klason lignin or acid insoluble lignin.

Selvendran et al.[15] hydrolyzed plant cell wall materials using 2 *M* TFA hydrolysis at 120°C for 2 hr and 1 *M* sulfuric acid hydrolysis at 100°C for 1, 2, 5, and 8 hr (Saeman hydrolysis) with or without a prior 72% H_2SO_4 step at 20°C for 3 hr. They found that Saeman hydrolysis with a 2 hr secondary hydrolysis liberated the most monosaccharides, although continuing the secondary hydrolysis for a total of 5 hr gave almost identical yields except for an increased yield of rhamnose and an appreciable decrease of galacturonic acid. A small amount of mannose was produced, possibly from C-2 epimerization of glucose, in the case of Saeman hydrolysis. TFA hydrolysis was not very effective in liberating glucose from cellulose, although the yield of other monosaccharides was similar to that of Saeman hydrolysis. This method has been used by others for the analysis of neutral detergent fiber (dietary fiber).[27]

Jeffery et al.[24] used 77% H_2SO_4 followed by 6.6% H_2SO_4 for 4 hr at 100°C, while Laver et al.[23] used 3% H_2SO_4 in their secondary hydrolysis for 4.5 hr which they claimed was better than the method of Jeffery et al. Oades used 72% H_2SO_4 followed by 0.5 *M* H_2SO_4 (approximately 5%) in his study of organic material in soils and peats.[25] Blakeney et al.[28] studied hydrolysis of plant cell walls with 72% H_2SO_4 followed by secondary hydrolysis with 1 *M* H_2SO_4 for (a) 1 hr at 121°C, (b) 2 hr at 100°C, and (c) 3 hr at 100°C. The highest recovery of neutral monosaccharides was for condition (c), accounting for 93% of the dry weight of cell walls.

There are several studies of hydrolysis of plant cell wall material using TFA hydrolysis.[19,26,30,31] Generally the method gives good yields of neutral monosaccharides, except for glucose from cellulose. In their investigation of hydrolysis of wood and pulp samples for the sugars arabinose, xylose, galactose, mannose, and glucose using TFA, Paice et al. studied 30, 50, and 80% (v/v) TFA at 100°C for up to 4.8 hr. The best conditions were with 80% TFA for 2 hr for 87% recovery of monosaccharides. Their method called for 5 mg of material dissolved in 1.6 mℓ of anhydrous TFA (a good cellulose solvent) for 2 days at room temperature. The acid was then diluted to the desired concentration, and the tubes sealed under vacuum.

2. Gums

Gums are used in foods such as ice creams, salad dressings, and soft cheese products since they bind water, improve flavor and texture, and are thickening agents. Gums are heteropolysaccharides. Guar gum is a galactomannan while others may also contain arabinose, xylose, rhamnose, and fucose. Varma et al. studied the neutral monosaccharide components of gums.[32] They hydrolyzed 2 mg dried gum using 1 mℓ of 75 m*M* H_2SO_4 in a sealed ampule at 95°C for 36 hr. They claimed that this method was the best compromise for a variety (unspecified) of hydrolysis conditions. In another study, the neutral monosaccharides of tragacanth, arabic, guar, carob, and other gums were investigated.[33] The gum identities and quantities of a variety of prepared foods were also analyzed. Hydrolysis was carried out with 0.5 *M* TFA at 100°C for 4 hr. Gums recoveries were determined based on the addition of known quantities of gums to some food samples. The recovery factors are therefore applicable for the entire method, not just sample hydrolysis recoveries. Guaran was also hydrolyzed in 1 *M* HCl at 80°C for 24 hr.[34]

3. Polyuronic Acids

As mentioned, 2 *M* TFA hydrolysis at 100°C may be a useful method for hydrolysis of polyuronic acids, though it has not been used specifically for this purpose. (Also, see Section II. D.3 of Chapter 1.) D-Mannuronic and L-guluronic acids have been prepared by acid hydrolysis of alginate samples.[35] Alginate (50 mg) was dissolved in 2.5 mℓ of 80% H_2SO_4 for 17 hr at 20°C. The H_2SO_4 was diluted to 1 *M* and hydrolysis was carried out at 100°C for 3 hr. The acid was neutralized with calcium carbonate and the sample was concentrated. Information on the hydrolysis of pectins may be found in a monograph.[36]

C. Glycoconjugates

Due to the numerous methods available for hydrolysis of various glycoconjugates (glycoproteins, mucins, mucopolysaccharides, gangliosides, and others), it is useful to break down these methods according to type of acid. Some studies compare a particular method of hydrolysis to several other methods; in these cases the comparison is described under the heading of the central method of such studies. For example, Neeser and Schweizer[21] developed 4 *M* TFA hydrolysis of glycoproteins and compared this method to other available methods; all of their results are described in the TFA section.

1. Sulfuric Acid

Sulfuric acid hydrolysis of glycoconjugates has not received much attention. Fox et al.,[37] however, tried various sulfuric acid hydrolysis conditions and found 1 *M* H_2SO_4 for 3 hr to be very useful for hydrolysis of bacterial cell walls. In this study, the sulfuric acid was neutralized with a 20% solution of *N,N*-dioctylmethylamine in chloroform.

2. Hydrochloric Acid

Niedermeier and Tomana used 0.5 to 6.0 *M* HCl at 100°C for 1 to 12 hr for the hydrolysis of human immunoglobulins and bovine submaxillary mucin while monitoring the liberated glucosamine and galactosamine.[38] Hydrolysis with 1 *M* HCl at 100° for 10 hr released all of the hexosamines from immunoglobulins, although the glucosamine liberated from bovine submaxillary mucin was only about 90% of that liberated by 3 *M* HCl for 3 hr, leading the authors to suggest that optimal hydrolysis conditions for each glycoprotein must be determined individually. In an earlier study, Niedermeier[16] used 1 *M* HCl at 100°C for 1, 4, and

10 hr to find the mildest conditions required to liberate certain monosaccharides. He found that IgM, 1 hr hydrolysis was suitable for galactose, 4 hr for mannose and glucosamine, and 10 hr for fucose, although for IgG 1 hr hydrolysis was suitable for fucose, 4 hr for mannose and galactose, and 10 hr glucosamine. The composition of IgM was proportional to the amount hydrolyzed for 1, 2, 3.5, and 5 mg in 5 mℓ of 1 *M* HCl indicating that reversion products are of no consequence at these concentrations.

Kannan et al.[39] also used 1 *M* HCl at 100°C for the hydrolysis of neutral glycolipids; 6 hr was sufficient for hydrolysis of galactolipids, but 12 hr was required for complete cleavage of the glucose-ceramide bond. The use of standard ceramide mono- and dihexosides subjected to 12 hr hydrolysis gave monosaccharide recoveries of over 94% in all cases. Torello et al.[17] used 1 *M* HCl at 100°C for 1 to 16 hr for the hydrolysis of human cerebral cortex ganglioside, GM_1; the highest yield of galactose was obtained after 4 hr, while the highest yield of glucose and glcNAC was obtained after 12 hr. There was appreciable degradation of the liberated monosaccharides under these conditions; purging the tubes with nitrogen (instead of air) did not significantly affect the yields. By using correction factors, corrected ratios of monosaccharides in various gangliosides were within 10 to 20% of the theoretical value.

Alpenfels et al.[40] studied the hydrolysis of glycoproteins and keratin fibers hydrolyzed with 1 or 2 *M* HCl at 100°C for various periods of time. These investigators found that the concentration of neutral monosaccharides from hard keratin reached a maximum after 2 hr of 2 *M* HCl hydrolysis at 100°C, and the yield of the neutral monosaccharides was linear up to 25 mg hair per mℓ HCl solution. The latter fact shows that a relatively large amount of protein does not interfere with the analysis of a relatively small amount of carbohydrate.

Griggs et al.[18] studied the hydrolysis of canine submaxillary mucin by 0.5, 3, and 6 *M* HCl for 1.5, 3, 4.5, 6, and 24 hr at 100°C. They found 3 *M* HCl for 3 hr to give the maximal release of neutral and amino monosaccharides with the minimal degradation of the liberated monosaccharides, although recoveries of neutral monosaccharides were only 76 to 88%. Guerrant and Moss[41] also used 3 *M* HCl hydrolysis, though at 75°C for 16 hr, for the hydrolysis of bacterial cell walls. Honda et al.[20] used 4 *M* HCl for 6 hr at 100°C for the hydrolysis of nondialyzable glycoconjugates when determining amino monosaccharides, but preferred 2 *M* TFA at 100°C for 6 hr when determining the neutral monosaccharides and uronic acids, since these compounds are subject to more severe degradation by 4 *M* HCl. They obtained complete hydrolysis with over 90% recovery of monosaccharides added prior to hydrolysis by using these two hydrolysis procedures.

3. Trifluoroacetic Acid

The use of 2 *M* TFA by Honda et al. was just discussed in the previous section. Eggert and Jones[42] used 2 *M* TFA at 105°C for various lengths of time for the hydrolysis of neutral sugars in glycoproteins. They found 6 to 10 hr of hydrolysis to be ideal for glycoproteins such as proteodermatan sulfate, salivary glycoproteins, and α-1-acid glycoprotein, while 4 hr of hydrolysis was sufficient for ovalbumin and fetuin; from this they recommend 8 hr hydrolysis for unknown glycoproteins and, ideally, hydrolysis for 6, 8, and 10 hr. They also found TFA hydrolysis to be superior to 2 *M* HCl hydrolysis at 105°C due to the decomposition of liberated monosaccharides with HCl hydrolysis. Neutralization of the HCl after hydrolysis, but prior to freeze-drying, reduced the decomposition somewhat, although there was still more decomposition than with TFA hydrolysis.

Recently Neeser and Schweizer introduced 4 *M* TFA at 121°C for 1 hr for hydrolysis of glycoproteins.[21] Both neutral and amino sugars were considered. They compared this method to hydrolysis with 0.6 *M* HCl at 100°C for 4 hr and 3 *M* HCl at 125°C for 0.75 hr. Hydrolysis of fetal calf serum fetuin, bovine submaxillary mucin, and horseradish peroxidase showed TFA hydrolysis to be superior.

4. *Other Methods*

Takemoto et al. used a mixture of equal amounts of 4 *M* TFA and 4 *M* HCl at 100°C for 6 hr for the measurement of neutral and amino monosaccharides of glycoconjugates.[43] Their rationale was that 2.5 *M* TFA at 100°C for 4 hr is useful for neutral sugars, but not severe enough for the liberation of amino sugars from glycoproteins (especially Asp-GlcNAc), while hydrolysis with 4 *M* HCl at 100°C for 6 hr is sufficient to release amino sugars from glycoproteins, but decomposes neutral sugars. The authors found that with their method, amino sugars were completely liberated even from Asn-GlcNAc with no decomposition of glucosamine. Neutral sugars were completely removed but with some decomposition. Correction factors were applied. This method was applied to α-1-acid glycoprotein, submaxillary gland mucin, and bovine brain gangliosides.

Some investigators have used ion-exchange resin in the acid form with a small amount of acid to hydrolyze glycoproteins.[44,45] Lehnhardt and Winzler[44] used 0.1 to 3.0 mg glycoprotein with 100 μℓ of a 20% suspension of Dowex 50 X2 (H^+) 200/400 mesh resin in 0.02 *M* HCl in a steam bath up to 40 hr. This method, compared to hydrolysis with 0.5 *M* H_2SO_4 in a steam bath, gave much less decomposition of the liberated neutral monosaccharides from orosomucoid. Since mannose was liberated much more slowly than galactose or fucose in both methods, there was some decomposition of galactose and fucose in the case of sulfuric acid hydrolysis before all of the mannose was liberated. Resin hydrolysis gave complete hydrolysis after 20 hr, while sulfuric acid hydrolysis gave complete hydrolysis after 10 hr. Porter[45] also used this resin hydrolysis method in his study of neutral and amino sugars of α-1-acid-glycoprotein and bovine luteinizing hormone.

D. Fructans

Polymers of fructose are important carbohydrate reserves in a number of plants. Inulins and levans are two major types which differ in structure. Fructans require only relatively mild conditions for their hydrolysis. Permethylated fructosans could be hydrolyzed with 2 *M* TFA at 60°C for 30 min.[46] Fructan oligosaccharides were hydrolyzed in dilute sulfuric acid at 70°C[47] or 95°C (0.1 *M*).[48] Fructans from timothy haplocorm (where they comprise 63% of the water-soluble carbohydrates) could be hydrolyzed in 0.01 *M* HCl at 98°C.[49]

V. FORMOLYSIS AND ACETOLYSIS

Formolysis and acetolysis are not common methods for cleavage of glycosidic linkages. They do have some unique applications, however. For instance, methylated polysaccharides are not generally soluble in hot water, so hydrolysis is preceded by formolysis under these circumstances. For example,[50] 5 mg of methylated polysaccharide is dissolved in 3 mℓ of 90% formic acid and kept at 100°C for 2 hr. The formic acid is removed by evaporation at 40°C. The residue is dissolved in 1 mℓ of 0.25 *M* sulfuric acid and heated at 100°C for 12 hr. The cooled solution is neutralized with barium carbonate, filtered, and concentrated by evaporation at 40°C. The formolysis procedure has also been used in the study of gangliosides.[43]

Acetolysis, with acetic anhydride/acetic acid/sulfuric acid, 10:10:1 (40°C for 2 to 13 hr) has been used for selective cleavage of glycosidic linkages in yeast mannans.[51] Selective cleavage of (1 → 6) linkages resulted in mixtures of relatively stable acetylated oligosaccharides containing (1 → 2) and (1 → 3) linkages. The composition of the oligosaccharide mixture is characteristic of the strain of yeast which is the source of the mannan. Acetolysis has also been used to form the acetates of cellobiose, cellotriose, and higher cellodextrins, but this reaction is not an analytical method.[52] Recently the application of acetolysis and formolysis were compared for hydrolysis of glycopeptides.[53] Both methods, used together, were recommended for analysis of samples containing both neutral and aminodeoxyhexitols. On-column injection into the GLC was also recommended.

VI. ENZYMATIC HYDROLYSIS

Enzymatic hydrolysis is very useful in identification of carbohydrate linkages[1c] as well as in hydrolysis of the labile sialic acids.[4] Neeser has developed a method where the sialic acids are enzymatically hydrolyzed and, simultaneously, enzymatically converted to stable 2-amino-2-deoxymannose derivatives.[4] This allows a ''one-pot'' determination of carbohydrate constituents of glycoproteins.

A number of selective glycosidases are now commercially available.[2] For glycoconjugates where the oligosaccharides are not released under alkaline conditions, enzymatic liberation is possible, the details of which are given.[1b] The basic approach is as follows: proteolytic digestion is first used followed by gel filtration to isolate the oligosaccharides from the free amino acids and short peptides; exoglycosidases may then be used to hydrolyze terminal sugar units from the oligo- or polysaccharides to elucidate the structure. The enzymes must be pure so that one does not obtain erroneous results, which has been a problem in the past.[1c] For relatively simple oligosaccharides, the use of a few exoglycosidases soon leaves a compound composed of a single monosaccharide linked to an amino acid or short peptide, which is the carbohydrate-peptide linkage. For more information on determination of structure with enzymes, there is an excellent review article available.[54]

VII. REDUCTIVE CLEAVAGE

One drawback of methylation analysis of polysaccharides is that the carbon atom involved with the acetal or ketal of a particular monosaccharide is not distinguished from linked positions after hydrolysis of the permethylated polysaccharide. For example, a 4-linked aldohexopyranose gives the same methylation product after acid hydrolysis as a 5-linked aldohexofuranose. By the application of a method which cleaves the glycosidic linkage with the addition of hydride, instead of the usual water, reduction of the anomeric carbon is achieved while still maintaining the ring structure.[55-57] If this method is applied after permethylation then the linked position is unequivocally identified. Added advantages of this method are that the anhydroalditol generated is stable and, for aldoses, a mixture of anomers is not formed since there is no stereochemistry at the anomeric carbon, although this means that the configuration of the linkages cannot be determined. The reductive-cleavage step is accomplished with boron trifluoride or trimethylsilyl trifluoromethanesulfonate-assisted organosilane reduction. This method has already been applied to structural determination of several complex polysaccharides[56-57] and promises to be an important method.

VIII. OTHER LITERATURE

A few general references of interest on this subject include the discussion of mechanisms in carbohydrate chemistry by Capon concentrating on reactions of the anomeric carbon,[58] the discussion on acid-catalyzed hydrolysis of glycosides by BeMiller,[59] the discussion of various specific degradation reactions of polysaccharides especially in regards to hydrolysis, acetolysis, periodate oxidation, deamination, and β-elimination[60], and cleavage of the glycosidic linkage.[61]

REFERENCES

1. **Sharon, N.,** *Complex Carbohydrates: Their Chemistry, Biosynthesis, and Functions,* Addison-Wesley, Reading, Mass., 1975, 54.

1a. **Sharon, N.,** *Complex Carbohydrates: Their Chemistry, Biosynthesis, and Functions,* Addison-Wesley, Reading, Mass., 1975, 65.

1b. **Sharon, N.,** *Complex Carbohydrates: Their Chemistry, Biosynthesis, and Functions,* Addison-Wesley, Reading, Mass., 1975, 84.

2. **Ogata, S.-I. and Lloyd, K. O.,** Mild alkaline borohydride treatment of glycoproteins — a method for liberating both *N*- and *O*-linked carbohydrate chains, *Anal. Biochem.,* 119, 351, 1982.

3. **Mawhinney, T. P.,** Simultaneous determination of *N*-acetylglucosamine, *N*-acetylgalactosamine, *N*-acetylglucosaminitol and *N*-acetylgalactosamintol by gas-liquid chromatography, *J. Chromatogr.,* 351, 91, 1986.

4. **Neeser, J.R.,** G.L.C. of *O*-methyloxime and alditol acetate derivatives of neutral sugars, hexosamines, and sialic acids: "one pot" quantitative determination of the carbohydrate constituents of glycoproteins and a study of the selectivity of alkaline borohydride reductions, *Carbohydr. Res.,* 138, 189, 1985.

5. **Mega, T. and Ikenaka, T.,** Methanolysis products of asparagine-linked *N*-acetylglucosamine and a new method for determination of *N*- and *O*-glycosidic *N*-acetylglucosamine in glycoproteins that contain asparagine-linked carbohydrates, *Anal. Biochem.,* 119, 17, 1982.

6. **Chambers, R. E. and Clamp, J. R.,** An assessment of methanolysis and other factors used in the analysis of carbohydrate-containing materials, *Biochem. J.,* 125, 1009, 1971.

7. **Jentoft, N.,** Analysis of sugars in glycoproteins by high-pressure liquid chromatography, *Anal. Biochem.,* 148, 424, 1985.

8. **Cheetham, N. W. H. and Sirimanne, P.,** Methanolysis studies of carbohydrates, using H.P.L.C., *Carbohydr. Res.,* 112, 1, 1983.

9. **Inoue, S. and Miyawaki, M.,** Quantitative analysis of iduronic acid and glucuronic acid in sulfated galactosaminoglycuronans by gas chromatography, *Anal. Biochem.,* 65, 164, 1975.

10. **Chaplain, M. F.,** A rapid and sensitive method for the analysis of carbohydrate components in glycoproteins using gas-liquid chromatography, *Anal. Biochem.,* 123, 336, 1982.

11. **Mononen, I.,** Quantitative analysis, by gas-liquid chromatography, of monosaccharides after methanolysis and deamination, *Carbohydr. Res.,* 88, 39, 1981.

12. **Brondz, I. and Olsen, I.,** Differentiation between *Acinobacillus acinomycetemcomitans* and *Haemophilus aphrophilus* based on carbohydrates in lipopolysaccharide, *J. Chromatogr.,* 310, 261, 1984.

13. **Bryn, K. and Jantzen, E.,** Analysis of lipopolysaccharides by methanolysis, trifluoroacetylation, and gas chromatography on a fused-silica capillary column, *J. Chromatogr.,* 240, 405, 1982.

13a. **Bryn, K. and Jantzen, E.,** Quantification of 2 Keto-3-deoxyoctanate in (lipo) polysaccharides by methanolytic release, trifluoroacetylation and capillary gas chromatography, *J. Chromatogr.,* 370, 103, 1986.

14. **Dudman, W. F. and Lacey, M. J.,** Identification of pyruvated monosaccharides in polysaccharides by gas-liquid chromatography-mass spectrometry, *Carbohydr. Res.,* 145, 175, 1986.

15. **Selvendran, R. R., March, J. F. and Ring, S. G.,** Determination of aldoses and uronic acid content of vegetable fiber, *Anal. Biochem.,* 96, 282, 1979.

16. **Niedermeier, W.,** Gas chromatography of neutral and amino sugars in glycoproteins, *Anal. Biochem.,* 40, 465, 1971.

17. **Torello, L. A., Yates, A. J., and Thompson, D. K.,** Critical study of the alditol acetate method for quantitating small quantities of hexoses and hexosamines in gangliosides, *J. Chromatogr.,* 202, 195, 1980.

18. **Griggs, L. E., Post, A., White, E. R., Finkelstein, J. A., Moeckel, W. E., Holden, K. G., Zarembo, J. E., and Weisbach, J. A.,** Identification and quantitation of alditol acetates of neutral and amino sugars from mucins by automated gas-liquid chromatography, *Anal. Biochem.,* 43, 369, 1971.

19. **Albersheim, P., Nevins, D. J., English, P. D., and Karr, A.,** A method for the analysis of sugars in plant cell-wall polysaccharides by gas-liquid chromatography, *Carbohydr. Res.,* 5, 340, 1967.

20. **Honda, S., Suzuki, S., Kakehi, K., Honda, A., and Takai, T.,** Analysis of the monosaccharide compositions of total nondialyzable urinary glycoconjugates by the dithioacetal method, *J. Chromatogr.,* 226, 341, 1981.

21. **Neeser, J.R. and Schweizer, T. F.,** A quantitative determination by capillary gas-liquid chromatography of neutral and amino sugars (as *O*-methyl oxime acetates), and a study of hydrolytic conditions for glycoproteins and polysaccharides in order to increase sugar recoveries, *Anal. Biochem.,* 142, 58, 1984.

22. **Saeman, J. F., Moore, W. E., Mitchell, R. L., and Millett, M. A.,** Techniques for the determination of pulp constituents by quantitative paper chromatography, *TAPPI,* 37, 336, 1954.

23. **Laver, M. L., Root, D. F., Shafizadeh, F., and Lowe, J. C.,** An improved method for the analysis of the carbohydrates of wood pulps through refined conditions of hydrolysis, neutralization, and monosaccharides separation, *TAPPI,* 50, 618, 1967.

24. **Jeffery, J. E., Partlow, E. V., and Polglase, W. J.,** Chromatographic estimation of sugars in wood cellulose hydrolyzates a routine reflectance method, *Anal. Chem.*, 32, 1774, 1960.
25. **Oades, J. M.,** Gas-liquid chromatography of alditol acetates and its application to the analysis of sugars in complex hydrolysates, *J. Chromatogr.*, 28, 246, 1967.
26. **Paice, M. G., Jurasek, L., and Desrochers, M.,** Simplified analysis of wood sugars, *TAPPI*, 65, 103, 1982.
27. **Slavin, J. L. and Marlett, J. A.,** Evaluation of high performance liquid chromatography for measurement of the neutral saccharides in neutral detergent fiber, *J. Agric. Food Chem.*, 31(3), 467, 1983.
28. **Blakeney, A. B., Harris, P. J., Henry, R. J., and Stone, B. A.,** A simple and rapid preparation of alditol acetates for monosaccharide analysis, *Carbohydr. Res.*, 113, 291, 1983.
29. **Pettersen, R. C., Schwandt, V. H., and Effland, M.,** An analysis of the wood sugar assay using HPLC: a comparison with paper chromatography, *J. Chromatogr. Sci.*, 22, 478, 1984.
30. **Jones, T. M. and Albersheim, P.,** A gas chromatographic method for the determination of aldoses and uronic acid constituents of plant cell wall polysaccharides, *Plant Physiol.*, 49, 926, 1972.
31. **Collings, G. F. and Yokoyama, M. T.,** Analysis of fiber components in feeds and forages using gas-liquid chromatography, *J. Agric. Food Chem.*, 27, 373, 1979.
32. **Varma, R., Varma, R. S., and Wardi, A. H.,** Separation of aldononitrile acetates of neutral sugars by gas-liquid chromatography and its application to polysaccharides, *J. Chromatogr.*, 77, 222, 1973.
33. **Lawrence, J. F. and Iyengar, J. R.,** Gas chromatographic determination of polysaccharide gums in foods after hydrolysis and derivatization, *J. Chromatogr.*, 350, 237, 1985.
34. **Thiem, J., Schwentner, J., Karl, H., Sievers, A., and Jeimer, J.,** Separation of peracetylated mono- and disaccharides and quantitative analysis of guaran by high-performance liquid chromatography on silica gel, *J. Chromatogr.*, 155, 107, 1978.
35. **Voragan, A. G. J., Schols, H. A., De Vries, J. A., and Pilnik, W.,** High-performance liquid chromatographic analysis of uronic acids and oligogalacturonic acids, *J. Chromatogr.*, 244, 327, 1982.
36. **Fishman, M. L. and Jen, J. J.,** Eds., *Chemistry and Function of Pectins, ACS Symp. Series 310,* American Chemical Soc., Washington, D.C., 1986.
37. **Fox, A., Morgan, S. L., Hudson, J. R., Zhu, Z. T., and Lau, P. Y.,** Capillary gas chromatographic analysis of alditol acetates of neutral and amino sugars in bacterial cell walls, *J. Chromatogr.*, 256, 429, 1983.
38. **Niedermeier, W. and Tomana, M.,** Gas chromatographic analysis of hexosamines in glycoproteins, *Anal. Biochem.*, 57, 363, 1974.
39. **Kannan, R., Seng, P. N., and Debuch, H.,** Evaluation of a gas chromatographic method for the quantitative estimation of hexoses from neutral glycolipids, *J. Chromatogr.*, 92, 95, 1974.
40. **Alpenfels, W. F., Mathews, R. A., Madden, D. E., and Newsom, A. E.,** The rapid determination of neutral sugars in biological samples by high-performance liquid chromatography, *J. Liq. Chromatogr.*, 5, 1711, 1982.
41. **Guerrant, G. O., and Moss, C. W.,** Determination of monosaccharides as aldononitrile, *O*-methyloxime, alditol, and cyclitol acetate derivatives by gas chromatography, *Anal. Chem.*, 56, 633, 1984.
42. **Eggert, F. M. and Jones, M.,** Measurement of neutral sugars in glycoproteins as dansyl derivatives by automated high-performance liquid chromatography, *J. Chromatogr.*, 333, 123, 1985.
43. **Takemoto, H., Hase, S., and Ikenaka, T.,** Microquantitative analysis of neutral and amino sugars as fluorescent pyridylamino derivatives by high-performance liquid chromatography, *Anal. Biochem.*, 145, 245, 1985.
44. **Lehnhardt, W. F. and Winzler, R. J.,** Determination of neutral sugars in glycoproteins by gas-liquid chromatography, *J. Chromatogr.*, 34, 471, 1968.
45. **Porter, W. H.,** Application of nitrous acid deamination of hexosamines to the simultaneous GLC determination of neutral and amino sugars in glycoproteins, *Anal. Biochem.*, 63, 27, 1975.
46. **Pollock, C. J., Hall, M. A., and Roberts, D. P.,** Structural analysis of fructose polymers by gas-liquid chromatography and gel filtration, *J. Chromatogr.*, 171, 411, 1979.
47. **Heyraud, A., Rinaudo, M., and Taravel, F. R.,** Isolation and characterization of oligosaccharides containing D-fructose from juices of the Jerusalem artichoke: kinetic constants for acid hydrolysis, *Carbohydr. Res.*, 128, 311, 1984.
48. **Wolf, D. D. and Ellmore, T. L.,** Automated hydrolysis of non reducing sugars and fructosans from plant tissue, *Crop Sci.*, 15, 775, 1975.
49. **Suzuki, M.,** Fructosan in the timothy haplocorm, *Can. J. Bot.*, 46, 1201, 1968.
50. **Lindberg, B.,** Methylation analysis of polysaccharides, *Methods Enzymol.*, 28, 178, 1972.
51. **Stewart, T. S. and Ballou, C. E.,** A comparison of yeast mannans and phosphomannans by acetolysis, *Biochemistry,* 7, 1855, 1968.
52. **Wolfrom, M. L. and Thompson, A.,** Acetolysis, *Methods Carbohydr. Chem.*, 3, 143, 1963.

53. **Conchie, J., Hay, A. J., and Lomax, J. A.,** A comparison of some hydrolytic and gas chromatographic procedures used in methylation analysis of the carbohydrate units of glycopeptides, *Carbohydr. Res.*, 103, 129, 1982.
54. **McCleary, B. V. and Matheson, N. K.,** Enzymic analysis of polysaccharide structure, *Adv. Carbohydr. Chem. Biochem.*, 44, 147, 1986.
55. **Rolf, D. and Gray, G. R.,** Reductive cleavage of glycosides, *J. Am. Chem. Soc.*, 104, 3539, 1982.
56. **Rolf, D. and Gray, G. R.,** Quantitative analysis of linkage positions in a complex D-glucan by the reductive-cleavage method, *Carbohydr. Res.*, 152, 343, 1986.
57. **Bennek, J. A., Rice, M. J., and Gray, G. R.,** Analysis of positions in 2-acetamido-2-deoxy-D-glucopyranosyl residues by the reductive-cleavage method, *Carbohydr. Res.*, 157, 125, 1986.
58. **Capon, B.,** Mechanism in carbohydrate chemistry, *Chem. Rev.*, 69(4), 407, 1969.
59. **BeMiller, J. N.,** Acid-catalyzed hydrolysis of glycosides, *Adv. Carbohydr. Chem.*, 22, 25, 1967.
60. **Lindberg, B., Lönngren, J., and Svensson, S.,** Specific degradation of polysaccharides, *Adv. Carbohydr. Chem. Biochem*, 31, 185, 1975.
61. **Bochkov, A. F. and Zaikov, G. E.,** *Chemistry of the O-Glycosidic Bond: Formation and Cleavage*, Pergamon Press, New York, 1979, 177.
62. **York, W. S., Darvill, A. G., McNeil, M., Stevenson, T. T., and Albersheim, P.,** Isolation and characterization of plant cell walls and cell wall components, *Methods Enzymol.*, 118, 3, 1986.

Chapter 4

SILYL ETHERS OF CARBOHYDRATES

Kazuaki Kakehi and Susumu Honda

TABLE OF CONTENTS

I. INTRODUCTION

Since carbohydrates are generally nonvolatile, due to the presence of highly polar hydroxyl groups, they have to be converted to volatile derivatives such as ethers (silyl, methyl, etc.) and esters (acetyl, trifluoroacetyl, etc.) before being submitted to gas chromatography (GC). Each of these derivatives has its own characteristic feature, but this chapter deals with silyl ethers.

Because of the simplicity of the derivatization procedure, silylation has been widely applied prior to analysis of various classes of carbohydrates by GC, including aldoses, ketoses, glycosides, alditols, uronic acids, deoxy sugars, and oligomers up to the tetrasaccharides. The first paper[1] on silylation of carbohydrates appeared in 1963, in which Sweeley et al. demonstrated excellent separation of a number of carbohydrates. This paper was epoch making, because it was also the first example of carbohydrate analysis by GC. Since this paper was published, GC has taken the place of paper chromatography, the then known, most reliable analytical method. Recent advancement in capillary technique has further improved separation, and development of mass spectrometry (MS) has realized more reliable identification of carbohydrates separated by GC.

There are a number of books and reviews[2-18] on GC of carbohydrates, and most of them include analysis of silyl ethers. Particular attention should be paid to the reviews by Dutton,[16,17] because they comprise almost all literature on GC of carbohydrates published before 1973.

Although there are various kinds of carbohydrates, ranging from monosaccharide to polysaccharide from the viewpoint of molecular size, the subject of GC is generally confined to monosaccharide because of limitation due to volatility. However, oligosaccharides having relatively low degrees of polymerization are also handled, but column temperatures must be raised to nearly their maximal points. Monosaccharides occur in nature in free state, but more abundantly in conjugated state. The monosaccharides in the latter state have to be released from the conjugates prior to GC analysis. When they are freed by alcoholysis, the products are usually glycosides having isomeric forms due to the difference in ring size and anomeric configuration. This diversity of monosaccharide species often makes chromatographic pattern complicated and determination of individual monosaccharides difficult, though decomposition of freed monosaccharides is minimized in alcoholysis. On the other hand, when they are liberated by hydrolysis, the problem of the presence of isomers can be solved by converting resultant monosaccharides to such derivatives as alditols and dithioacetals. Oximation is also useful, though the problem of the presence of *syn* and *anti* isomers is left unsolved. Acid hydrolysis is known to be accompanied by a considerable degree of nonhydrolytic degradation, which leads to loss of component monosaccharides. In addition, separation of derivatized products is more difficult than that of glycosides, due to reduction of steric variability by losing molecular rigidity. In any case GC, even in capillary mode, cannot afford complete separation of all known monosaccharides. Therefore, appropriate conditions should be selected to adapt to the combination of monosaccharides to be analyzed.

This chapter describes details of silylation procedure and GC condition for analysis of silylated derivatives, taking as many examples as possible from the literature.

II. GENERAL CONSIDERATIONS

A. Characteristics and Advantages of Silylation

Silylation involves substitution of the hydrogen atom in the hydroxyl group by the silyl group to give an ether, as shown in Equation 1.

$$R\text{–}OH + R'_3Si\text{–}X \rightarrow R\text{–}O\text{–}Si\text{–}R'_3 + HX \quad (1)$$

Table 1
SILYLATING REAGENTS USED FOR DERIVATIZATION OF CARBOHYDRATES

Reagent	Abbreviation	Structure
N, O-Bis-(trimethylsilyl)-acetamide	BSA	$CH_3—C(OSi(CH_3)_3)=NSi(CH_3)_3$
Hexamethyldisilazane	HMDS	$(CH_3)_3SiNHSi(CH_3)_3$
Trimethylchlorosilane	TMCS	$(CH_3)_3SiCl$
Trimethylsilylimidazole	TMSI	$(H_3C)_3Si—N$ (imidazole ring)
N,O-Bis-(Trimethylsilyl)-trifluoroacetamide	BSTFA	$CF_3—C(OSi(CH_3)_3)=N—Si(CH_3)_3$

Carbohydrates, which are polyhydroxyl compounds, of course undergo this type of reaction to give persilylated derivatives. The sulfhydryl and amino groups may also be substituted to give a thioether and an iminoether, respectively.

The silylation reaction of the hydroxyl group is very rapid and proceeds under mild conditions. Silylation endows derivatives with a volatile nature, permitting distillation under reduced pressure. The substituent group R′ may be varied from alkyl to aryl groups, but the methyl group is the most popular for GC. Trimethylsilylation (TMS) increases sensitivity to the detection by flame ionization. Bromomethyldimethylsilylation has an advantage that the derivatives can be selectively detected, with extremely high sensitivity, by electron-capture detectors. Therefore, bromomethyldimethylsilylation is useful when only very small amounts of carbohydrates are present in the sample.

B. Procedures for Silylation

In the first paper of silylation of carbohydrates for GC, Sweeley et al.[1] described the use of a combination of tyrimethylchlorosilane (TMCS) and hexamethyldisilazane (HMDS). This method is still one of the important methods for silylation, however, much improvement has been made. Table 1 lists the silylating reagents now in use for derivatization of carbohydrates. These reagents are used not only alone, but also in mixtures.

In addition to these reagents, *N*-trimethylsilylacetamide (TMSA), *N*-methyl-*N*-trimethylsilyltrifluoroacetamide (MSTFA) and *N*-trimethylsilyldiethylamine (TMSDEA) are also known to be useful for TMS of the hydroxyl group, but they are rarely used for derivatization of carbohydrates.

In some cases, silylation of carbohydrates is performed without solvent, but in most cases pyridine is used as a reaction solvent. Typical procedures for silylation are given below.

1. General Procedure for O-Silylation by Use of a Combination of TMCS and HMDS

This is a historical procedure devised by Sweeley et al.,[1] but is still evaluated as a general procedure for *O*-silylation. The amino group is not at all, or only partially, silylated by this procedure. Reducing sugars generally give plural peaks due to ring-structural and configurational isomers.

Dissolve a carbohydrate sample (<10 mg) in anhydrous pyridine (1 mℓ), which is prepared by reflux of commercial pyridine of analytical grade with barium oxide, followed by distillation, and add TMCS (0.1 mℓ) and HMDS (0.2 mℓ) to the resultant solution. When the carbohydrate sample is difficult to dissolve, heating to 75 to 85°C facilitates dissolution. Shake the mixture vigorously for 30 sec, then allow it to stand for more than 5 min. Centrifuge the mixture for 5 min at 3000 rpm to remove the white precipitates and submit the supernatant to GC. When analysis is repeated by using a flame ionization detector, the surface of the electrodes is covered by a solid mass of resultant siloxane and the sensitivity is decreased. Therefore, the surface of the electrodes should be cleaned mechanically after a series of analyses.

When only a small amount of a sample is available, the TMS derivatives in the supernatant can be concentrated by the following series of operations: addition of chloroform (3 mℓ) and water (3 mℓ), shaking the mixture, evaporation of the chloroform layer to dryness *in vacuo*, and redissolution of the residue in a small volume of chloroform.[19] Experience has indicated that carbohydrate quantities, as small as 1 μg (1 to 10 ng as the injected amount) per a single sample, can be determined by a slight modification of this concentration technique without any interference by the peaks of solvent and reagents. The operation should be performed as quickly as possible at low temperature to avoid partial decomposition by hydrolysis of the derivatives.

2. *A Simplified Procedure for O-Silylation by Use of BSA*[20]

This procedure is simpler than the Sweeley et al. procedure (Section II.B.1), but is as effective as the latter.

A carbohydrate sample (0.5 to 1 mg) is dissolved in anhydrous pyridine (70 μℓ) in a small vial. Add BSA (130 μℓ) to this solution, allow the mixture to stand for 30 min with a stopper tightly closed, and submit a portion to GC.

This procedure allows pertrimethylsilylation of mono- as well as oligosaccharides, but the derivatives in the reaction mixture are relatively unstable when stored as they are. The peaks of degradation products from BSA often disturb the analysis by overlap on the peaks of samples.

The velocity of silylation with BSA is slower than that with the Sweeley et al. reagent, but this procedure has an advantage that no solid of siloxane is formed on detector electrodes. Englmaier[21] succeeded in accelerating the reaction by adding 5% TMCS to BSA.

3. *Procedure for Mercaptalation of Reducing Monosaccharides with Ethanethiol, Followed by Trimethylsilylation with TMCS-HMDS*[22,23]

This is one of the devices to eliminate peak splitting due to the presence of isomers. This procedure is applicable to aldoses, uronic acids, and *N*-acetylated amino sugars. Free amino sugars are converted to 2,5-anhydroaldoses by the action of nitrous acid, and the latter can be mercaptalated and trimethylsilylated.

For neutral reducing sugars[22] — Add an 0.1% aqueous solution (10 to 100 μℓ) of 3-*O*–methylglucose (internal standard) to a solution of a sample (1 to 100 μg) contained in a small vial with a polyethylene or PTFE stopper and evaporate the mixture to dryness under reduced pressure by placing the vial in a desiccator. Add a 2:1 (by volume) mixture (20 μℓ) of ethanethiol and trifluoroacetic acid to the residue, dissolve the residue completely by swirling the vial, and allow the solution to stand for 10 min at 25°C. Add anhydrous pyridine (50 μℓ), HMDS (100 μℓ) and TMCS (50 μℓ) in this order, and shake the mixture immediately. Keep the mixture for 30 min at 50°C, then centrifuge it. Submit a portion to GC.

For ultramicro analysis, add a series of operations for concentration, i.e., solvent extraction, evaporation, and redissolution in a smaller volume of solvent, as described for the *O*-trimethylsilylation with TMCS-HMDS (Section II.B.1). n-Hexane is preferable to chloroform in this case.

For free amino sugars[23] — Although amino sugars are difficult to convert to dithioacetals in an intact state, the 2,5-anhydroaldoses derived thereof can be mercaptalated and trimethylsilylated with the same ease as that for neutral reducing sugars. Add an 0.1% aqueous solution (1 to 100 μℓ) of 3-*O*-methylglucose (internal standard) to a sample (1 to 100 μℓ) contained in a small vial with a polyethylene or PTFE stopper, and evaporate the mixture to dryness under reduced pressure by placing the vial in a desiccator. Cool the vial in an ice bath, add successively chilled solutions of 0.1 *M* barium nitrite (100 μℓ) and 0.1 *M* sulfuric acid (75 μℓ), gently shake the mixture, and allow it to stand for 1 hr in an ice bath. Evaporate the mixture to dryness as soon as possible in a similar manner as that mentioned above for preparation of a sample containing the internal standard. Add a 2:1 (by volume) mixture of ethanethiol and trifluoroacetic acid to the residue, and treat the mixture in the same manner as that described for mercaptalation and TMS of neutral reducing sugars.

In these procedures, mercaptalation and TMS can be conveniently performed in a single vial. However, the operation should be carried out in a hood to avoid the vile smell of ethanethiol. It is recommended to rinse the pipette and the reaction vial with a dilute solution of sodium hypochlorite or a bleach solution immediately after use. The remaining ethanethiol is easily converted to nonsmelling ethanesulfonate.

4. Procedure for O-Methyloximation of Aldoses and Ketoses, Followed by Trimethylsilylation with BSTFA[24]

Evaporate to dryness a sample solution of a reducing sugar or a mixture of reducing sugars (10 to 100 μg) contained in a small vial under reduced pressure. Add a 1% pyridine solution (500 μℓ) of *O*-methylhydroxylamine hydrochloride, and heat the mixture for 2 hr at 80°C. Subsequently add BSTFA (150 μℓ) to the mixture, and keep it at 80°C for another 15 min. After it is cooled, submit an aliquot to GC.

This is a simple procedure for derivatization of reducing sugars. However, blank peaks often interfere with the determination of reducing sugars, unless *O*-methylhydroxylamine hydrochloride of high purity is used. The derivatized products are reported to be stable for hours.

C. Chromatographic Conditions

In early works of GC of carbohydrate derivatives, stainless steel columns were usually employed. However, the derivatives, especially those of amino sugars and sialic acids, are partially decomposed during analysis by contact with the inner wall of columns, causing loss of samples or appearance of secondary peaks. For this reason glass columns should be used. Glass-to-glass connections at the inlets and the outlets of the columns are also desirable.

Stationary phases have been much improved, and a number of commercial samples having high and consistent quality are now available. Proper choice can lead to satisfactory results.

Sweeley et al.[1] compared silicone SE-52 and ethyleneglycol succinate polymer, selected as representatives of nonpolar and polar phases, respectively. Sawardeker and Sloneker[25] studied quantification of carbohydrates, and indicated that the former phase gave symmetric peaks of TMS derivatives, but separation was not satisfactory. On the contrary, the latter phase caused tailing due to adsorption, though separation was good. Eventually they succeeded in separation and quantification of the TMS derivatives of galactose, glucose, and mannose by using a column of Carbowax® 20 *M*, a moderately polar material. Separation of pentoses and deoxyhexoses was not successful even with this stationary phase. On the other hand, Ellis[26] stated that moderately polar stationary phases such as silicone XE-60, XF-1150, and QF-1 gave good separation among 15 candidates of stationary phases. Bhatti et al.[27] reported separation of a wide range of carbohydrates on silicone SE-30 and UC-W98. They recommended the use of Diatoport S as a supporting material.

Reducing sugars are often analyzed as alditol and diethyldithioacetal derivatives to eliminate the problem of peak splitting due to isomers. Separation of these derivatives on packed columns is difficult because they reduce steric variability by ring opening. However, they can be separated by using capillary columns. Separation on capillary columns is further improved by application of temperature programs.

Table 2 summarizes the analytical conditions reported for separation of TMS derivatives of carbohydrates. It comprises almost all conditions in recent literature published since 1974 that contain chromatograms, together with details of derivatization procedures.

III. SILYLATED DERIVATIVES OF VARIOUS CLASSES OF CARBOHYDRATES

The procedure for silylation is common to all kinds of hydroxyl groups in carbohydrates, but each class of carbohydrates has its own characteristic features (due to other functional group[s]). Therefore, total derivatization procedures and analytical conditions are varied among classes.

A. Aldose

Since the aldose is the carbohydrate of the most fundamental class, a great number of papers concerning GC of carbohydrates of this class have appeared. Because the aldose has isomers based on ring size and anomeric configuration, direct silylation give plural peaks, as already pointed out. Notwithstanding this peak splitting, many papers deal with direct silylation. On the other hand, peak unification by borohydride reduction or mercaptalation, prior to silylation, has been investigated.

Borohydride reduction is described in the chapter on alditol acetates (Chapter 5), since the separation of acetylated alditols is easier than that of silylated alditols. Oximation or alkyloximation is another method for simplification of peaks, though *syn* and *anti* isomers are formed.

1. Analysis as Pertrimethylsilylated Derivatives

Direct TMS of aldoses with either TMCS-HMDS or BSA affords pertrimethylsilylated derivatives of pyranose anomers. Some aldoses also give the derivatives of furanoses. Figure 1 shows a chromatogram presented by Sweeley et al.,[1] which shows a number of peaks arising from a mixture of monosaccharides, including tetroses, pentoses, and hexoses, together with oligosaccharides.

The analysis was performed on a column of silicone SE-52 with a temperature program. Separation was fairly good, 18 sugars giving a total of 20 peaks. Figure 2 gives another example of the separation of TMS derivatives of aldoses on a column of Carbowax® 20 *M* at a constant temperature, reported by Sawardeker and Sloneker.[25]

On the other hand, Bhatti et al.[27] tabulated retention times of TMS derivatives of 37 monosaccharides on a column of silicone SE-30. All these data are complicated by the presence of too many peaks. Some aldoses gave peaks superimposed on those of other aldoses. However, analysis of aldoses as TMS derivatives is useful in cases where the number of aldose species is relatively small.

2. Analysis as Trimethylsilylated Diethyldithioacetals

Mercaptalation of aldoses with a mixture of ethanethiol and trifluoroacetic acid, followed by TMS with TMCS-HMDS, quantitatively yields trimethylsilylated aldose diethyldithioacetals under mild conditions, and these derivatives are separable on a capillary column coated with silicone SF-96, as shown in Figure 3.

The derivatives of all naturally occurring aldoses were separated in the presence of the derivatives of all naturally occurring uronic acids and *N*-acetylhexosamines. The details of the derivatization procedure were described in the preceding paragraph (Section II.B.3).

Table 2
CONDITIONS USED FOR GAS CHROMATOGRAPHY OF TMS ETHERS OF CARBOHYDRATES AND THEIR DERIVATIVES (1974—1986)

Carbohydrates analyzed	Stationary phase	Supporting material	Column size	Column temp. (°C)	Carrier gas	Flow rate (mℓ/min)	Analysis time (min)	Method for detection	Magnitude of separation	Ref.
Neutral Saccharides										
Aldoses (ribose, galactose, glucose, mannose)	15% Carbowax® 20 *M*	Chromosorb® W (HMDS — treated)	12 ft, 0.25 in. I.D.	170	Helium	100	80	FID	Complete	25
Aldoses (fucose, galactose, glucose, mannose)	Pentasil	Stainless steel capillary	1000 ft, 0.03 in. I.D.	170	Helium	10	45	FID	Complete	67
Aldoses (arabinose, ribose, xylose, 2-deoxy-ribose) as O-methyloximes	Methyl silicone	Glass capillary	25 m	120→260 at 30/min for initial 2 min, followed by 11/min	Helium	45	10	FID	Complete	61
Aldoses (arabinose, galactose, glucose, mannose) treated with 2-aminoethanethiol	3% OV-1	Gas Chrom Q	1 m, 3 mm I.D.	210	Nitrogen	55	8	FID	Partial	64
	3% OV-17	Chromosorb® W	2 m, 3 mm I.D.	195	Nitrogen	55	10	FPD (for sulfur)	Partial	64
Enantiomers of aldoses (D,L-arabinose, D,L-xylose, D,L-galactose, D,L-glucose, D,L-mannose, D,L-fucose, D,L-rhamnose) as (−)-2-butyl glycosides	SE-30	Glass capillary	25 m, 0.31 mm I.D.	135→200 at 1/min	Nitrogen	1	50	FID	Complete	33
Reducing monosaccharides (arabinose, lyxose, ribose, xylose, galactose, glucose, mannose, rhamnose, fructose, sorbose, and tagatose)	OV-101	Glass capillary	16m, 0.17 mm I.D.	160	Nitrogen		50	FID	Complete	69
	XE-60	Glass capillary	28 m, 0.17mm I.D.	160	Nitrogen		50	FID	Complete	69

Table 2 (continued)
CONDITIONS USED FOR GAS CHROMATOGRAPHY OF TMS ETHERS OF CARBOHYDRATES AND THEIR DERIVATIVES (1974—1986)

Carbohydrates analyzed	Stationary phase	Supporting material	Column size	Column temp. (°C)	Carrier gas	Flow rate (mℓ/min)	Analysis time (min)	Method for detection	Magnitude of separation	Ref.
Neutral Saccharides (continued)										
Reducing monosaccharides (galactose, glucose, and fructose) as *O*-methyloximes	SP-2100	Fused-silica capillary	50 m, 0.2 mm I.D.	180	Helium	1	25	EI-MS (selective ion monitoring)	Complete	62
Reducing monosaccharides (glucose, fructose) as oximes and a nonreducing disaccharide (sucrose)	SP-2250	Supelcoport (80—100 mesh)	6 ft, 0.125 in. I.D.	170→300 at 10/min	Helium	30	10	FID	Complete	46
Reducing mono- and disaccharides (glucose, fructose, lactose) as oximes and a nonreducing disaccharide (sucrose)	3% SP-2250	Supelcoport (80—100 mesh)	6 ft, 0.125 in. O.D.	170→300 at 10/min	Helium	30	15	FID	Complete	30
Reducing monosaccharides (glucose, mannose, fructose, 1-deoxyglucose), alditols (ribitol, glucitol, mannitol) and a cyclitol (inositol)	OV-101	Glass capillary	20 m, 0.25 mm I.D.	160→200 at 0.5/min	Nitrogen	0.25	60	FID EI-MS	Complete	60
3 Aldoses, 2 ketoses, 17 alditols, and 5 inositols	OV-101	Glass capillary	30 m, 0.25 mm I.D.	120→260 at 3/min	Nitrogen		50	EI-MS	Complete	59
Cyclitols (D-inositol, L-inositol, *epi*-inositol, *muco*-inositol, *neo*-inositol, *scyllo*-inositol) and a	3% SE-30 15% PEGS	Gas Chrom P (AW-DMCS, 80—100 mesh)	6 ft, 0.125 in. I.D.	175	Argon	15—18 psi	27	FID	Partial	57

derivative (*myo*-inosone)		Gas Chrom P (AW-DMCS, 80—100 mesh)	6 ft, 0.125 in. I.D.	150	Argon	15—18 psi	30	FID	Partial	57
Reducing monosaccharides (glucose, mannose, fructose), alditols (arabinitol, ribitol, glucitol, 1,5-anhydroglucitol) and a cyclitol (*myo*-inositol)	SE-52	Glass capillary	50 m, 0.25 mm I.D.	140→240 at 2/min	Nitrogen		70	FID	Complete	58
Reducing monosaccharides (arabinose, galacotse, glucose, rhamnose, 2-deoxy-glucose, fructose) as *O*-methyloximes and an alditol (glucitol)	OV-1	Glass capillary	50 m, 0.3 mm I.D.	175	Argon	0.7	45	FID	Complete	31
Partially methylated aldoses as diethyldithioacetals	SF-96	SCOT capillary	50 m, 0.28 mm I.D.	225	Nitrogen	2	40	FID	Complete	66
Aldoses (galactose, glucose, fucose), deaminated amino sugars (2,5-anhydromannitol from glucosamine, 2,5-anhydrotalitol from galactosamine, deaminated neuraminic acid), an alditol (mannitol) and a cyclitol (*myo*-inositol)	OV-101	Glass capillary	25 m, 0.25 mm I.D.	120→250 at 4/min			30	FID	Complete	36
	2.2% SE-30	Gas Chrom Q (100—120 mesh)	2 m, 2 mm I.D.	120→250 at 4/min			30	EI-MS	Partial	36
Aldoses (ribose, glucose), a ketose (fructose), an alditol (erythritol), a cyclitol (inositol), and oligosaccharides (sucrose, raffinose, stachyose)	3% Dexsil® 300 GC	Chromosorb® W (AW-DMCS, 80—100 mesh)	6 ft, 2 mm I.D.	160→350 at 10/min for initial 8 min, followed by 30/min	Nitrogen	20	23	FID	Complete	20
Aldoses (erythrose, arabinose, lyxose, ribose, xy-	Cross-linked silicone	Fused-silica capillary	25 m, 0.2 mm I.D.	120→295 at 3/min for initial 45	Helium	0.6	50	FID	Complete	51

Table 2 (continued)
CONDITIONS USED FOR GAS CHROMATOGRAPHY OF TMS ETHERS OF CARBOHYDRATES AND THEIR DERIVATIVES (1974—1986)

Carbohydrates analyzed	Stationary phase	Supporting material	Column size	Column temp. (°C)	Carrier gas	Flow rate (mℓ/min)	Analysis time (min)	Method for detection	Magnitude of separation	Ref.
Neutral Saccharides (continued)										
lose, allose, galactose, glucose, gulose, mannose, talose, fucose, rhamnose), ketoses (ribulose, fructose, sorbose, tagatose, glucoheptulose, mannoheptulose), alditols (erythritol, threitol, arabinitol, ribitol, xylitol, galactitol, glucitol, mannitol, perseitol), a cyclitol (*myo*-inositol), and oligosaccharides (maltose, cellobiose, melibiose, trehalose, sucrose)				min, followed by 10/min						
Disaccharides (maltose, cellobiose, lactose, melibiose, gentiobiose, trehal-	10% OV-17	Gas Chrom Q (80—100 mesh)	2.7 m, 6 mm O.D.	255	Argon	55	21	FID	Partial	50
ose, palatinose, sucrose, turanose, lactulose)	3% OV-1	Gas Chrom Q (80—100 mesh)	1.5 m, 6 mm O.D.	255	Argon	55	20	FID	Partial	50
Reducing disaccharides (kojibiose, nigerose, maltose, cellobiose, gentiobiose, lactose,	1.5% SE-52	Chromosorb® W (AW-DMCS, 60—80 mesh)	2 m, 3 mm I.D.	215	Nitrogen	40	29	FID	Partial	63
neolactose, melibiose,	1.5% OV-17	Shimalite W	2 m, 3	215	Nitrogen	40	32	FID	Partial	63

lactulose) as oximes and nonreducing disaccharides (trehalose, sucrose, turanose)		(80—100 mesh)	mm I.D.							
Oligosaccharides of raffinose family (sucrose, raffinose, manninotriose, stachyose, verbascose)	3% Dexsil® 300	Chromosorb® W (AW-DMCS, 80—100 mesh)	0.46 m, 3.2 mm I.D.	130→370 at 3/min	Helium	40	30	FID	Complete	49
Acidic saccharides										
Uronic acids and their lactones (galacturonic acid, glucuronic acid, gulonic acid, iduronic acid, mannuronic acid)	10% SE-30	Celite® (100—120 mesh)	1.52 m, 4 mm I.D.	200	Nitrogen	40		FID	Partial	37
	4% XE-60	Celite® (100 — 120 mesh)	1.52 m, 4 mm I.D.	175	Nitrogen	40		FID	Partial	37
Hexuronic acids (glucuronic acid, iduronic acid) as methyl glycoside methyl esters	3.5% SE-30	Gas Chrom Q (80—100 mesh)	2 m, 4 mm I.D.	140→200 at 1/min	Nitrogen	30	30	FID	Partial	39
Aldonic acids (erythronic acid, ribonic acid) and an uronic acid (galacturonic acid)	10% OV-17	Gas Chrom Q (80—100 mesh)	3 m, 2 mm I.D.	120 at 1/min	Nitrogen	40	60	FID	Partial	65
	3% OV-225	Veraport 30 (100—120 mesh)	2 m, 2 mm I.D.	150 at 2/min	Nitrogen	40	60	FID	Partial	65
Acidic monosaccharides (5 uronic acids, 7 glycaric acids and their methyl esters)	2.5% SE-52	Chromosorb® G (AW-DMCS, 80—100 mesh)	2 m, 3 mm I.D.	190	Nitrogen	30	25	FID	Partial	38
	1% SE-30	Chromosorb® G (AW-DMCS, 80—100 mesh)	2 m, 3 mm I.D.	170	Nitrogen	30	25	FID	Partial	38
12 Sialic acids	3% OV-17	Gas Chrom Q (100—120 mesh)	1.6 m, 2 mm I.D.	196	Nitrogen	50	30	FID	Partial	40

Table 2 (continued)
CONDITIONS USED FOR GAS CHROMATOGRAPHY OF TMS ETHERS OF CARBOHYDRATES AND THEIR DERIVATIVES (1974—1986)

Carbohydrates analyzed	Stationary phase	Supporting material	Column size	Column temp. (°C)	Carrier gas	Flow rate (mℓ/min)	Analysis time (min)	Method for detection	Magnitude of separation	Ref.
Acidic saccharides (continued)										
	1% OV-1	Anachrom S (100—120 mesh)	1.6 m, 2 mm I.D.	206	Nitrogen	50	30	FID	Partial	40
N-Acetyl-, *N*,*O*-diacetyl and *N*-glycolylmannosamines derived from *N*-acetyl, *N*,*O*-diacetyl-, and *N*-glycolylneuraminic acids, respectively, as diethyldithioacetals	2% OV-1	Chromosorb® W (AW-DMCS, 80—100 mesh)	1 m, 3 mm I.D.	190	Nitrogen	30	50	FID	Complete	43
Basic saccharides including *N*-acyl derivatives										
Amino sugars (12 glucosamine and galactosamine derivatives)	2.2% SE-30	Gas Chrom S (100—120 mesh)	2 m, 3 mm I.D.	187	Argon	15		FID EI-MS	Partial	56
	3% QF-1	Gas Chrom Q (80—100 mesh)	2 m, 3 mm I.D.	140, 200	Argon	15			Partial	56
Amino sugars and related compounds (galactosamine, glucosamine, mannosamine, methyl *N*-acetylgalactosaminide, methyl *N*-acetylglucosaminide, methyl *N*-acetyl-	3% SE-30	Chromosorb® W HP (80—100 mesh)	6 ft, 2 mm I.D.	140 for 30 min, followed by temperature rise at 0.5/min to 200	Nitrogen	30	38	FID	Complete	35
				180	Nitrogen	30	8	EI-MS		
				243	Nitrogen	30	14			
				245	Nitrogen	30	11			

mannosaminide, chitobiose, Glc-NAcl→6GlcNAc, N-acetyl-glucosaminylasparagine)										
21 Amino sugars and related compounds	3% SP-2100	Supelcoport (80—100 mesh)	6 ft, 2 mm I.D.	180	Nitrogen	40	20	FID	Partial	18
	3% Poly A-103	Gas Chrom Q (100—120 mesh)	6 ft, 2 mm I.D.	200, 210	Nitrogen	40	20	EI-MS	Partial	18
N-Acetylhexosamines as O-methyloximes	3% OV-225	Supelcoport	1.86 m, 3 mm I.D.	170	Helium	30	25	FID	Partial	68
				Miscellaneous						
Glucose, fructose, pinitol, *myo*-inositol, sucrose, shikimic acid and quinic acid	3% OV-1	Chromosorb® W	1.5 m, 4 mm I.D.	145→250 at 4/min	Nitrogen	25	30	FID	Partial	45
Xylose, galactose, glucose, rhamnose, perseitol, *N*-acetylglucosamine, and *N*-acetylmuramic acid	3% SE-30	Chromosorb® W HP (100—200 mesh)	1.83 m, 4 mm I.D.	80→250 at 2/min	Nitrogen	25		FID	Complete	48
19 Aldoses, 6 deoxy sugars, 12 alditols, 15 aldonolactones, 13 methyl glycosides, 6 hexosamines, *N*-acetylneuraminic acid, and 14 oligosaccharides	3% SE-52	Chromosorb® W (AW-silanized, 80—100 mesh)	6 ft, 0.25 in. O.D.	140 125→250 at 2.3/min	Argon	75—150	50—130	FID	Partial	1
	15 % PEGS polyester	Chromosorb® W (80—100 mesh)	8 ft, 0.25 in. O.D.	140, 150, 170,	Argon	75—150			Partial	1
4 Pentoses, 2 *N*-acetylpentosamines, 7 *N*-acetylhexosamines, 2 *N*-acetylhexosaminitols, 2 hexuronic acids, 6 methyl ester of hexuronic acids,	3—3.8% SE-30 or UC-W98	Diatoport S (80—100 mesh)	2.5 m, 3.2 mm I.D.	140→200 at 0.5/min 220 300 350	Nitrogen	50	20—120	FID	Complete	27

Table 2 (continued)
CONDITIONS USED FOR GAS CHROMATOGRAPHY OF TMS ETHERS OF CARBOHYDRATES AND THEIR DERIVATIVES (1974—1986)

Carbohydrates analyzed	Stationary phase	Supporting material	Column size	Column temp. (°C)	Carrier gas	Flow rate (mℓ/min)	Analysis time (min)	Method for detection	Magnitude of separation	Ref.
				Miscellaneous (continued)						
13 disaccharides, 4 trisaccharides, 4 tetrasaccharides, and 3 pentasaccharides										
Reducing monosaccharides (arabinose, ribose, xylose, galactose, glucose, mannose, fucose, rhamnose, glucuronic acid, *N*-acetylgalactosamine, *N*-acetylglucosamine, *N*-acetylmannosamine, *N*-acetylneuraminic acid) as *O*-methyl glycosides	SE-30	Glass capillary	30 m, 0.3 mm I.D.	150→350 at 5/min	Helium	2	26	FID	Complete	32
Reducing monosaccharides (xylose, galactose, glucose, mannose, fucose, *N*-acetylgalactosamine, *N*-acetylglucosamine, *N*-acetylneuraminic acid, 9-*O*-acetyl-*N*-acetylneuraminic acid) as *O*-methyl glycosides	SE-30	Chromosorb® W (AW-DMCS, 80—100 mesh)	2 m, 4 mm I.D.	135→200 at 0.5/min	Nitrogen	40	120	FID	Complete	41
Reducing monosaccharides (xylose, galactose, mannose, fucose, *N*-acetylgalactosamine,	CPsil5 WCOT	Fused silica capillary	25 m, 0.32 mm I.D.	130→220 at 2/min	Nitrogen	35	50	FID	Complete	42

N-acetylglucosamine, *N*-acetylneuraminic acid, *N,O*-diacetylneuraminic acid) as methyl glycosides or methyl ester methyl glycosides										
Reducing monosaccharides (glyceraldehyde, erythrose, arabinose, ribose, xylose, galactose, glucose, mannose, fucose, rhamnose, galacturonic acid, glucuronic acid, *N*-acetylgalactosamine, *N*-acetylglucosamine) as diethyldithioacetals	SF-96	SCOT capillary	50 m, 0.28 mm I.D.	225	Nitrogen	1.5	60	FID	Complete	22
Reducing monosaccharides (glyceraldehyde, erythrose, arabinose, ribose, xylose, galactose, glucose, mannose, galacturonic acid, glucuronic acid, *N*-acetyl galactosamine, *N*-acetylglucosamine, 2,5-anhydromannose derived from glucosamine, 2,5-anhydrotalose derived from galactosamine) as diethyldithioacetals	SF-96	SCOT capillary	50 m, 0.28 mm I.D.	225	Nitrogen	2	80	FID	Complete	23
Reducing monosaccharides (ribose, deoxyribose, glucose, deoxyglucose, fucose, fructose, glucuronic acid, N-acetylgalactosamine, *N*-acetylglucosamine) as	3% SP-2250	Supelcoport (100—120 mesh)	6 ft, 0.125 in. O.D.	120→220 at 5/min	Nitrogen	18	20	FID	Complete	29

Table 2 (continued)
CONDITIONS USED FOR GAS CHROMATOGRAPHY OF TMS ETHERS OF CARBOHYDRATES AND THEIR DERIVATIVES (1974—1986)

Carbohydrates analyzed	Stationary phase	Supporting material	Column size	Column temp. (°C)	Carrier gas	Flow rate (mℓ/min)	Analysis time (min)	Method for detection	Magnitude of separation	Ref.
Miscellaneous (continued)										
O-methyloximes, alditols (*N*-acetylgalactosaminitol, *N*-acetylglucosaminitol) and a cyclitol (*myo*-inositol)										

Note: FID: flame ionization detection; FPD: flame photometric detection; EI-MS: electron-impact mass spectometry.

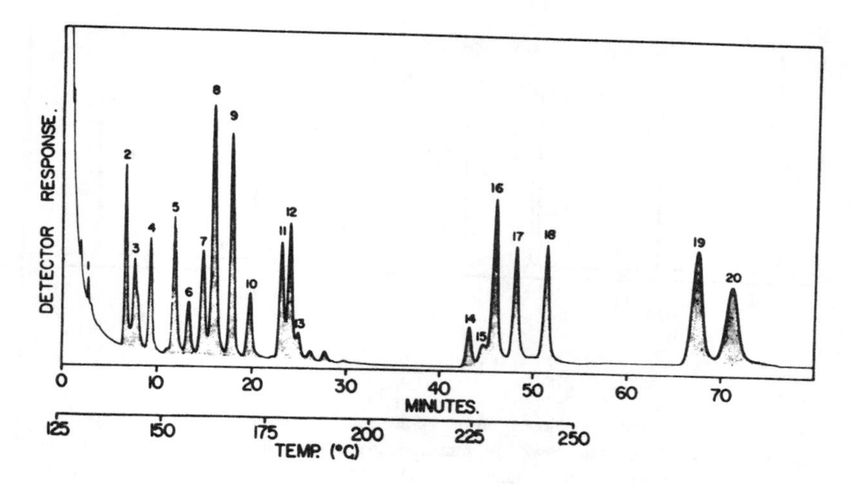

FIGURE 1. Linearly temperature-programmed separation of trimethylsilyl ethers of various carbohydrates. Peak assignment: 1, erythrose; 2, β-arabinose; 3, ribose; 4, α-xylose; 5, methyl α-mannoside; 6, α-glucose; 7, α-galactose; 8, α-glucose; 9, ascorbic acid; 10, β-glucose; 11, *N*-acetylglucosamine; 12, *N*-acetylgalactosamine; 13, D-*glycero*-D-*gluco*-heptonolactone; 14, sucrose; 15, α-maltose; 16, β-maltose; 17, β-cellobiose; 18, gentiobiose; 19, raffinose; 20, melezitose. Chromatographic conditions: column, 3% silicone SE-52 (8 ft, 0.25 in O.D.); column temperature, 125→250°C at 2.3 °C/min; carrier, argon (75 mℓ/min). (Reprinted with permission from Sweeley, C. C., Bentley, R., Makita, M., and Wells, W. W., *J. Am. Chem. Soc.*, 85, 2497, 1963. Copyright 1963 American Chemical Society.)

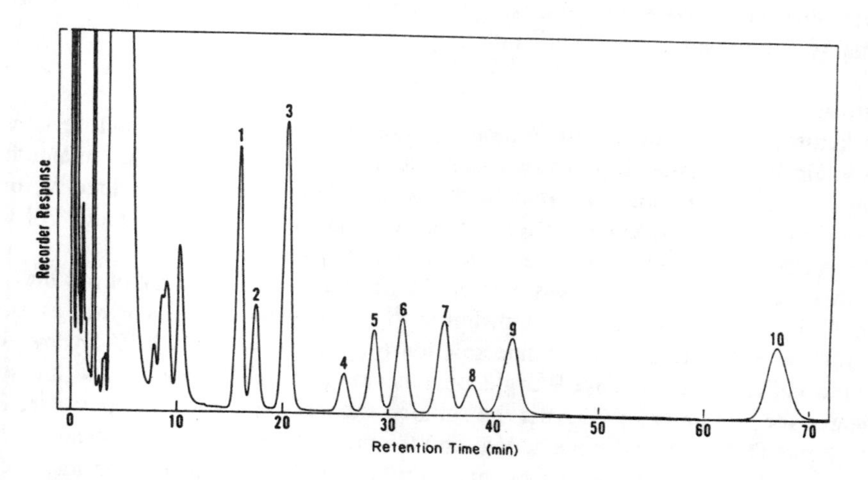

FIGURE 2. Separation of trimethylsilyl ethers of various carbohydrates. Peak assignment: 1,2, ribose; 3, α-mannose; 4, γ-galactose; 5, α-galactose; 6, methyl α-mannopyranoside (internal standard); 7, α-glucose; 8, β-mannose; 9, β-galactose; 10, β-glucose. Chromatographic conditions: column, 15% Carbowax® 20*M* on Chromosorb W (treated with HMDS, 12 ft, 0.25 in I.D.); column temperature, 170°C; carrier, helium (100 mℓ/min). (Reprinted with permission from Sawardeker, J. S. and Sloneker, J. H., *Anal. Chem.*, 37, 945, 1965. Copyright 1965 American Chemical Society.)

3. *Analysis as Trimethylsilylated Oximes or O-Alkyloximes*

TMS derivatives of aldose oximes, as well as ketose oximes, are often utilized for analysis of carbohydrates in food stuffs. The use of derivatives of oximes and the derivatization procedure were first described by Sweeley et al.,[1] followed by Brobst and Lott.[28] Later this method was extended to the *O*-alkyloxime methods. These methods have, however, a draw-

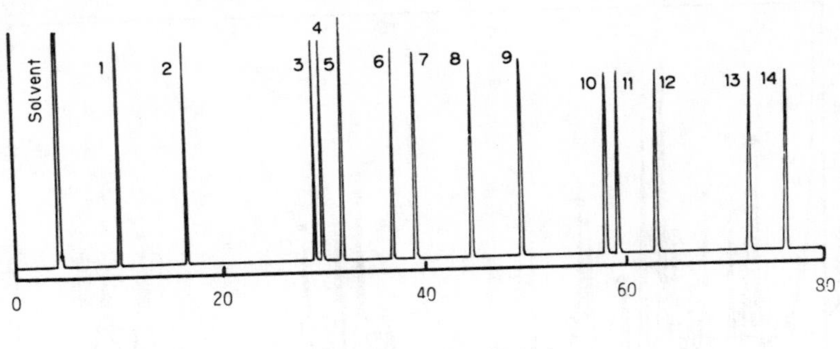

FIGURE 3. Separation of aldoses, uronic acids, and *N*-acetylhexosamines as trimethylsilylated diethyldithioacetals. Peak assignment: 1, glyceraldehyde; 2, erythrose; 3, xylose; 4, arabinose; 5, inositol (internal standard); 6, rhamnose; 7, fucose; 8, glucuronic acid; 9, galacturonic acid; 10, glucose; 11, mannose; 12, galactose; 13, *N*-acetylglucosamine; 14, N-acetylgalactosamine. Chromatographic conditions: column, a SCOT capillary column (50 m, 0.28 mm I.D.) coated with silicone SF-96; column temperature, 225°C; carrier, nitrogen (1.5 mℓ/min). (From Honda, S., Yamauchi, N., and Kakehi, K., *J. Chromatogr.*, 169, 287, 1979. With permission.)

back — the products have the *syn* and *anti* isomers. Mahwhinney[29] pointed out this deficiency by demonstrating separation of these isomers, but stated that these methods were still useful for class separation of amino sugar, aldohexose, and 6-deoxyaldohexose. An example of the separation of trimethylsilylated aldose *O*-methyloximes is shown in Figure 4. It is observed that each ribose and glucose gave a minor peak, together with a major one, presumably due to isomers.

B. Ketose

The ketose, which is important in food analysis, is fructose. When fructose is reduced with sodium borohydride, it gives two isomeric alditols, i.e., glucitol and mannitol. For this reason fructose is not converted to the alditol derivatives, but is directly, or after oximation as well as *O*-alkyloximation, trimethylsilylated. A typical example[30] of the derivatization procedure to the oxime derivatives is given below.

Granola cereal (0.1 to 0.5 g) was extracted with aqueous 80% methanol (40 mℓ), and a portion (0.5 mℓ) of the extract was transferred to a vial. After the solvent was evaporated, the residue was dissolved in a pyridine solution (0.5 mℓ) containing 2.5% hydroxylamine hydrochloride and 0.2% phenyl β-D-glucopyranoside (internal standard), and the solution was heated for 30 min at 75°C. After the solution was cooled to room temperature, HMDS (0.5 mℓ) and TFA (4 drops) were added, and the mixture was allowed to stand for 30 min (silylation method according to Brobst and Lott[28]). An aliquot was injected into a column of 3% silicone SP-2250 and analyzed with temperature programming. The chromatogram in Figure 5 obtained by this procedure indicates good separation of the fructose derivative from the derivatives of accompanying sugars (glucose, sucrose, and lactose).

Zegota[31] used a capillary column coated with silicone OV-101 to obtain excellent separation, as shown in Figure 6, but fructose, as well as rhamnose, galactose, and glucose, gave two peaks due to the *syn*/*anti* isomers.

C. Glycoside

In analysis of carbohydrates, glycosides are important as alcoholysis products of glycoconjugates. Although each glycoside has ring-structural and/or configurational isomers, it is relatively stable in an alcoholysis mixture, compared to the parent monosaccharide in a

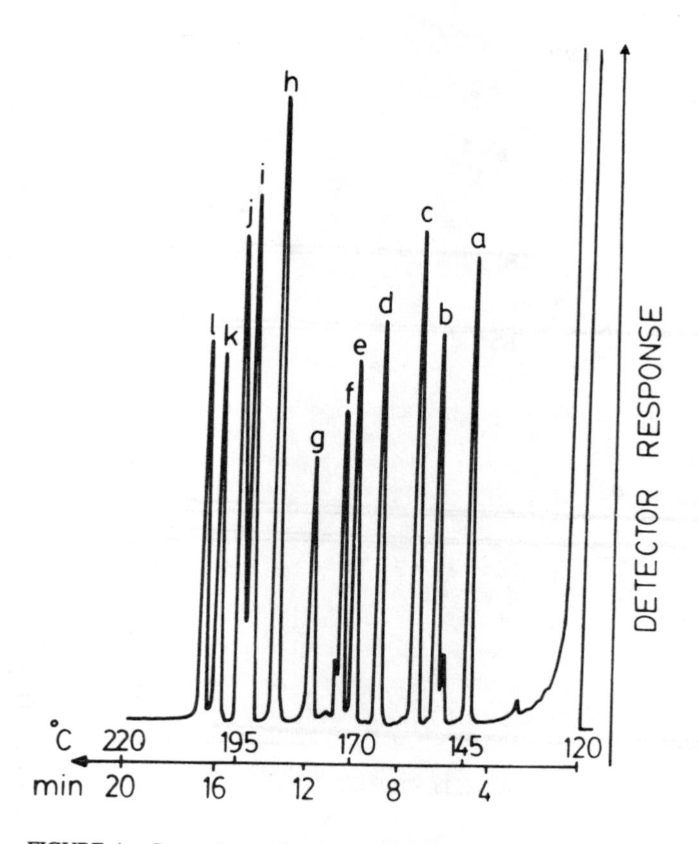

FIGURE 4. Separation of the trimethylsilylated *O*-methyloximes of: a, 2-deoxyribose; b, ribose; c, fucose; d, 2-deoxyglucose; e, fructose; f, glucose; g, glucuronic acid; i, *N*-acetylglucosaminitol; j, *N*-acetylgalactosaminitol; k, *N*-acetylglucosamine; l, *N*-acetylgalactosamine; and h, TMS *myo*-inositol (internal standard). Unlabeled small peaks at the base of the (b) ribose and (f) glucose peaks represent the resolved *syn* and *anti* isomers. Chromatographic conditions: column, a glass column (6 ft, 0.125 in I.D.) packed with 3% silicone SP-2250 on Supelcoport (100 to 120 mesh); column temperature, 120→220°C at 5°C/min; carrier, nitrogen (18 mℓ/min); sample amount, 1 μg each. (From Mahwhinney, T. P., *J. Chromatogr.*, 351, 91, 1986. With permission.)

hydrolysis mixture. With regard to methanolysis, Bhatti et al.[27] reported details on experimental conditions. Figure 7 shows a chromatogram for a standard mixture of trimethylsilylated aldoses, N-acetylhexosamines, and *N*-acetylneuraminic acid, separated on a column of silicone SE-30 in the gradient temperature mode.

The following procedure is a methanolysis procedure described by Bombardelli et al.[32] for the determination of the monosaccharide composition of gastric proteoglycan.

To a lyophilized sample (5 to 10 mg) of a mucin obtained from the rat stomach, contained in a screw-capped vial, 0.5 *M* hydrogen chloride in anhydrous methanol (0.5 mℓ) was added. The vial was flushed with nitrogen and heated for 20 hr at 50°C with the stopper tightly closed. The reaction solution was cooled, and anhydrous pyridine (150 μℓ) and acetic anhydride (100 μℓ) were added. The mixture was allowed to stand for 30 min, then evaporated to dryness by flushing with nitrogen, followed by placing the vial in a desiccator under reduced pressure. To the residue was added a mixture (0.5 mℓ) of BSTFA-TMSI-pyridine (10:1:0.4, by volume) and a mixture of authentic hydrocarbons as internal standards. After shaking the mixture, followed by letting it stand for a few minutes, a 2 μℓ-sample

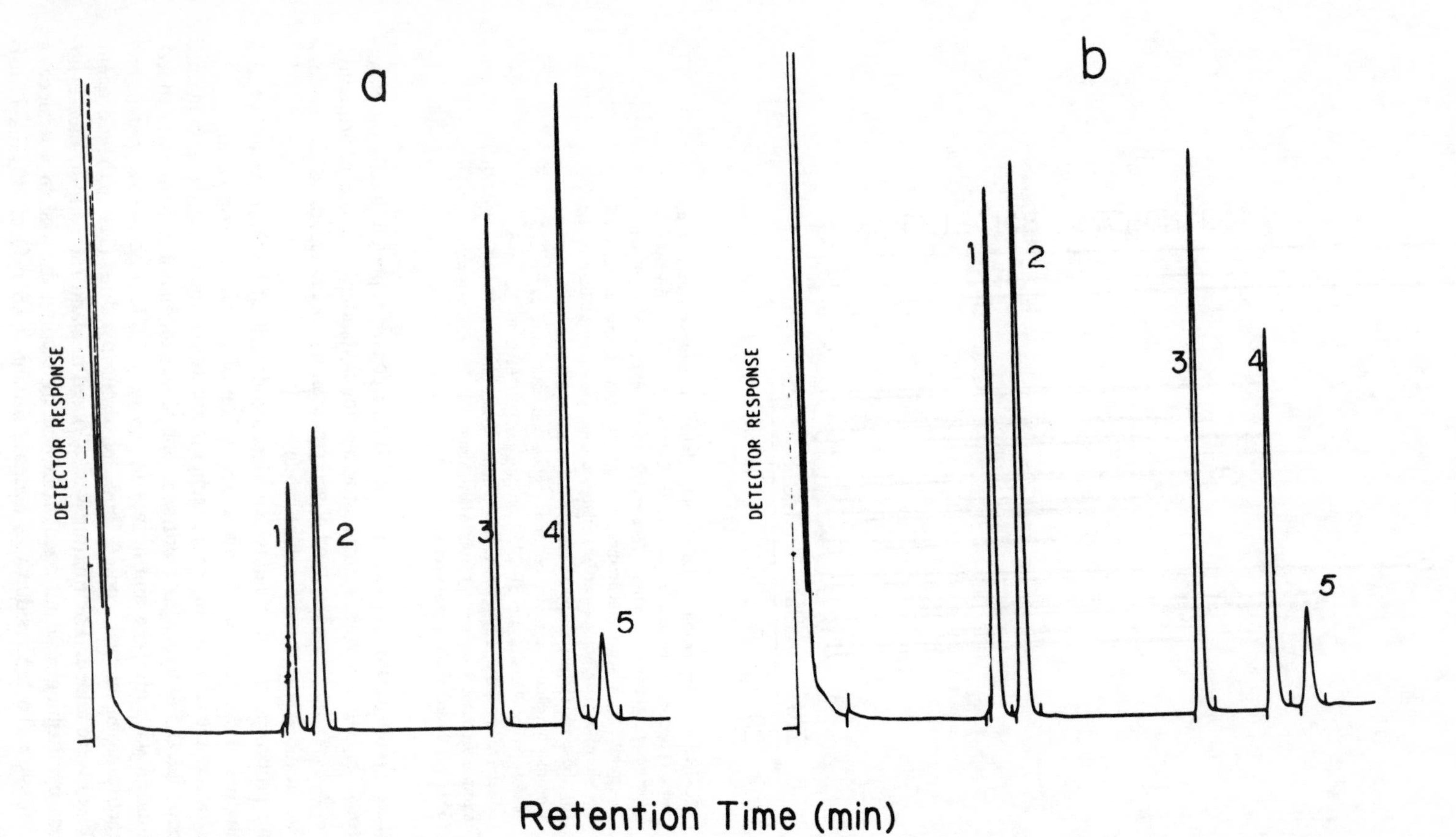

FIGURE 5. Separation of the trimethylsilylated oximes of reducing carbohydrates and trimethylsilyl ethers of nonreducing carbohydrates in (a) a methanol extract and (b) a water extract of a granola cereal. Peak assignment: 1, fructose; 2, glucose; 3, phenyl β-glucopyranoside (internal standard); 4, sucrose; 5, lactose. Chromatographic conditions: column, a stainless steel column (6 ft, 0.125 in I.D.) packed with 3% silicone SP-2250 on Supelcoport (80 to 100 mesh); column temperature, 170→300°C at 10°C/min; carrier, helium (30 mℓ/min). (Reprinted from *Journal of Food Science* 1981. 46:425-427, Copyright © by Institute of Food Technologists.)

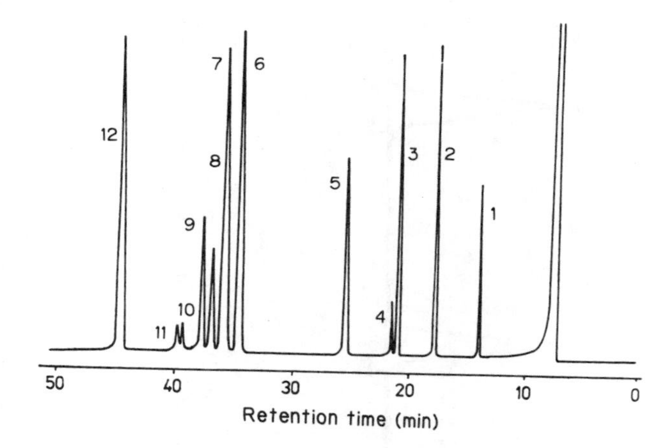

FIGURE 6. Separation of trimethylsilylated *O*-methyloximes of monosaccharides. Peak assignment: 1, 2-deoxyribose; 2, arabinose; 3,4, rhamnose; 5, 2-deoxyglucose; 6,7, fructose; 8,10, galactose; 9,11, glucose; 12, glucitol. Chromatographic conditions: column, a glass capillary column (50 m, 0.3 mm I.D.) coated with silicone OV-1; column temperature, 175°C; carrier, argon (0.7 mℓ/min). (From Zegota, H., *J. Chromatogr.*, 192, 446, 1980. With permission.)

was injected on the GC column. Use of a capillary column (Figure 8) greatly improved separation of TMS derivatives of methyl glycosides, as compared to that on a packed column (Figure 7).

It is notable that D- and L-enantiomers of a monosaccharide could be resolved by coupling them with an optically active alcohol. The following paragraph describes a procedure developed by Gerwig et al.[33]

A sample (0.5 mg) of a monosaccharide was dissolved in (−)-2-butanol containing 1 *M* hydrogen chloride (0.5 mℓ). The solution was flushed with nitrogen and heated for 6 hr at 60°C. The reaction solution was treated with an excess amount of silver carbonate and the mixture was centrifuged. The supernatant was evaporated to dryness at 45°C, and the residue was dried in a desiccator containing phosphorus pentoxide. The resultant syrup was submitted to TMS with a 1:1:5 (by volume) mixture (0.1 mℓ) of TMCS-HMDS-pyridine for 30 min at room temperature. Figure 9 shows the separation of D/L-enantiomers.

Most aldoses gave plural peaks of ring-structural and/or configurational isomers, and some of them overlapped each other. Therefore, simultaneous determination of enantiomers was possible for simple mixtures of aldoses such as the D-fucose-L-mannose-D-glactose and the L-fucose-D-mannose-L-galactose systems. However, when a number of aldoses were present, accurate determination of enantiomers became difficult due to overlap of peaks, though identification of peaks was possible.

D. Amino Sugar

Amino sugars are distributed in nature as constituents of glycoconjugates and exist in *N*-acetylated forms. Therefore, in determination of the monosaccharide composition of glycoconjugate sample it is hydrolyzed with acid (Chapter 3). Under such condition, the *N*-acetyl groups are removed and the component amino sugar(s) are liberated. Hurst[34] studied TMS of amino sugars and found that *N*-trimethylsilylation is incomplete by the method of Sweeley et al.,[1] which uses a mixture of TMCS and HMDS. For this reason, the free amino sugars are usually re-*N*-acetylated before TMS. Coduti and Bush[35] reported a result of separation of *N*-acetylgalactosamine, *N*-acetylglucosamine, methyl *N*-acetylgalactosaminide,

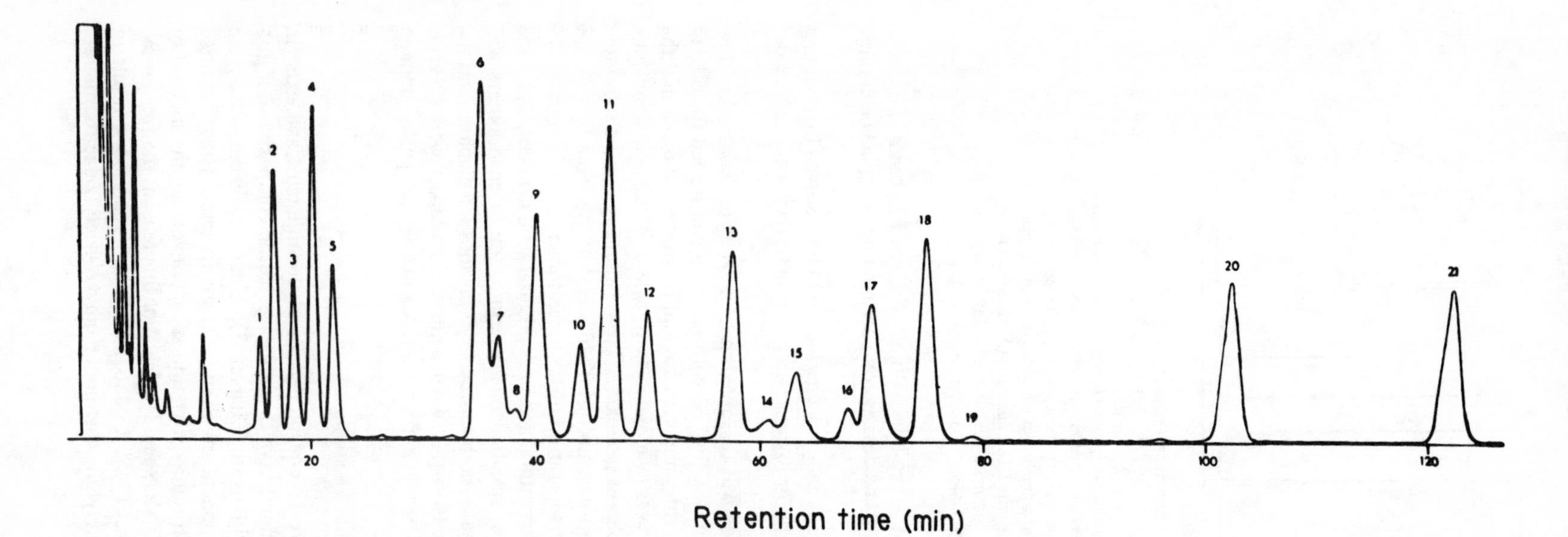

FIGURE 7. Separation of a standard mixture of monosaccharides as the trimethylsilylated methyl glycosides. Peak assignment: 1,2,3, fucose; 4,5, xylose; 6,8, mannose; 7,9,10, galactose; 11,12, glucose; 13, mannitol (internal standard); 14,16,18,19, *N*-acetylglucosamine; 15,17, *N*-acetylgalactosamine; 20, perseitol (internal standard); 21, *N*-acetylneuraminic acid. Chromatographic conditions: column, a glass column (2.5 m, 3.2 mm I.D.) packed with 3.8% silicone SE-30 on Diatoport S; column temperature, 140 → 200°C at 0.5°C/min; carrier, nitrogen (50 mℓ/min). (From Bhatti, T., Chambers, R. E., and Clamp, J. R., *Biochim. Biophys. Acta*, 222, 339, 1970. With permission.)

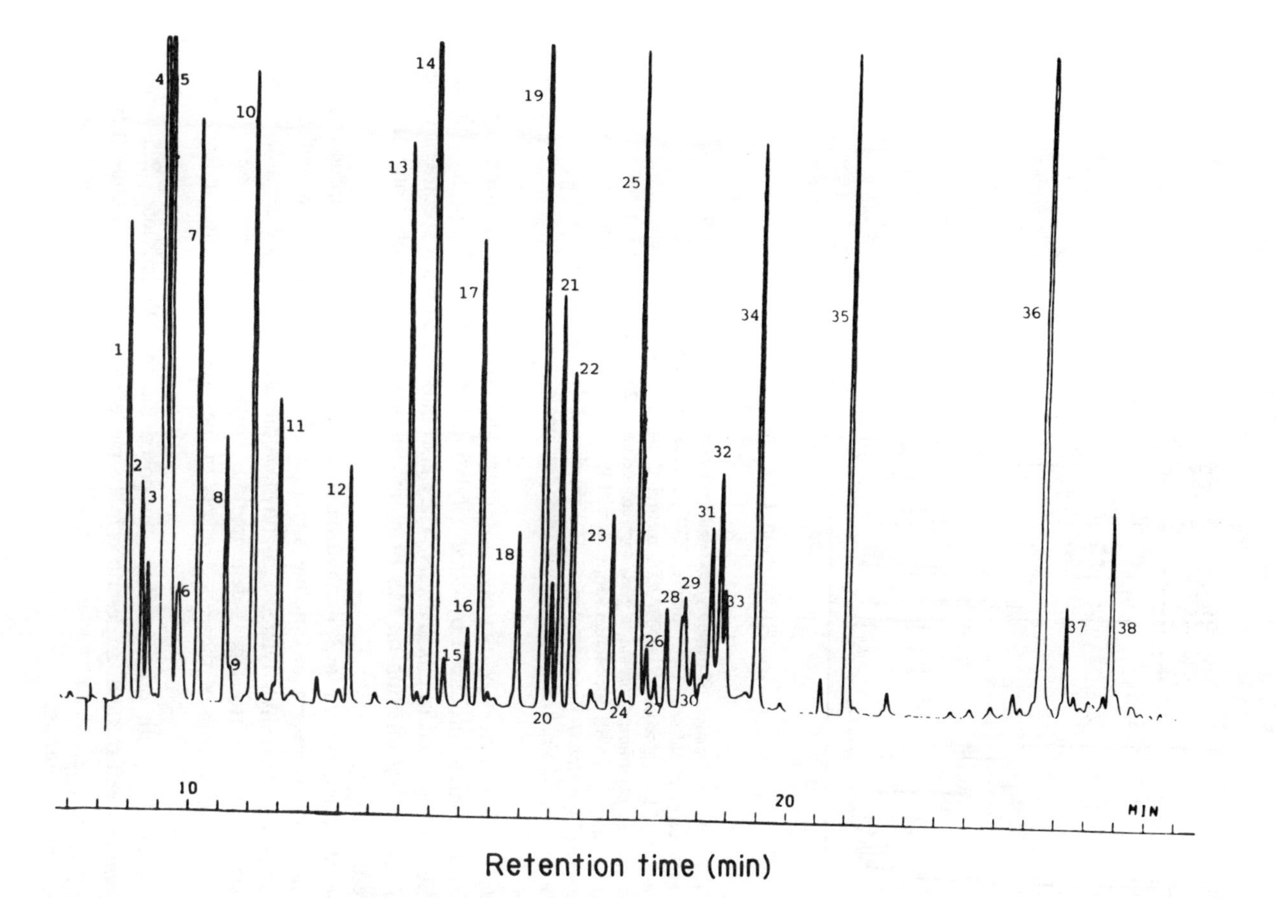

FIGURE 8. Separation of the trimethylsilylated methylglycosides of standard carbohydrates. Peak assignment: 1,2 arabinose; 3, ribose; 4, ribose, C_{16}; 5,6, rhamnose; 7,8,9, fucose; 10,11, xylose; 12,20,21, glucuronic acid; 13, C_{18}; 14,16,18, mannose; 15,17,18, galactose; 19, 22, glucose; 23,34,30, *N*-acetylmannosamine; 25, C_{20}; 26,27,28,29,32, *N*-acetylgalactosamine; 30,31,33,34, *N*-acetylglucosamine; 35, C_{22}; 36, C_{24}; 37,38, *N*-acetylneuraminic acid. Chromatographic conditions: column, a glass capillary column (30 m, 0.3 mm I.D.) coated with silicone SE-30; column temperature, 150→350°C at 5°C/min; carrier; helium (2.0 m /min). (From Bombardelli, E., Conti, M., Magistretti, M. J., and Martinelli, E. M., *J. Chromatogr.*, 279, 593, 1983. With permission.)

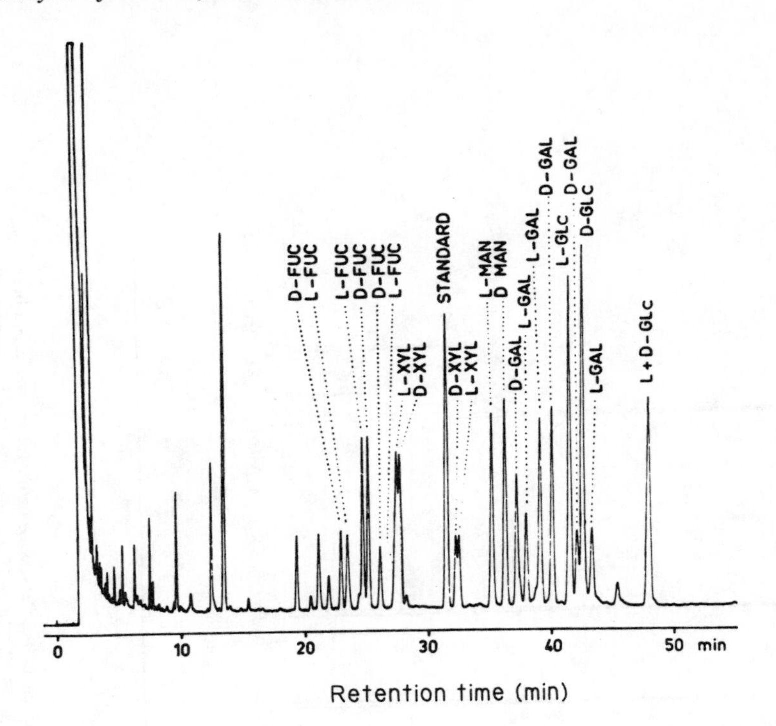

FIGURE 9. Separation of trimethylsilylated (−)-2-butyl glycosides of D- and L-enantiomers of aldoses. FUC, fucose; XYL, xylose; MAN, mannose; GAL, galactose; GLC, glucose; STANDARD, methyl β-D-galactopyranoside. Chromatographic conditions: column, a glass capillary column (25 m, 0.31 mm I.D.) coated with silicone SE-30; column temperature, 135→200°C at 1°C/min; carrier, nitrogen (1.0 mℓ/min). (From Gerwig, G. J., Kamerling, J. P., and Vliegenthart, J. F. G., *Carbohydr. Res.*, 62, 349, 1978. With permission.)

methyl *N*-acetylglucosaminide, *N,N'*-diacetylchitobiose, and *N*-acetylglucosaminyl asparagine as the TMS derivatives produced by direct silylation with TMCS-HMDS. A column of 3% silicone SE-30 was used in this analysis. They also interpreted mass spectra of these compounds. For further information on MS of amino sugars, the paper of Wong et al.[18] should be viewed.

Derivatization of amino sugars to 2,5-anhydroaldoses by treatment with nitrous acid is another useful method to prevent incomplete silylation. Mononen[36] devised a method to analyze amino sugars as trimethylsilylated 2,5-anhydroalditols. This method requires three consecutive reactions, i.e., nitrous acid deamination, borohydride reduction, and TMS. Figure 10 shows the separation of these derivatives in the presence of some aldoses and other sugars.

The authors,[23] together with other workers, proposed the trimethylsilylated 2,5-anhydroaldose diethyldithioacetal method, as described in the preceding paragraph (Section II.B.3). This method is also convenient and allows separation of the derivatives of galactosamine, glucosamine, and mannosamine.

E. Uronic Acid

Substitution of the hydroxymethyl group in a molecule of aldose by the carboxyl group gives rise to a uronic acid. In aqueous solutions, uronic acids may exist as an equilibrium mixture of the free acid and the intramolecular lactone, and the proportion of free acid and the lactone proportion are dependent on pH. Kennedy et al.[37] studied the separation of the

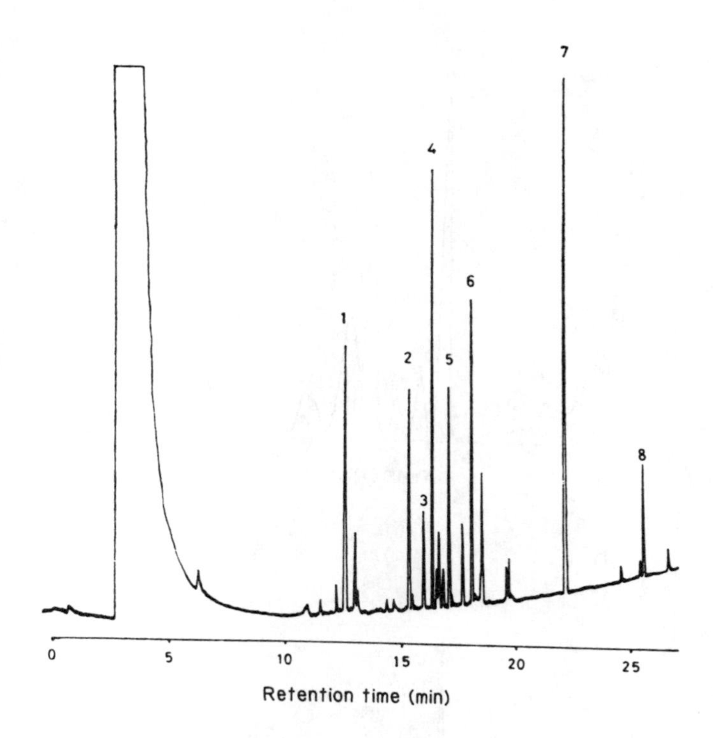

FIGURE 10. Separation of 2,5-anhydroalditols in the presence of other carbohydrates as trimethylsilyl ethers. Peak assignment: 1, fucose; 2, 2,5-anhydromannitol derived from glucosamine; 3, 2,5-anhydrotalitol derived from galactosamine; 4, mannitol; 5, galactose; 6, glucose; 7, *myo*-inositol; 8, deaminated sialic acid. Chromatographic conditions: column, a glass capillary column (25 m, 0.25 mm I.D.) coated with silicone OV-101; column temperature, 120→250°C at 4°C/min. (From Mononen, I., *Carbohydr. Res.*, 88, 39, 1981. With permission.)

free acids, sodium salts, and lactones of uronic acids as the TMS derivatives. Raunhardt et al.[38] also reported the results of similar experiments.

Galacturonic and mannuronic acids are important as constituents of plant mucilages, and glucuronic and iduronic acids are contained in animal proteoglycans. Since they are all in conjugated state, they should be released before analysis. Hydrolysis at elevated temperature causes partial decomposition of uronic acids. Methanolysis is preferable to hydrolysis because it reduces decomposition, but under the conditions for methanolysis, the carboxyl group is simultaneously converted to the methyl ester. Eventually uronic acids are analyzed as trimethylsilylated methyl glycosides of methyl esters. The following paragraph describes a detailed procedure for the analysis of uronic acids in proteoglycans.[39]

To a proteoglycan sample (200 μg) contained in a small vial was added an appropriate volume of a solution of mannuronolactone or mannitol (internal standard), and the mixture was evaporated to dryness under reduced pressure in a desiccator containing phosphorus pentoxide. The residue was dissolved in 1 *M* hydrogen chloride in methanol (0.5 mℓ), and the vial was sealed. The solution was heated for 20 hr at 100°C, then evaporated to dryness in a similar manner as that for sample preparation. The residue was dissolved in anhydrous pyridine (100 μℓ), and trimethylsilylated with TMCS (10 μℓ) and HMDS (20 μℓ) for 30 min at room temperature. The reaction mixture was centrifuged, and the supernatant was evaporated to dryness. The residue was dissolved in *n*-hexane (50 μℓ), and an aliquot was injected on the GC column. Figure 11 shows the result obtained from pig skin dermatan sulfate, together with that from a mixture of authentic uronic acids.

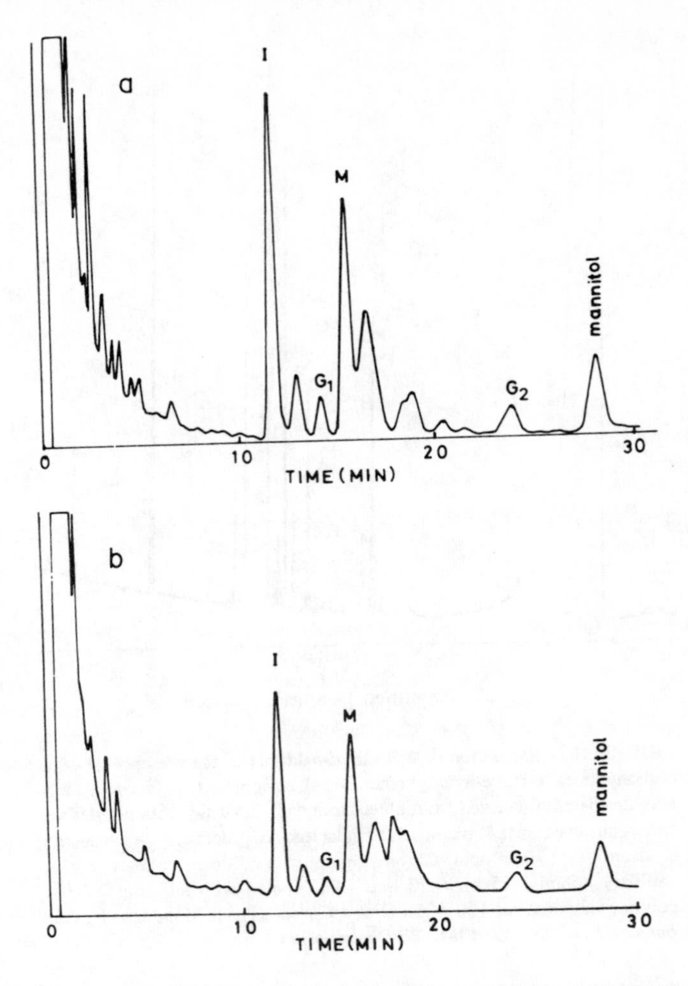

FIGURE 11. Separation of uronic acids as the trimethylsilylated methyl glycosides (of methyl esters or lactones): (a) standard mixture, (b) methanolysate of pig skin dermatan sulfate. I, iduronic acid; G_1 and G_2, glucuronic acid; M, mannuronic acid. Chromatographic conditions: column, a glass column (2 m, 4 mm I.D.) packed with 3.5% silicone SE-30 on Gas-Chrom Q (80 to 100 mesh); column temperature, 140→200°C at 1°C/min; carrier, nitrogen (30 mℓ/min). (From Inoue, S. and Miyawaki, M., *Anal. Biochem.*, 65, 164, 1975. With permission.)

The highest peak was assigned to a methyl glycoside of iduronic acid methyl ester, and two minor peaks were considered to have arisen from anomeric methyl glycosides of glucuronic acid methyl ester. Although qualitative analysis was possible, accurate determination of these uronic acids was difficult under these conditions, due to overlap on unidentified small peaks.

F. Sialic Acid

Sialic acids are acyl derivatives of neuraminic acid (5-amino-3,5-dideoxy-D-*glycero*-D-*galacto*-nonulosonic acid). All sialic acids are either *N*-acetylated or *N*-glycolylated, and some members have *O*-acetyl group(s) in addition to an *N*-acyl group. In nature, sialic acids are almost exclusively bound to other saccharides in glycoconjugates, hence necessitate hydrolytic liberation prior to GC analysis. Acid may be used to catalyze hydrolysis, but the

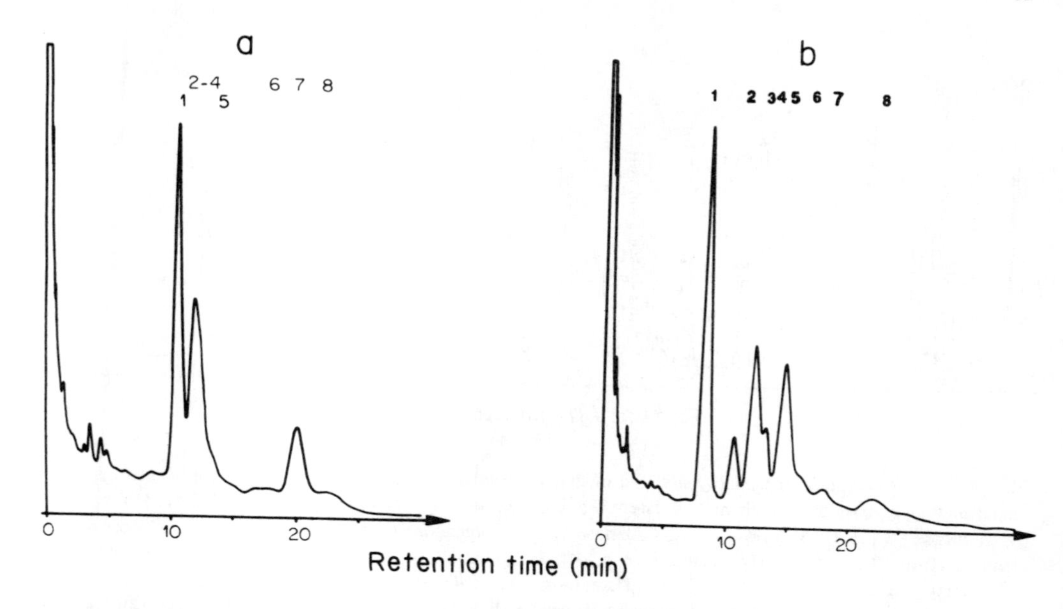

FIGURE 12. Separation of the sialic acids obtained from bovine submandibular gland glycoproteins as trimethylsilyl ethers. Peak assignment: 1, NANA; 2, NANA-9 (?)-OGl; 3, NANA-7 (?)-OGl; 4, NANA-7-OAc. 5, NGNA; 6, unknown sialic acid; 7, NANA-7,9-di-OAc; 8, NGNA-9-OAc, NGNA-7 (?)-OAc. Chromatographic conditions: column, (a) a glass column (2 m, 2 mm I.D.) packed with 1% silicone OV-1 on Anachrom S or (b) a glass column (2 m, 2 mm I.D.) packed with 3% silicone OV-17 on Gas Chrom Q; column temperature, (a) 206°C or (b) 196°C; carrier, nitrogen (40 mℓ/min). (From Casals-Stenzel, J., Buscher, H. P., and Schauer, R., *Anal. Biochem.*, 65, 507, 1975. With permission.)

liberated sialic acids are unstable to acid. In addition, some of the *O*-acetyl groups are removed during acid treatment. Therefore, the acid should be as dilute as possible, and the reaction temperature as low as possible. Heating for 1 hr at 80°C in 0.05 *M* sulfuric acid is the standard procedure. Even under these conditions some decomposition cannot be avoided. Liberated sialic acids can be analyzed as TMS derivatives. A typical procedure is as follows.[40]

A solution containing a sialic acid or a mixture of sialic acids (10 to 100 μg) was lyophilized in a small vial, and TMSI (25 μℓ) was added to the residue. After allowing the mixture to stand for 15 min, an aliquot was submitted to GC. A 1:2:10 (by volume) mixture of TMCS-HMDS-pyridine may be used, but the reaction time should be prolonged to more than 30 min. BSA, BSTFA, TMSDEA, and a 3:3:2 (by volume) mixture of TMSI-BSA-TMCS were also attempted for TMS of sialic acids, all giving satisfactory results. Figure 12 shows a chromatogram of trimethylsilylated sialic acids from bovine submandibular gland glycoproteins on columns of silicone OV-1 and OV-17. The latter column gave better resolution.

Treatment of sialic acids in methanolic hydrogen chloride, followed by TMS, yields trimethylsilylated methyl esters of sialic acids. Since reducing sugars give trimethylsilylated methyl glycosides under the identical conditions, the peaks of sialic acids must be separated from those of the methyl glycosides from accompanying reducing sugars. Figure 13 shows an example of good separation of these peaks on a column of silicone SE-30.[41]

The peaks of sialic acid derivatives were eluted slower than other peaks and were well separated from each other. Better separation was obtained by use of a CPsil WCOT silica capillary column.[42]

A novel method was developed by the authors,[43] which involved liberation of sialic acids with neuraminidase, followed by cleavage of the liberated sialic acids into pyruvate and *N*-acylmannosamine derivatives with *N*-acetylneuraminate pyruvate lyase. The latter com-

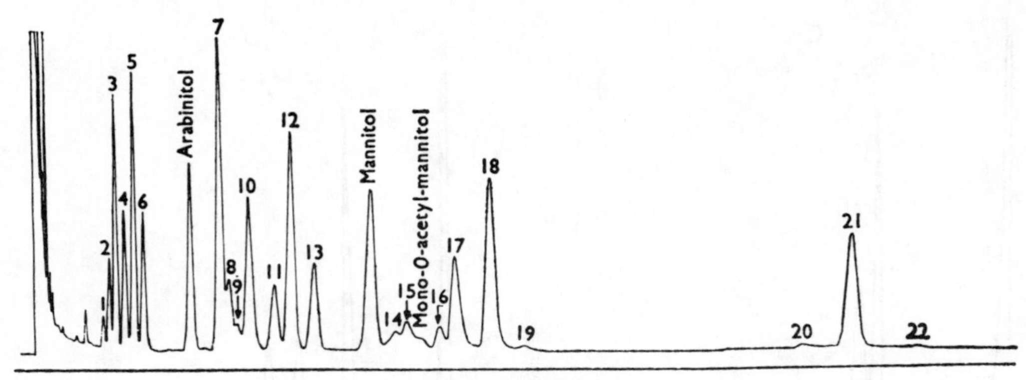

FIGURE 13. Separation of trimethylsilylated methyl glycosides. Sialic acids were analyzed as the trimethylsilylated methyl glycosides of their methyl esters. Peak assignment: 1, xylose (β-f); 2, fucose (β-f); 3, fucose (α-p); 4, fucose (β-p,α-f); 5, xylose (α-p); 6, xylose (β-p); 7, mannose (α-p); 8, galactose (β-f), glucose (f); 9, mannose (β-p); 10, mannose (f), galactose (α-p,α-f); 11, galactose (β-p); 12, glucose (α-p); 13, glucose (β-p); 14, *N*-acetylglucosamine (α-f); 15, *N*-acetylgalactosamine (α, β-f); 16, *N*-acetylglucosamine (β-p); 17, *N*-acetylgalactosamine (α,β-p); 18, *N*-acetylglucosamine (β-p); 19, *N*-acetylglucosamine (free p); 20, *N*-acetylneuraminic acid (α-p); 21, *N*-acetylneuraminic acid (β-p); 22, 9-*O*-acetyl-*N*-acetylneuraminic acid (β-p). f, furanoside; p, pyranoside. Gas chromatographic conditions: column, a glass column (2 m, 4 mm I.D.) packed with 3.8% silicone SE-30 on Chromosorb W (AW-DMCS, 80 to 100 mesh); column temperature, 135→200°C at 0.5°C/min; carrier, nitrogen (40 mℓ/min). (Reprinted by permission from *Biochem. J.*, 151, 491, Copyright © 1975 The Biochemical Society, London.)

pounds could be analyzed as their TMS diethyldithioacetals. Figure 14 shows good separation of the acylmannosamines derived from sialic acids in bovine submaxillary mucin.

Both enzymic reactions proceeded in the same medium, hence these reactions were performed in a single vessel. However, the enzyme specificities to sialic acids having *O*-acetyl group(s) were left unexamined. Figure 14 also gives the electron-impact (EI) mass spectra of these derivatives.

G. Oligosaccharide

Generally oligosaccharides are not suitable subjects of GC. However, those with a low degree of polymerization can be analyzed as TMS derivatives, though their analysis requires high temperatures. In many cases, oligosaccharides are accompanied by monosaccharides and have to be analyzed together, with the monosaccharides eluting at relatively low temperatures. Therefore, a program with a wide temperature range is necessary for analysis of such mixtures. Figure 15 gives an example of separation of oligosaccharides of the raffinose family up to the tetraose, together with monosaccharides.[20] By using a column of Dexsil® with a wide temperature range program (160 to 350°C), excellent separation was obtained.

IV. ANALYSIS OF CARBOHYDRATES IN VARIOUS SAMPLES AS SILYLATED DERIVATIVES

A. Plant Materials

In analyzing carbohydrates from plants, the extracts are sometimes obtained as a syrup containing a considerable amount of water. Therefore, methods for derivatization are required that are applicable even in the presence of water.

Ford[44] reexamined the method for TMS with TMSI of carbohydrates in extracts of tissues of plants such as stachyforpheta and mint, and stated that water content ranging from 19% to 23% did not interfere with the derivatization. Ericsson et al.[45] studied a similar subject

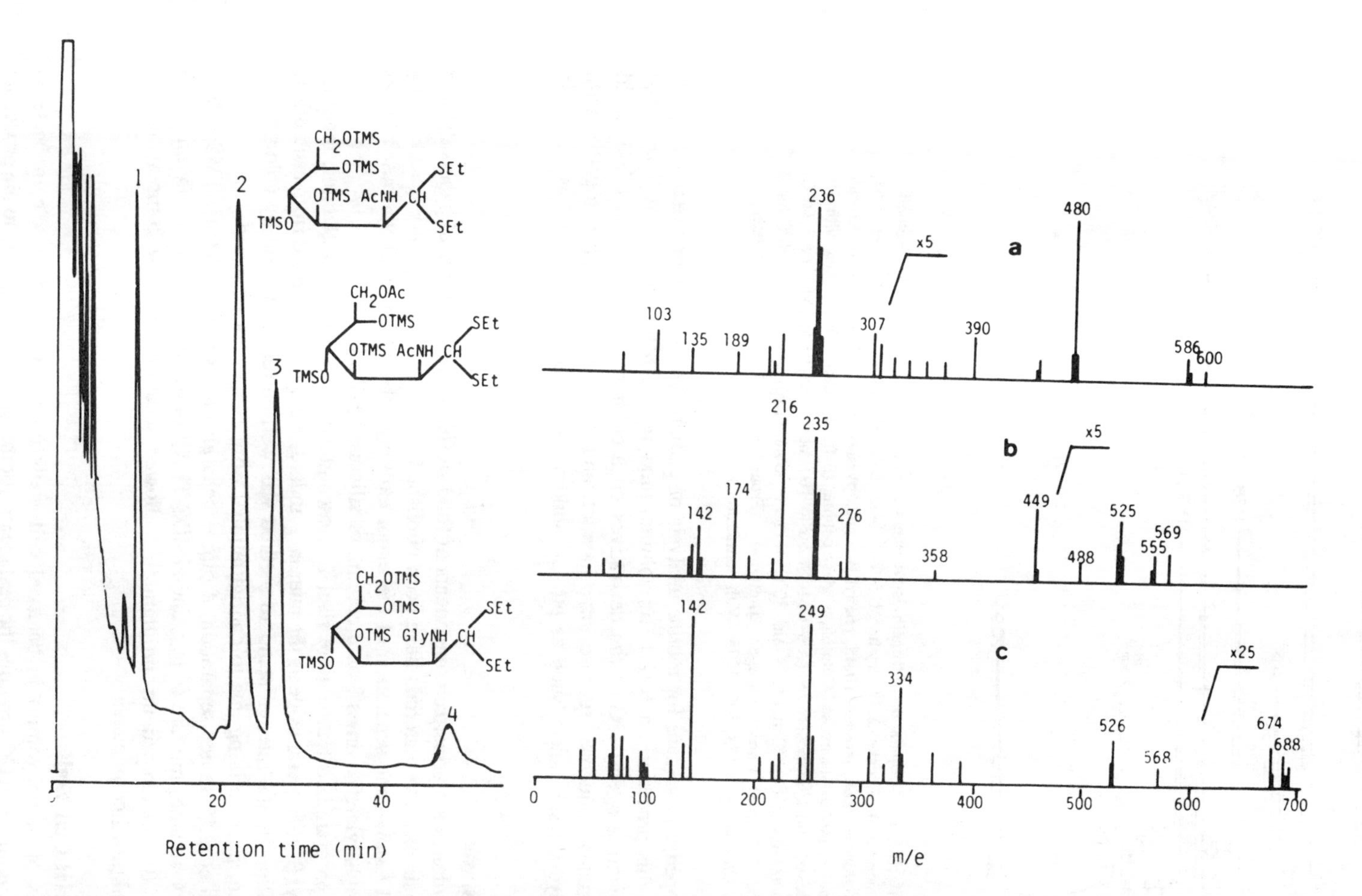

FIGURE 14. Separation of the trimethylsilylated diethyldithioacetals of *N*-acylmannosamines derived from the sialic acids in bovine submaxillary mucin by the action of *N*-acetylneuraminate pyruvate lyase. Peak assignment: 1, 3-*O*-methylglucose (internal standard); 2, *N*-acetylmannosamine; 3, *N*,*O*-diacetylmannosamine; 4, *N*-glycolylmannosamine. The mass spectra a, b, and c represent those of the compounds of Peaks 2, 3, and 4, respectively. Chromatographic conditions: column, a glass column (1 m, 3 mm I.D.) packed with 2% silicone OV-1 on Chromosorb W (AW-DMCS, 80 to 100 mesh); column temperature, 190°C; carrier, nitrogen (30 mℓ/min). (From Kakehi, K., Maeda, K., Teramae, M., Honda, S., and Takai, T., *J. Chromatogr.*, 272, 1, 1983. With permission.)

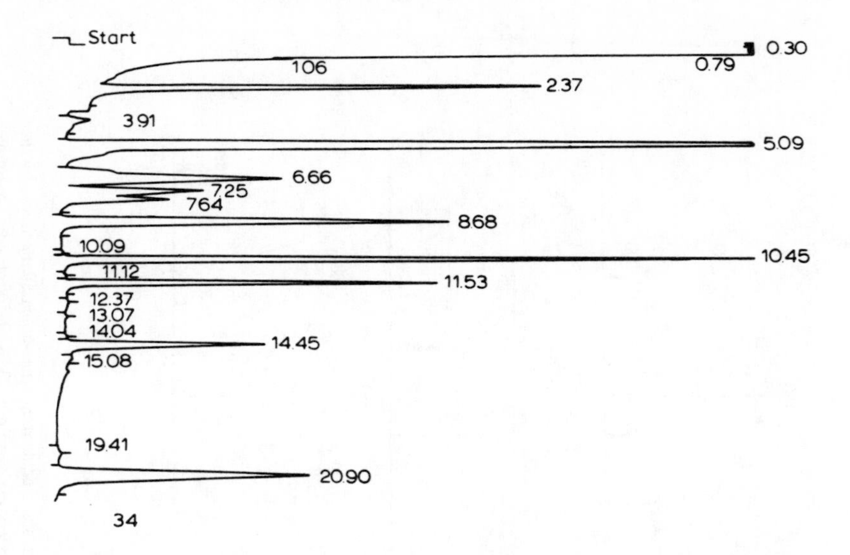

FIGURE 15. Separation of trimethylsilyl ethers of mono- and oligosaccharides. Retention times (min) of peaks: 2.37, erythritol; 5.09, ribose; 6.66, fructose; 7.25, α-glucose; 7.64, β-glucose; 8.68, inositol; 10.45, phenyl β-D-glucopyranoside; 11.53, sucrose; 14.45, raffinose; 20.90, stachyose. Column, a glass column (6 ft, 2.0 mm I.D.) packed with 3% Dexsil® on Chromosorb W (AW-DMCS, 80 to 100 mesh). Initial temperature, 160°C; increasing rate, 10°C/min for 8 min, 30°C/min for 3.66 min; final temperature, 350°C (held for 11.34 min). Carrier, nitrogen (20 mℓ/min). (From Janauer, G. A., and Englmaier, P., *J. Chromatogr.*, 153, 539, 1978. With permission.)

and developed a procedure for routine analysis of carbohydrates in water extracts of plant tissues. In this procedure, a 10:2:1 (by volume) mixture of DMF-HMDS-TMCS was used for the silylating agent. GC of the derivatives on a column of 3% silicone OV-1 could identify fructose, glucose, sucrose, *myo*-inositol, and its methyl derivative, together with noncarbohydrate substances such as pinitol, shikimic acid, and quinic acid, as shown in Figure 16.

B. Food Stuffs

Carbohydrates are principal components of food stuffs, hence a number of papers dealing with analysis of these materials have been published.

Figure 17 shows the separation of the major carbohydrates in juices of various fruits as the trimethylsilylated oximes[46] on a column of silicone SP-2250.

With a programmed temperature rise, fructose, glucose, sucrose, and glucitol, together with phenyl β-glucopyranoside as the internal standard, were well separated from each other. Quantification by the internal standard method was well correlated to that by colorimetry. In this case, peak splitting due to configurational isomers was not observed.

Sosulski et al.[47] studied separation of oligosaccharides in extracts of beans. TMS with a 30:10:1 (by volume) mixture of pyridine-HMDS-TFA, followed by GC on a column of 3% Dexsil® 300 with temperature programming allowed separation of nine components, including unknown compounds.

C. Bacterial Cell Walls

Analysis of carbohydrates in bacterial cell walls plays an important role in bacterial chemotaxonomy, and GC provides the best means for this subject. According to the procedure of Aluyi and Drucker,[48] cell walls were methanolyzed, and the resultant methyl glycosides

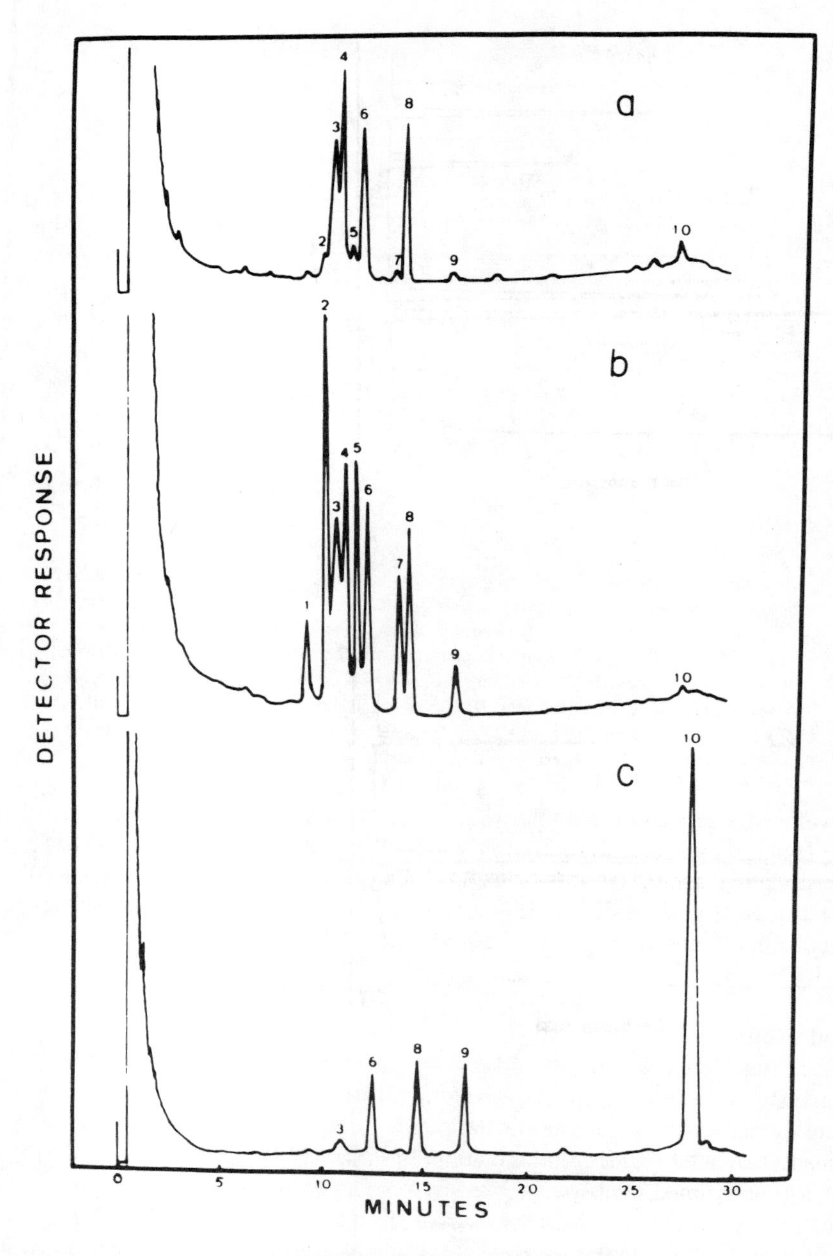

FIGURE 16. Gas chromatograms of the TMS derivatives of carbohydrates in crude extracts of (a) needles of a 20-year-old pine tree, (b) shoots of 2-month-old pine seedings, and (c) 11-day-old peas. Peak assignment: 1, unknown; 2, shikimic acid; 3, fructose; 4, pinitol; 5, quinic acid; 6, fructose, glucose; 7, unknown *O*-methyl-inositol; 8, glucose; 9, *myo*-inositol; 10, sucrose. Chromatographic conditions: column, a glass column (1.5 m, 4 mm I.D.) packed with 3% silicone OV-1 on Chromosorb W (AW-DMCS, 100 to 120 mesh); column temperature, 145°C for 1 min followed by heating to 250°C at 4°C/min; carrier, nitrogen (25 mℓ/min). (From Ericsson, A., Hansen, J., and Dalgaard, L., *Anal. Biochem.*, 86, 552, 1978. With permission.)

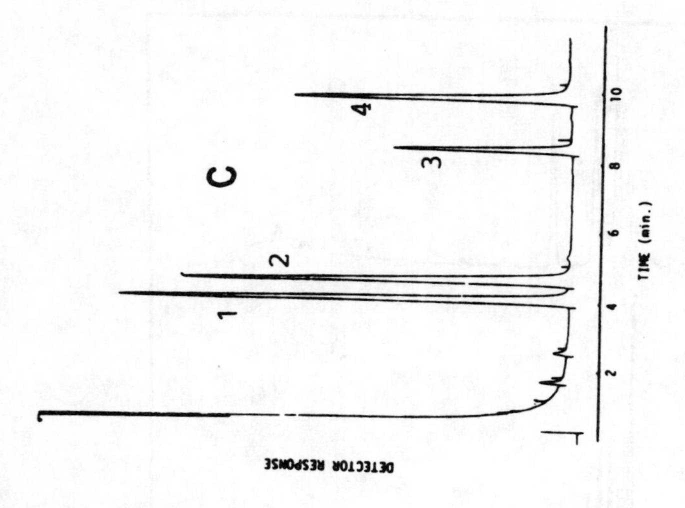
c
1
2
3
4
DETECTOR RESPONSE
TIME (min.)
2
4
6
8
10

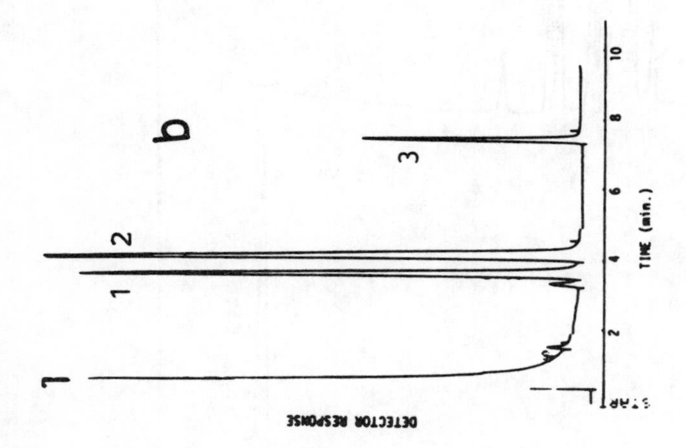
b
1
2
3
DETECTOR RESPONSE
TIME (min.)
2
4
6
8
10

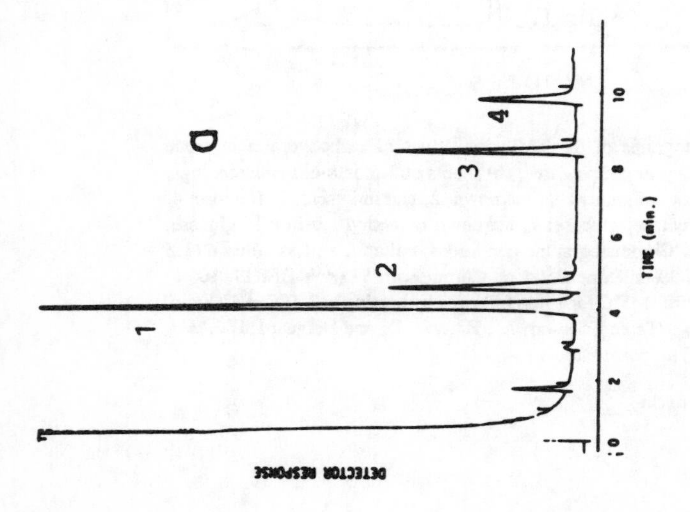
a
1
2
3
4
DETECTOR RESPONSE
TIME (min.)
0
2
4
6
8
10

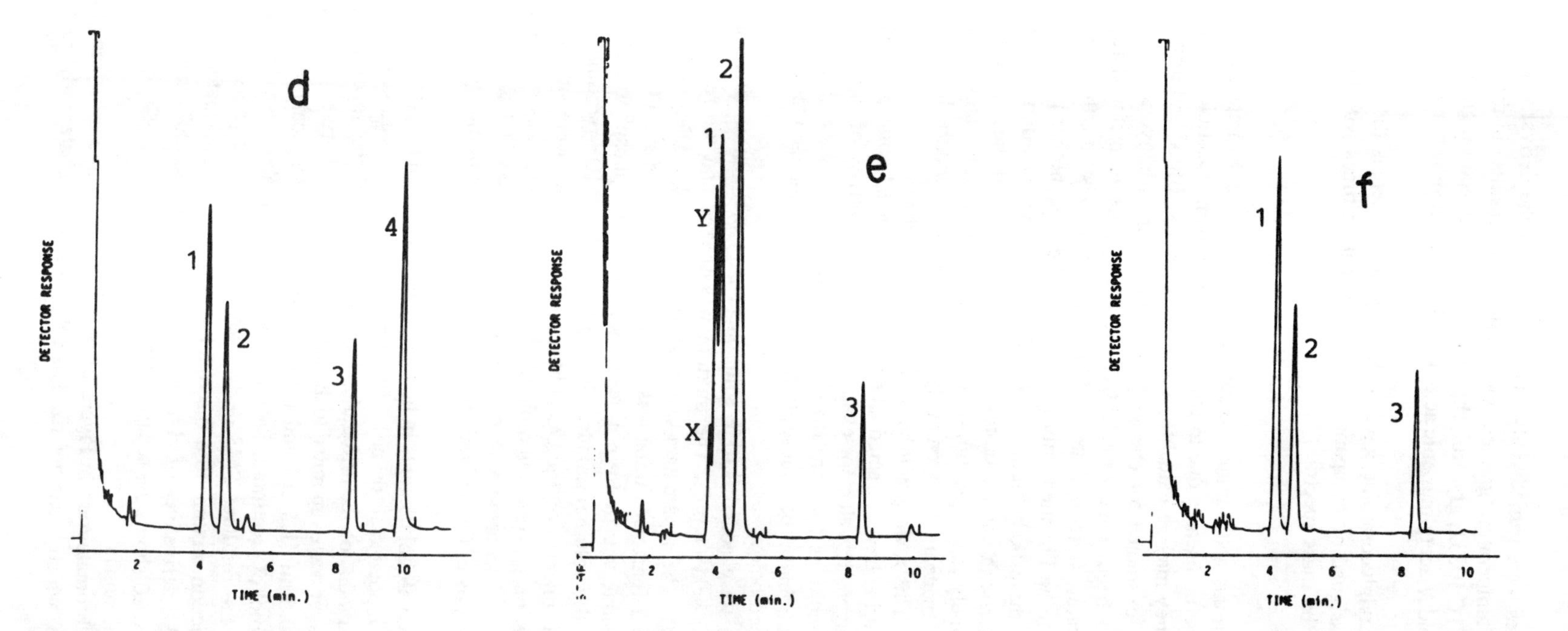

FIGURE 17. Separation of the trimethylsilylated oximes of reducing carbohydrates in (a) apple juice, (b) cranberry juice, (c) grape juice, (d) orange juice, (e) prune juice, and (f) tomato juice. Peak assignment: 1, fructose; 2, glucose; 3, phenyl-β-D-glucopyranoside; 4, sucrose; X, unknown; Y, glucitol. Chromatographic conditions: column, a stainless steel column (6 ft, 0.125 in I.D.) packed with silicone SP-2250 on Supelcoport (80 to 100 mesh); column temperature, 170 → 300°C at 10°C/min; carrier gas, helium (30 mℓ/min). (Reprinted from *Journal of Food Science* 1983. 48:633-635. Copyright © by Institute of Food Technologists.)

of the component monosaccharides were trimethylsilylated. Analysis of the derivatives on a column of silicone SE-30 with temperature gradient gave satisfactory results. In the cell walls of *Streptococcus mutans* NCTC 10832, glycerol, rhamnose, xylose, galactose, glucose, perseitol, *N*-acetylglucosamine, and *N*-acetylmuramic acid were detected, as shown in Figure 18.

Galactose, glucose, and *N*-acetylglucosamine gave multiple peaks, whereas the others gave single peaks. Peaks were assigned by comparison of their retention times with those of authentic specimens, and also by mass spectra.

D. Clinical Samples

Tissues and body fluids of animals and humans contain a variety of carbohydrates, and some of them increase or decrease in diseases due to abnormal metabolism or deficiency of particular enzymes. Therefore, they may be valuable markers of diseases. Blood and urine samples are the most frequently examined for checking abnormality of carbohydrate metabolism, and great efforts have been made to establish rapid and reliable methods for determination of such carbohydrates. Carbohydrates in blood and urine exist in either free or bound form. The latter need to be liberated prior to analysis. There is another problem of cleanup of samples, because these body fluids, especially blood, contain a number of proteinaceous substances together with carbohydrates. In addition, many of these accompanying substances are in extremely large quantities, compared to the carbohydrates to be analyzed. Recently the authors[49] made a review on analysis of carbohydrates in body fluids with special interest in cleanup. This review will be useful for this problem.

Disaccharides in plasma and urine have attracted attention from a clinical point of view, because some of them have diagnostic value for examination of absorptivity of digestive organs. Laker[50] investigated conditions for derivatization and GC of cellobiose, lactose, maltose, melibiose, sucrose, gentiobiose, palatinose, trehalose, turanose, and lactulose. The established procedure is described below.

To a plasma sample (1 mℓ) were added 0.01% solution (1 mℓ) of turanose (internal standard) and 7% sulfosalicylic acid (1 mℓ). The mixture was shaken and the resultant precipitates were centrifuged off. The supernatant was deionized with Eolite DM-F (about 0.5 mℓ) in batch mode, and the deionized solution was evaporated to dryness at 50°C under a nitrogen atmosphere. The residue was dried in a desiccator containing phosphorus pentoxide, and treated with a 1:1:2 (by volume) mixture (50 μℓ) of TMCS-HMDS-pyridine for 20 min at 60°C. An aliquot was injected on the GC column. Urine samples were similarly processed, but the deproteinization process was omitted and the TMS agent was replaced by a 1:1:5 (by volume) mixture of the same reagents. Columns of silicone OV-1 and OV-17 were appropriate for the analysis of the derivatives. Figure 19 shows a chromatogram of authentic samples.

Jansen et al.[51] reported data on alditol contents in human urine, plasma, and erythrocytes. Urine samples were directly evaporated to dryness. Plasma and erythrocyte samples were subject to cleanup by adding methanol, centrifugation, and extraction of the supernatant with *n*-hexane. The methanol layer was evaporated to dryness. The volumes of urine, plasma, and erythrocyte samples were 0.5 mℓ, 1.5 mℓ, and 1.5 mℓ, respectively. The scale of urine samples was equivalent to 1 μmol of creatinine. Each residue obtained as above was treated with TMSI-BSA-TMCS, and the resultant TMS derivatives of alditols were analyzed by injecting an aliquot of the reaction mixture to a fused-silica capillary column coated with cross-linked methylsilicone. All derivatives of alditols were well separated and quantified from their peak heights. Figure 20 shows the separation of authentic alditols, together with possibly accompanying neutral sugars.

On the other hand the top chromatogram in Figure 21 was obtained from a urine sample of a subject with galactosemia, which is characterized as galactose-1-phosphate uridyltransferase deficiency.

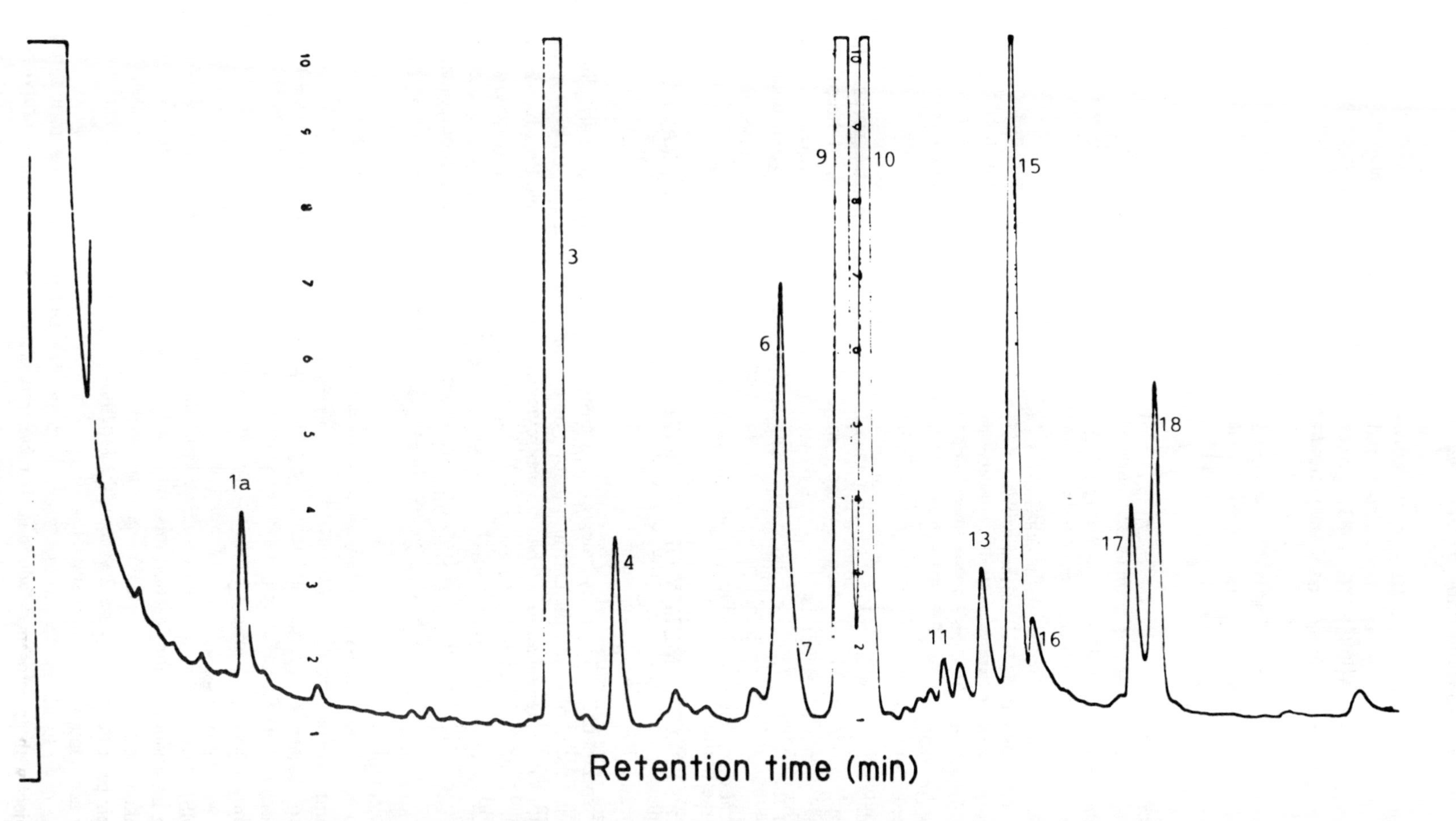

FIGURE 18. Trimethylsilylated-sugar fingerprint of *Streptococcus mutans* NCTC 10823. Peak assignment: 1a, glycerol; 3, rhamnose; 4, xylose; 6,7, galactose; 9,10, glucose; 11,13,15,16, *N*-acetylglucosamine; 17, *N*-acetylmuramic acid; 18, perseitol. Chromatographic conditions: column, a glass column (1.83 m, 4 mm I.D.) packed with 3% silicone SE-30 on Chromosorb® W HP (100 to 120 mesh); column temperature, 80 → 250°C at 2°C/min; carrier gas, nitrogen (25 mℓ/min). (From Aluyi, H. A. S. and Drucker, D. B., *J. Chromatogr.*, 178, 209, 1979. With Permission.)

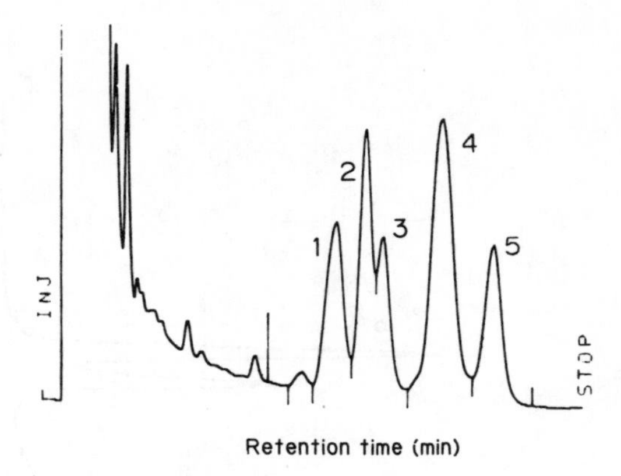

FIGURE 19. Separation of disaccharides found in human urine as TMS derivatives. Peak assignment: 1, lactulose; 2, sucrose; 3,5, lactose; 4, turanose (internal standard). Chromatographic conditions: a glass column (2.7 m, 6 mm O.D.) packed with 10% silicone OV-17; column temperature, 255 °C; carrier gas, nitrogen (55 mℓ/min). (From Laker, M. F., *J. Chromatogr.*, 163, 9, 1979. With permission.)

The galactitol peak was strikingly higher than those in the other chromatograms. The middle chromatogram obtained from a urine sample of a subject with Type 1 hereditary tyrosinemia, noted for fumarylacetoacetate deficiency, showed a marked increase of the heights of the glucitol, fructose, and sucrose peaks. The bottom chromatogram obtained from a urine sample of an infant patient, who was under gastric drip feeding because of neurological disorder, indicated abnormal increase of the sucrose level.

V. MASS SPECTROMETRY OF TRIMETHYLSILYLATED DERIVATIVES OF CARBOHYDRATES

EI mass spectra of TMS carbohydrates were extensively studied by DeJongh et al.[52] Since the EI mass spectra offer multilateral information on carbohydrate structure, they serve for peak assignment. Fragmentation mechanism was reviewed by Kochetkov and Chizov.[53] Here the outline of the mechanism for trimethylsilyl ethers of simple sugars is briefly described.

Figure 22 gives the spectra of the trimethylsilyl ethers of (a) α-glucopyranose, (b) methyl α-D-glucopyranoside, and (c) ethyl β-galactofuranoside, obtained at an ionization potential of 70 eV, and Scheme 1 proposes possible routes of fragmentation.[52]

(a)

$[\text{CH}_2\text{OTMS}, \text{OTMSi}, \text{TMS O}, \text{OTMS}, \text{OTMS}]^{\bullet +}$ *m/e* 540 → *m/e* 422 + O=$\overset{1}{\text{C}}$HOTMS

m/e 422 → *m/e* 333 + TMS O•

m/e 333 → *m/e* 243 + TMS OH

m/e 422 → *m/e* 332 + TMS OH

m/e 540 → *m/e* 407 + $CH_3^{\bullet}$

m/e 407 → *m/e* 317 + TMS OH

Scheme Ia

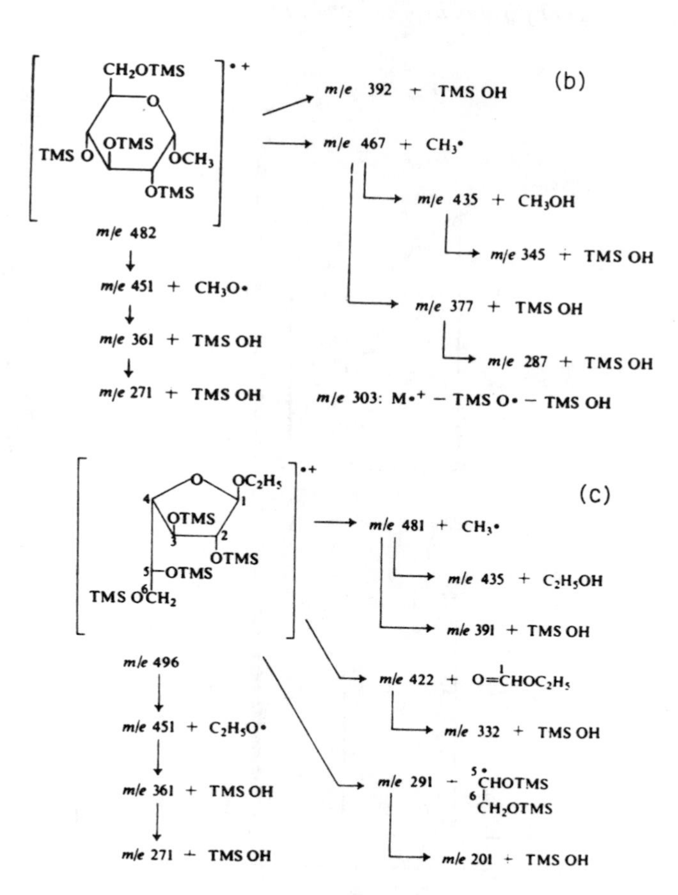

Scheme 1. Possible pathways of fragmentation of trimethylsilylated (a) α-glucose, (b) methyl α-glucopyranoside, and (c) ethyl β-galactofuranoside in EI-MS. (Reprinted with permission from DeJongh, D. C., Radford, T., Hribar, J. D., Hanessian, S., Bieber, M., Dawson, G., and Sweeley, C. C., *J. Am. Chem. Soc.*, 91, 1728, 1969. Copyright 1969 American Chemical Society.)

Under such conditions, the molecular ions of these derivatives could hardly be detected, because the initial fragmentation around the C-1 carbon was extremely drastic. A reasonable mechanism of fragmentation of the trimethylsilyl ether of α-glucopyranose is the one initiated by elimination of O = ^{1}CHOTMS to give the m/e 422 ion. This ion further undergoes degradation via three different pathways. The first pathway involves successive fragmentation to the m/e 333 and the m/e 243 ions by the release of the TMSO· radical and TMSOH, respectively. The second pathway is the degradation to the m/e 332 ion and TMSOH. The third pathway gives, consecutively, the m/e 407 and 317 ions by the removal of the CH_3· radical and TMSOH, respectively. However, only the 332 and 243 peaks were visible in the spectrum. Other mechanisms involving recombination of these fragment ions should be taken into consideration.

The trimethylsilyl ether of methyl α-glucopyranoside is considered to be cleaved via three routes. The first route leads to the m/e 451 ion by the removal of the CH_3O· radical. The m/e 451 ion is successively decomposed to the m/e 361 and 271 ions by leaving a TMSOH group, followed by another one. The second route is the direct degradation to the m/e 397 ion with the release of TMSOH. The third route gives the m/e 467 ion by the removal of the CH_3· radical, and this ion is subsequently degraded by two processes, one to the m/e

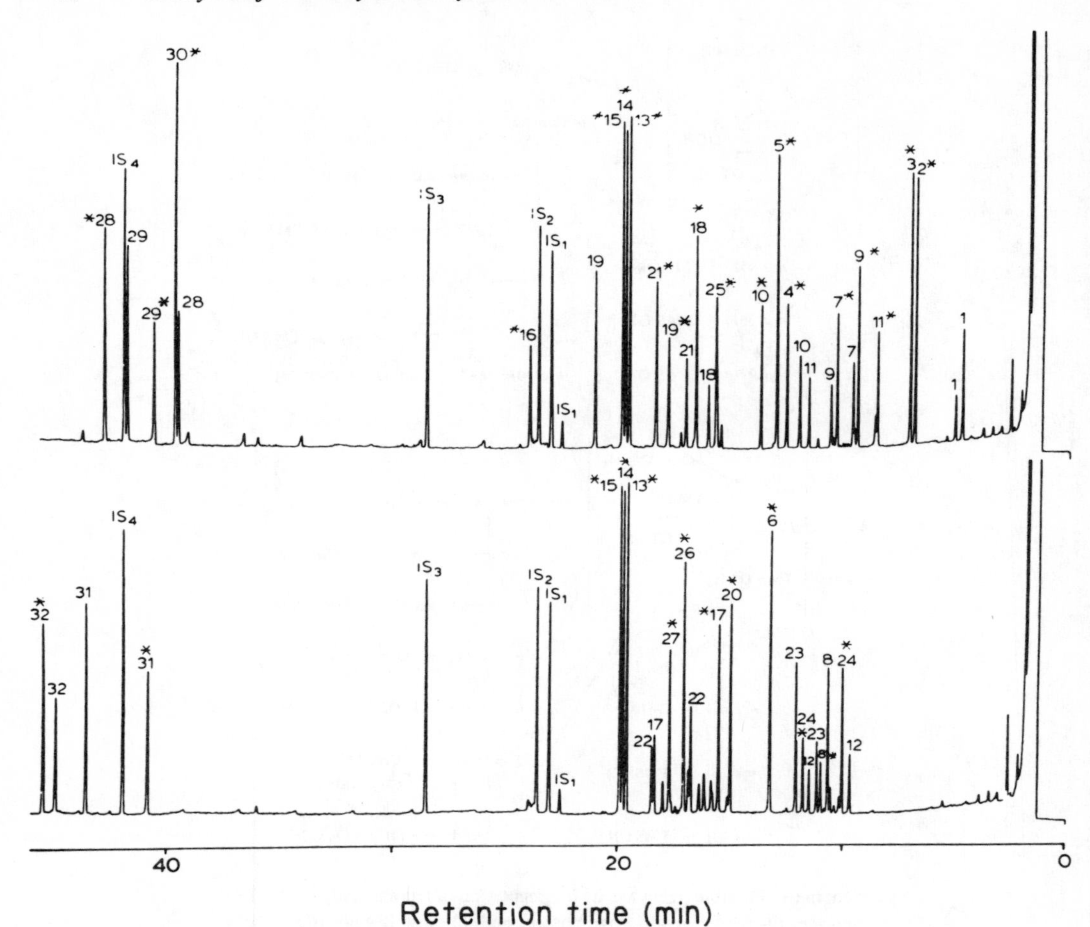

FIGURE 20. Separation of authentic trimethylsilylated sugars, used for the identification and quantification of sugars and polyols in urine. Peak assignment: 1, erythrose (two peaks); 2, threitol; 3, erythritol; 4, xylitol; 5, arabinitol; 6, ribitol; 7, arabinose (two peaks); 8, ribose (two peaks); 9, lyxose (two peaks); 10, xylose (two peaks); 11, xylulose (two peaks); 12, ribulose (two peaks); 13, mannitol; 14, glucitol; 15, galactitol; 16, *myo*-inositol; 17, mannose (two peaks); 18, allose (two peaks); 19, glucose (two peaks); 20, gulose; 21, galactose (two peaks); 22, talose (two peaks); 23, fucose (two peaks); 24, rhamnose (two peaks); 25, fructose; 26, sorbose; 27, tagatose; 28, lactose (two peaks); 29, maltose (two peaks); 30, sucrose; 31, cellobiose (two peaks); 32, melibiose (two peaks); IS_1, glucoheptulose (two peaks); IS_2, mannoheptulose; IS_3, perseitol; IS_4, trehalose. Chromatographic conditions: a capillary column (25 m, 0.2 mm I.D.) coated with cross-linked methyl silicone; column temperature, 120→265°C at 3°C/min, then to 295°C at 10°C/min. (From Jansen, G., Muskiet, F. A. J., Schierbeek, H., Berger, R., and van der Slik, W., *Clin. Chim. Acta*, 157, 277, 1986. With permission.)

345 ion via the m/e 435 ion and the other to the m/e 287 ion via the m/e 377 ion. The eliminated species in the first process are probably CH_3OH, followed by TMSOH. Those in the second process are commonly considered to be TMSOH. All these fragment ions, except for those in the initial stage, i.e., m/e 451 and 392 ions, were detected in the spectrum; the m/e 377 peak was especially intense. A minor fragment ion of m/e 303 is considered to be produced by elimination of TMSO· and TMS–OH from the molecular ion.

The fragmentation of the TMS ethers of ethyl β-galactofuranoside is more complex. It can be degraded by four different pathways by the elimination of $C_2H_5O\cdot$, $CH_3\cdot$, $O{=}^1CHOC_2H_5$ and $\cdot^5CHOTMS - {}^6CH_2OTMS$. The resultant fragment ions with m/e 451, 481, 422, and 291 are further decomposed to the ions with m/e 271 (via 361), 435, 332, and 201, re-

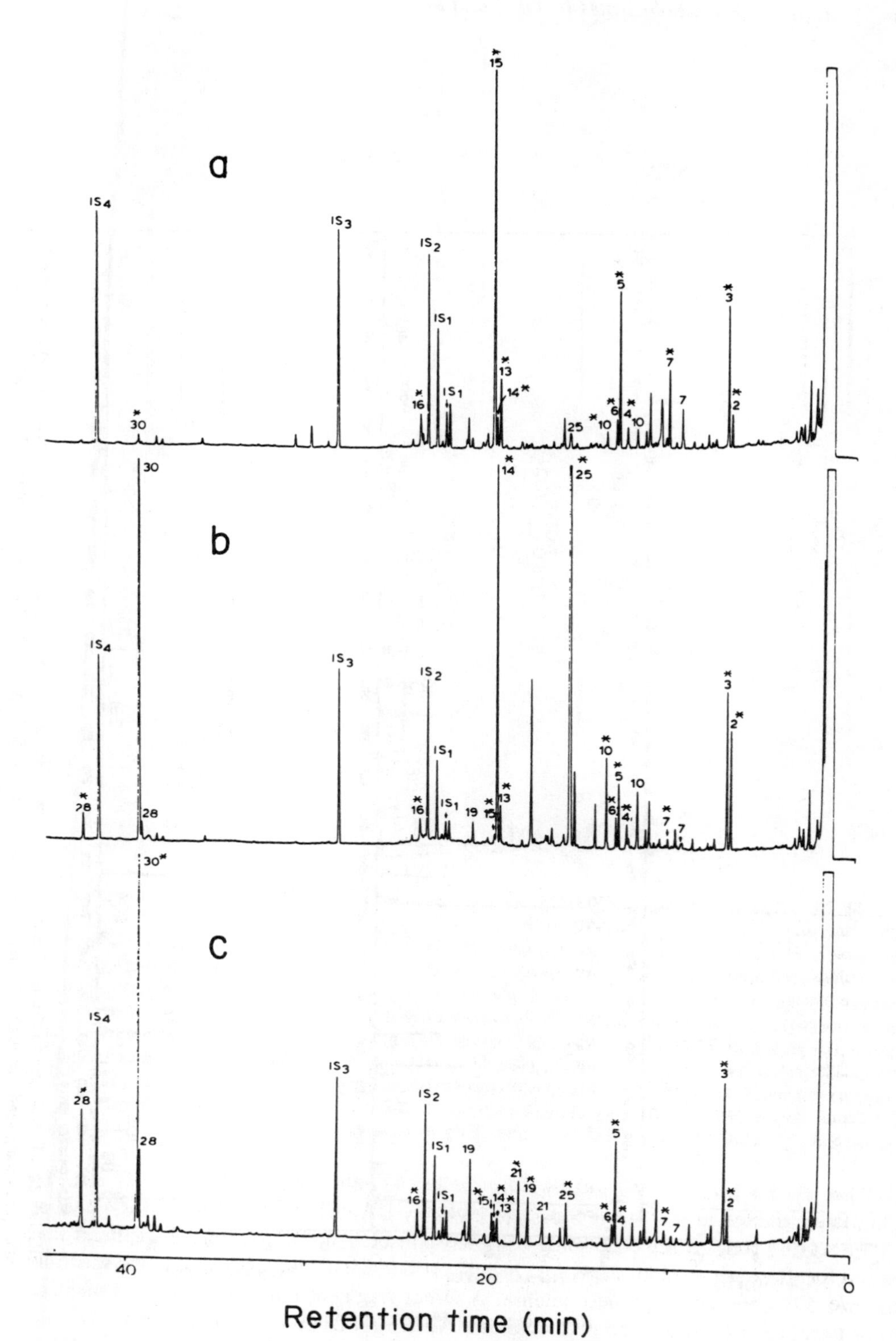

FIGURE 21. Analysis of carbohydrates in human urine of infant patients as TMS derivatives: (a) a 1-year-old patient with galactosaemia; (b) a 6-year-old patient with type 1 hereditary tyrosinemia; (c) 4-year-old patient with gastric drip feeding due to neurological disorder. For peak identification and chromatographic conditions see the caption for Figure 20. (From Jansen, G., Muskiet, F. A. J., Schierbeek, H., Berger, R., and van der Slik, W., *Clin. Chim. Acta*, 157, 227, 1986. With permission.)

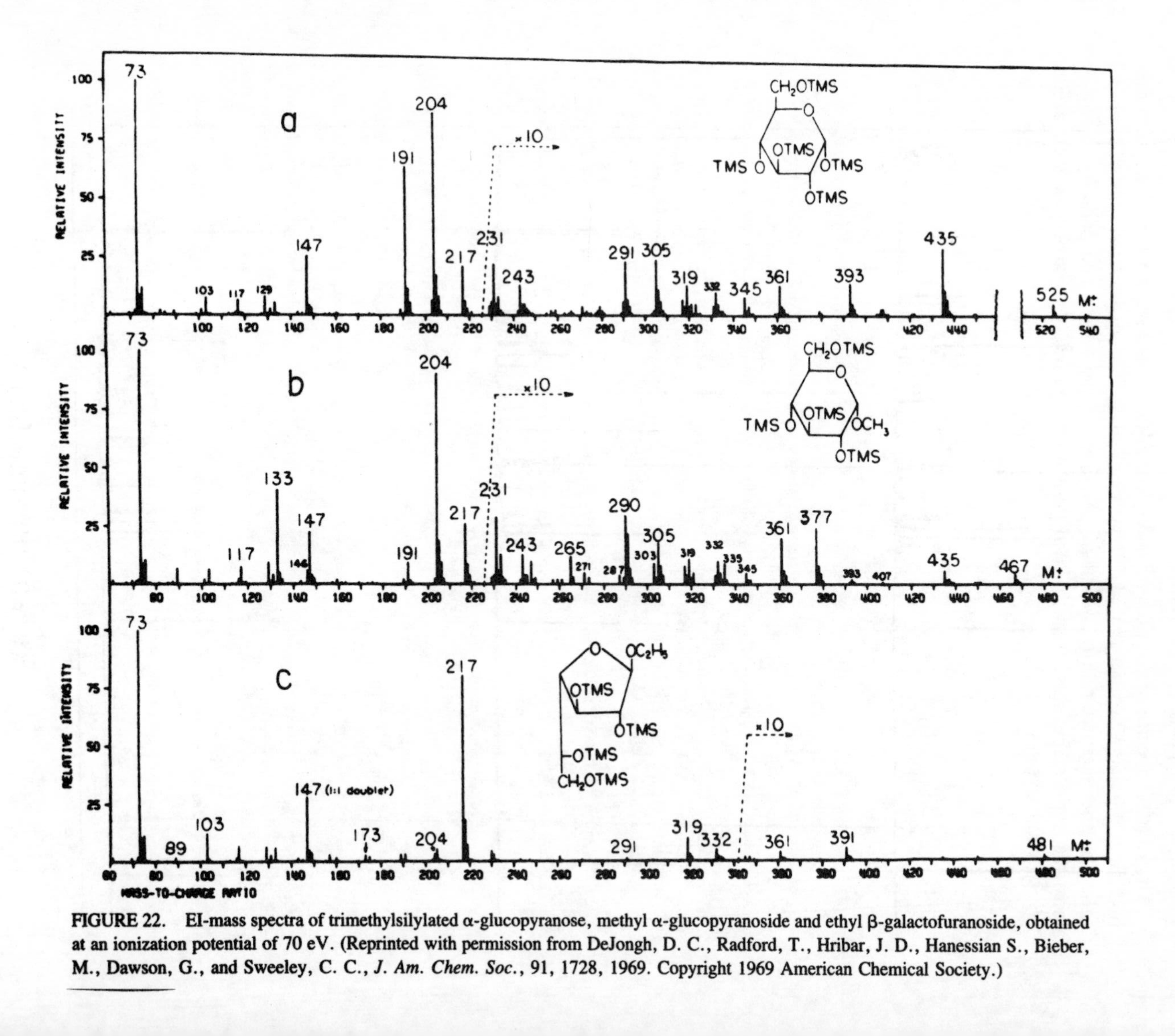

FIGURE 22. EI-mass spectra of trimethylsilylated α-glucopyranose, methyl α-glucopyranoside and ethyl β-galactofuranoside, obtained at an ionization potential of 70 eV. (Reprinted with permission from DeJongh, D. C., Radford, T., Hribar, J. D., Hanessian S., Bieber, M., Dawson, G., and Sweeley, C. C., *J. Am. Chem. Soc.*, 91, 1728, 1969. Copyright 1969 American Chemical Society.)

spectively. The m/e 481 ion was also led to the m/e 391 ion. The m/e 481, 391, 361, 332, and 291 ions were detected in the spectrum.

The pyranoside has an intense peak at m/e 204, but the m/e 319 peak is almost negligible. Vice versa is the furanoside. This is a useful means to discriminate furanosides from pyranosides.

The EI mass spectra of the trimethylsilyl derivatives of sialic acids were studied in detail by Kamerling and Vliegenthart[54] and Haverkamp et al.[55] Many papers[35,56] have been published on the interpretation of EI mass spectra of amino sugars and partially methylated monosaccharides, but their trimethylsilyl derivatives are less important than their acetates and trifruoroacetates.

REFERENCES

1. **Sweeley, C. C., Bentley, R., Makita, M., and Wells, W. W.,** Gas-liquid chromatography of trimethylsilyl derivatives of sugars and related substances, *J. Am. Chem. Soc.*, 85, 2497, 1963.
2. **Kircher, H. W.,** Gas-liquid partition chromatography of sugar derivatives, in *Methods in Carbohydrate Chemistry,* Vol. 1, Whistler, R. L. and Wolfrom, M. L., Eds., Academic Press, New York, 1962, 13.
3. **Bishop, C. T.,** Separation of carbohydrate derivatives by gas-liquid partition chromatography, in *Methods of Biochemical Analysis,* Vol. 10, Glick, D., Ed., Interscience, New York, 1962, 1.
4. **Bishop, C. T.,** Gas-liquid chromatography of carbohydrate derivatives, in *Advances in Carbohydrate Chemistry,* Vol. 19, Wolfrom, M. L. and Tipson, R. S., Eds., Academic Press, New York, 1964, 95.
5. **Wells, W. W., Sweeley, C. C., and Bentley, R.,** Gas chromatography of carbohydrates, in *Biochemical Application of Gas Chromatography,* Vol. 1, Szymanski, H. A., Ed., Plenum Press, New York, 1964, 169.
6. **Sweeley, C. C., Wells, W. W., and Bentley, R.,** Gas chromatography of carbohydrates, in *Methods in Enzymology,* Vol. 8, Neufeld, E. F. and Ginsburg, V., Eds., Academic Press, New York, 1966, 95.
7. **Sloneker, J. H.,** Gas chromatography of carbohydrates, in *Biochemical Application of Gas Chromatography,* Vol. 2, Szymanski, H. A., Ed., Plenum Press, New York, 1968, 27.
8. **Clamp, J. R., Bhatti, T., and Chambers, R. E.,** The determination of carbohydrate in biological materials by gas-liquid chromatography, in *Methods of Biochemical Analysis,* Vol. 19, Glick, D., Ed., Interscience, New York, 1971, 229.
9. **Clamp, J. R., Bhatti, T., and Chambers, R. E.,** The examination of carbohydrate in glycoproteins by gas-liquid chromatography, in *Glycoproteins,* Gottshalk, A., Ed., Elsevier, Amsterdam, 1972, 300.
10. **Laine, R. A., Esselman, W. J. and Sweeley, C. C.,** Gas-liquid chromatography of carbohydrates, in *Methods in Enzymology,* Vol. 28, Ginsburg, V., Ed., Academic Press, New York, 1972, 159.
11. **Brobst, K. M.,** Gas-liquid chromatography of trimethylsilyl derivatives, in *Methods in Carbohydrate Chemistry,* Vol. 6, Whistler, R. L. and BeMiller, J. N., Eds., Academic Press, New York, 1972, 3.
12. **Sweeley, C. C. and Tao, R. V. P.,** Gas chromatographic estimation of carbohydrates in glycosphingolipids, in *Methods in Carbohydrate Chemistry,* Vol. 6, Whistler, R. L. and BeMiller, J. N., Eds., Academic Press, New York, 1972, 8.
13. **Loewus, F. and Shah, R. H.,** Gas-liquid chromatography of trimethylsilyl ethers of cyclitols, in *Methods in Carbohydrate Chemistry,* Vol. 6, Whistler, R. L. and BeMiller, J. N., Eds., Academic Press, New York, 1972, 14.
14. **Sloneker, J. H.,** Gas-liquid chromatography of alditol acetates, in *Methods in Carbohydrate Chemistry,* Vol. 6, Whistler, R. L. and BeMiller, J. N., Eds., Academic Press, New York, 1972, 20.
15. **Jones, H. G.,** Gas-liquid chromatography of methylated sugars, in *Methods in Carbohydrate Chemistry,* Vol. 6, Whistler, R. L. and BeMiller, J. N., Eds., Academic Press, New York, 1972, 25.
16. **Dutton, G. G. S.,** Applications of gas-liquid chromatography to carbohydrates: Part 1, in *Advances in Carbohydrate Chemistry and Biochemistry,* Vol. 28, Tipson, R. S. and Horton, D., Eds., Academic Press, New York, 1973, 11.
17. **Dutton, G. G. S.,** Applications of gas-liquid chromatography to carbohydrates: Part 2, in *Advances in Carbohydrate Chemistry and Biochemistry,* Vol. 30, Tipson, R. S. and Horton, D., Eds., Academic Press, New York, 1974, 9.
18. **Wong, C. G., Sung, S. J., and Sweeley, C. C.,** Analysis and structural characterization of amino sugars by gas-liquid chromatography and mass spectrometry, in *Methods in Carbohydrate Chemistry,* Vol. 8, Whistler, R. L. and BeMiller, J. N., Eds., Academic Press, New York, 1980, 55.

19. **Yamakawa, T., Ueta, N., and Ishizuka, I.,** Gas chromatography of trimethylsilylated monosaccharides, *Jpn. J. Exp. Med.*, 34, 231, 1964.
20. **Janauer, G. A. and Englmaier, P.,** Multi-step time program for the rapid gas-liquid chromatography of carbohydrates, *J. Chromatogr.*, 153, 539, 1978.
21. **Englmaier, P.,** Identification and quantitative estimation of plant cyclitols and polyols by gas chromatography, *Fresenius Z. Anal. Chem.*, 324, 338, 1986.
22. **Honda, S., Yamauchi, N., and Kakehi, K.,** Rapid gas chromatographic analysis of aldoses as their diethyldithioacetal trimethylsilylates, *J. Chromatogr.*, 169, 287, 1979.
23. **Honda, S., Kakehi, K., and Okada, K.,** Convenient method for the gas chromatographic analysis of hexosamines in the presence of neutral monosaccharides and uronic acids, *J. Chromatogr.*, 176, 367, 1979.
24. **Pelletier, O. and Cadieux, S.,** Glass capillary or fused-silica gas chromatography-mass spectrometry of several monosaccharides and related sugars: improved resolution, *J. Chromatogr.*, 231, 225, 1982.
25. **Sawardeker, J. S. and Sloneker, J. H.,** Quantitative determination of monosaccharides by gas liquid chromatography, *Anal. Chem.*, 37, 945, 1965.
26. **Ellis, W. C.,** Liquid phases and solid supports for gas-liquid chromatography of trimethylsilyl derivatives of monosaccharides, *J. Chromatogr.*, 41, 335, 1969.
27. **Bhatti, T., Chambers, R. E., and Clamp, J. R.,** The gas chromatographic properties of biologically important *N*-acetylglucosamine derivatives, monosaccharides, disaccharides, trisaccharides, tetrasaccharides and pentasaccharides, *Biochim. Biophys. Acta*, 222, 339, 1970.
28. **Brobst, K. M. and Lott, C. E., Jr.,** Determination of some components in corn syrup by gas-liquid chromatography of the trimethylsilyl derivatives, *Cereal Chem.*, 43, 35, 1966.
29. **Mahwhinney, T. P.,** Simultaneous determination of *N*-acetylglucosamine, *N*-acetylgalactosamine, *N*-acetylglucosaminitol and *N*-acetylgalactosaminitol by gas-liquid chromatography, *J. Chromatogr.*, 351, 91, 1986.
30. **Li, B. W. and Schuhmann, P. J.,** Gas chromatographic analysis of sugars in granola cereals, *J. Food Sci.*, 46, 425, 1981.
31. **Zegota, H.,** Separation and quantitative determination of fructose as the *O*-methyloxime by gas liquid chromatography using glass capillary columns, *J. Chromatogr.*, 192, 446, 1980.
32. **Bombardelli, E., Conti, M., Magistretti, M. J., and Martinelli, E. M.,** Computer-aided evaluation of gastric proteoglycans by high-resolution gas chromatography, *J. Chromatogr.*, 279, 593, 1983.
33. **Gerwig, G. J., Kamerling, J. P., and Vliegenthart, J. F. G.,** Determination of the D and L configuration of neutral monosaccharides by high-resolution capillary G. L. C., *Carbohydr. Res.*, 62, 349, 1978.
34. **Hurst, R. E.,** The trimethylsilylation reactions of hexosamines, and gas-chromatographic separation of the derivatives, *Carbohydr. Res.*, 30, 143, 1973.
35. **Coduti, P. L. and Bush, C. A.,** Structure determination of *N*-acetyl amino sugar derivatives and disaccharides by gas chromatography and mass spectroscopy, *Anal. Biochem.*, 78, 21, 1977.
36. **Mononen, I.,** Quantitative analysis, by gas-liquid chromatography and mass fragmentography, of monosaccharides after methanolysis and deamination, *Carbohydr. Res.*, 88, 39, 1981.
37. **Kennedy, J. F., Robertson, S. M., and Stacey, M.,** G. L. C. of the *O*-trimethylsilyl derivatives of hexuronic acids, *Carbohydr. Res.*, 49, 243, 1976.
38. **Raunhardt, O., Schmidt. H. W. H., and Neukom, H.,** Gas-chromatographische Untersuchungen an Uronsäuren and Uronsäure derivaten, *Helv. Chim. Acta*, 50, 1267, 1967.
39. **Inoue, S. and Miyawaki, M.,** Quantitative analysis of iduronic acid and glucuronic acid in sulfated galactosaminoglycuronans by gas chromatography, *Anal. Biochem.*, 65, 164, 1975.
40. **Casals-Stenzel, J., Buscher, H. P., and Schauer, R.,** Gas-liquid chromatography of *N*- and *O*-acylated neuraminic acids, *Anal. Biochem.*, 65, 507, 1975.
41. **Kamerling, J. P., Gerwig, G. J., Vliegenthart, J. F. G., and Clamp, J. R.,** Characterization by gas-liquid chromatography-mass spectrometry and proton-magnetic-resonance spectroscopy of pertrimethylsilyl methyl glycosides obtained in the methanolysis of glycoproteins and glycopeptides, *Biochem. J.*, 151, 491, 1975.
42. **Kamerling, J. P. and Vliegenthart, J. F. G.,** Gas-liquid chromatography and mass spectrometry of sialic acids, in *Sialic Acids, Chemistry, Metabolism and Function*, Schauer, R., Ed., Springer-Verlag, New York, 1982, 95.
43. **Kakehi, K., Maeda, K., Teramae, M., Honda, S., and Takai, T.,** Analysis of sialic acids by gas chromatography of the mannosamine derivatives released by the action of *N*-acetylneuraminate lyase, *J. Chromatogr.*, 272, 1, 1983.
44. **Ford, C. W.,** Semimicro quantitative determination of carbohydrates in plant material by gas-liquid chromatography, *Anal. Biochem.*, 57, 413, 1974.
45. **Ericsson, A., Hansen, J., and Dalgaard, L.,** A routine method for quantitative determination of soluble carbohydrates in small samples of plant material with gas-liquid chromatography, *Anal. Biochem.*, 86, 552, 1978.

46. **Li, B. W. and Schuhmann, P. J.,** Sugar analysis of fruit juices: content and method, *J. Food Sci.*, 48, 633, 1983.
47. **Sosulski, F. W., Elkowicz, L., and Reichert, R. D.,** Oligosaccharides in eleven legumes and their air-classified protein and starch fractions, *J. Food Sci.*, 47, 498, 1982.
48. **Aluyi, H. A. S. and Drucker, D. B.,** Fingerprinting of carbohydrates of *Streptococcus Mutans* by combined gas-liquid chromatography-mass spectrometry, *J. Chromatogr.*, 178, 209, 1979.
49. **Kakehi, K. and Honda, S.,** Profiling of carbohydrates, glycoproteins and glycolipids, *J. Chromatogr.*, 379, 27, 1986.
50. **Laker, M. F.,** Estimation of disaccharides in plasma and urine by gas-liquid chromatography, *J. Chromatogr.*, 163, 9, 1979.
51. **Jansen, G., Muskiet, F. A. J., Schierbeek, H., Berger, R., and van der Slik, W.,** Capillary gas chromatographic profiling of urinary, plasma and erythrocyte sugars and polyols as their trimethylsilyl derivatives, preceded by a simple and rapid prepurification method, *Clin. Chim. Acta.*, 157, 277, 1986.
52. **DeJongh, D. C., Radford, T., Hribar, J. D., Hanessian, S., Bieber, M., Dawson, G., and Sweeley, C. C.,** Analysis of trimethylsilyl derivatives of carbohydrates by gas chromatography and mass spectrometry, *J. Am. Chem. Soc.*, 91, 1728, 1969.
53. **Kochetkov, N. K. and Chizov, O. S.,** Mass spectrometry of carbohydrates, in *Methods in Carbohydrate Chemistry*, Vol. 6, Whistler, R. L. and BeMiller, J. N., Eds., Academic Press, New York, 1972, 540.,
54. **Kamerling, J. P. and Vliegenthart, J. F. G.,** Mass spectrometry of pertrimethylsilyl neuraminic acid derivatives, *Carbohydr. Res.*, 33, 297, 1974.
55. **Haverkamp, J., Schauer, R., Wember, M., Kamerling, J. P., and Vliegenthart, J. F. G.,** Synthesis of 9-*O*-acetyl- and 4,9-di-*O*-acetyl derivatives of the methyl ester of *N*-acetyl-α-D-neuraminic acid methyl glycoside: their use as models in periodate oxidation studies, *Hoppe-Seyler's Z. Physiol. Chem.*, 356, 1575, 1975.
56. **Karkkainen, J. and Vihko, R.,** Characterization of 2-amino-2-deoxy-D-glucose, 2-amino-2-deoxy-D-galactose, and related compounds, as their trimethylsilyl derivatives by gas-liquid chromatography-mass spectrometry, *Carbohydr. Res.*, 10, 113, 1969.
57. **Wells, W. W., Pittman, T. A., and Well, H. J.,** Quantitative analysis of *myo*-inositol in rat tissue by gas-liquid chromatography, *Anal. Biochem.*, 10, 450, 1965.
58. **Shalon, L. S., Novotony, M., and Karmen, A.,** Evaluation of certain polyols in the cerebrospinal fluid of patients with multiple sclerosis, *J. Chromatogr.*, 336, 351, 1984.
59. **Niwa, T., Yamada, K., Ohki, T., Saito, A., and Mori, M.,** Identification of 6-dexoyallitol and 6-deoxyglucitol in human urine. Electron impact mass spectra of eight isomers of 6-deoxyhexitol, *J. Chromatogr.*, 336, 345, 1984.
60. **Yoshioka, S., Saitoh, S., Seki, S., and Seki, K.,** Concentrations of non-glucose polyols in serum and cerebrospinal fluid from apparently healthy adults and children, *Clin. Chem.*, 30, 188, 1984.
61. **Johnson, S. L. and Mayersohn, M.,** Quantitation of xylose from plasma and urine by capillary column gas chromatography, *Clin. Chim. Acta*, 137, 13, 1984.
62. **Pelletier, O. and Cadieux, S.,** Quantitative determination of glucose in serum by isotope dilution mass spectrometry following gas liquid chromatography with fused silica column, *Biomed. Mass Spectr.*, 10, 130, 1983.
63. **Toba, T. and Adachi, S.,** Gas-liquid chromatography of trimethylsilylated disaccharide oximes, *J. Chromatogr.*, 135, 411, 1977.
64. **Yuki, H., Yamaji, A., Takai, H., Nagamura, K., Bando, R., and Kawasaki, H.,** Analysis of aldoses by gas chromatography as their 2-amino-ethanethiol derivatives, *Yakugaku Zasshi*, 98, 77, 1978.
65. **Kringstad, R. and Bakke, I. L. F.,** Identification of hydroxy acids by gas-liquid chromatography, *J. Chromatogr.*, 144, 209, 1977.
66. **Honda, S., Nagata, M., and Kakehi, K.,** Rapid gas chromatographic analysis of partially methylated aldoses as trimethylsilylated diethyldithioacetals, *J. Chromatogr.*, 209, 299, 1981.
67. **Swaninathan, N., Apon, B., and Aladjem, F.,** Use of capillary columns for the analysis of monosaccharides by gas-liquid chromatography, *Anal. Biochem.*, 75, 646, 1976.
68. **Orme, T. W., Boone, C. W., and Roller, P. P.,** The analysis of 2-acetamido-2-deoxy-aldose derivatives by gas-liquid chromatography and mass spectrometry, *Carbohydr. Res.*, 37, 261, 1974.
69. **Tesarik, K.,** The separation of some monosaccharides by capillary column gas chromatography, *J. Chromatogr.*, 65, 295, 1972.

Chapter 5

PREPARATION OF ALDITOL ACETATES AND THEIR ANALYSIS BY GAS CHROMATOGRAPHY (GC) AND MASS SPECTROMETRY (MS)

Alvin Fox, Stephen L. Morgan, and James Gilbart

TABLE OF CONTENTS

I. INTRODUCTION

The alditol acetate derivatization procedure for the gas chromatographic (GC) analysis of carbohydrates is a popular method for analysis of the sugar composition of macromolecules.[1,2] A major advantage of this method in comparison to others such as trimethylsilylation TMS[3], or trifluoroacetylation[4], is that a single peak is produced for each derivatized sugar. Only a limited number of other methods, such as the aldononitrile acetate[5,6] or *O*-methyl oxime acetate procedures, also produce single peaks.[7] A detailed discussion of these other methods is outside the scope of this chapter, but is given elsewhere (see Chapters 4, 6, 7, and 8). Even though high-resolution capillary columns can adequately resolve complex mixtures, multiple peaks may confound qualitative identification and quantitative measurement. The production of multiple peaks for each single sugar of interest also may adversely affect the limit of detection. A further advantage of alditol acetates is that, once formed, they are extremely stable, thus allowing postderivatization cleanup, as well as storage of treated samples for extended periods. This method, however, requires a large number of manual processing steps which make the procedure time consuming and tedious to perform.[1,2,8,9] Another problem of concern to our laboratories in using the alditol acetate method for trace analysis has been the appearance in the chromatogram of contaminating extraneous background peaks that originate from side reactions during the hydrolysis and/ or the derivatization processes.

The objective of this chapter is to provide a practical guide to the use of the alditol acetate derivatization procedure and to review the relevant methods-development literature. Several more general reviews on the chromatographic analysis of carbohydrates have appeared recently.[10-12] A detailed description of a modified alditol acetate method used in our laboratories for trace analysis of carbohydrates in microbiological applications and photographs of specialized derivatization equipment is included. Representative chromatograms of sugar mixtures and mass spectra of various common and some uncommon sugars are also presented for reference. Finally, we make some suggestions for future work to further simplify and increase the utility of the alditol acetate procedure.

II. THE ALDITOL ACETATE METHOD — A GENERAL DISCUSSION

A. Overview

The alditol acetate method was originally described by Gunner, et al.[1] in 1961 and popularized by a number of researchers including Sawardeker.[2] The derivatization reaction itself involves the reduction and acetylation of monomeric sugars in a series of steps. Table 1 summarizes the four basic steps of the method. Figure 1 shows the chemical reactions involved in the conversion of an aldose to an alditol acetate using the conversion of glucosamine to glucosaminitol hexaacetate as an example.

The first step in the chemical derivatization is a sodium borohydride reduction of the aldose to an alditol. Aldoses exist in aqueous solution in dynamic equilibrium between ring and straight chain forms, whereas alditols only occur as straight chains. If the reduction is not carried out, acetylation of the ring form of aldoses complicates the chromatogram. Usually two or four multiple peaks are produced from each sugar (two from pyranose anomers and two from furanose anomers). As a result, interpretation of the peaks in the chromatogram and quantitation of specific sugar amounts becomes more difficult. Several other common methods for carbohydrate analysis, including direct TMS or trifluoroacetylation of the unreduced sugar, also produce this problem.[3,4]

The second step of the original method involves multiple evaporations with acidic methanol to remove borate generated in the reduction step.[1,2] The presence of borate dramatically inhibits the acetylation reaction to follow because borate forms strong complexes with polyhydroxyl compounds.

Table 1
SUMMARY OF THE ORIGINAL ALDITOL ACETATE METHOD

Step 1	Reduction of aldose to alditol using sodium borohydride
Step 2	Removal of borate generated in first step by multiple evaporations with acidic methanol
Step 3	Acetylation with acetic anhydride using a catalyst such as pyridine
Step 4	Gas chromatographic analysis

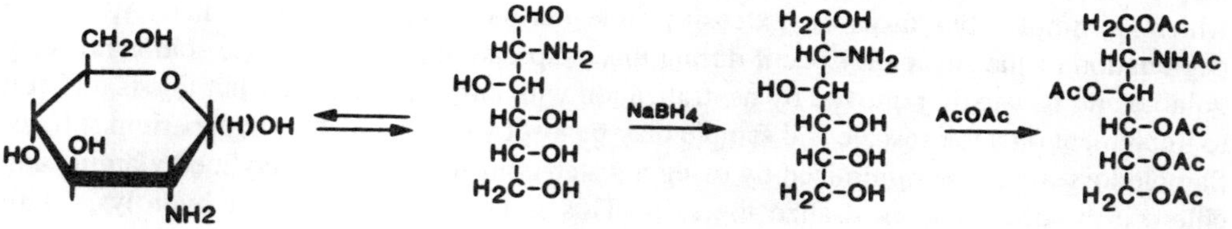

FIGURE 1. An example of the alditol acetate derivatization reaction. Glucosamine, an aminoaldose which exists predominantly in the ring form, is reduced with sodium borohydride to glucosaminitol, which exists entirely as a linear molecule. After drying the sample, hydroxyl groups are acetylated with acetic anhydride to produce glucosaminitol hexaacetate.

Hydrogen bonding of hydroxyl groups on the alditol tends to raise the boiling point so high that they decompose before becoming volatile. The third step in the classical alditol acetate method is acetylation of the hydroxyl and amino groups on the alditol to form the corresponding esters and amides.[1,2] This acetylation step eliminates hydrogen bonding and the resulting alditol acetate is sufficiently volatile to make the GC analysis of the derivative possible. Furthermore, the acetylation of free hydroxyl groups in the alditol decreases interactions with other components of the sample and the GC system.

The final step in the analysis of alditol acetates is the GC analysis. Early work with these derivatives employed packed columns and nonpolar stationary phases or even phases of only moderate polarity such as Carbowax®-20M or ethylene glycol succinate.[2]

Since the introduction of the alditol acetate method, numerous investigators have introduced modifications designed to handle specific application problems and difficulties in the analysis. The original method can be time consuming because a considerable number of manual operations are involved, especially during removal of the borate. Simultaneous analysis of a large batch of samples is difficult by most published procedures. The original methods assumed the investigator was starting with mixtures of monomeric sugars and did not consider the problem of analyzing the constituents of polymers containing carbohydrates. Subsequent methods provide a hydrolysis step to release monomeric sugars from polysaccharides; however, the destruction of monomeric sugars during this acidic hydrolysis was not adequately considered.[9] Many researchers have encountered problems in neutralization or evaporation of the acid prior to derivatization. Uncontrolled side reactions during acid hydrolysis, removal of the acid, or undesired reactions during the derivatization may produce extraneous peaks in the resulting chromatogram.[13] Impure reagents also can contaminate samples and increase the background in chromatograms. Another issue in the analysis of complex carbohydrate mixtures involves cleanup of the sample to reduce or eliminate undesirable contaminating peaks in the chromatogram. Finally, the use of modern capillary GC columns and selective detectors have added a new dimension to the selectivity and sensitivity with which sugars can be separated and identified.

B. Hydrolysis

The first step in the determination of the carbohydrate composition of a polysaccharide is hydrolysis of the polymer to release its monomeric constituents. Sugars are notoriously unstable when heated to high temperatures in strong acid solutions. Any such degradation of the sample is undesirable because it results in erroneously low values for the monomeric sugars and may introduce additional peaks in the chromatogram. Two commonly used acids for release of sugars from polymers are hydrochloric acid and sulfuric acid. Hydrochloric acid has been found to introduce extraneous peaks in the chromatogram, either due to side reactions or sugar degradation.[13] The use of sulfuric acid and the careful exclusion of air during hydrolysis can minimize these effects.[14] Steps taken to remove the acid after hydrolysis may cause losses of some sugars. Hydrochloric acid is usually removed by lyophilization which is simple, but takes an extensive time period (usually about 24 hr). Additional degradation of the sugar may occur during this evaporation period. Sulfuric acid is not very volatile and is usually removed by neutralization with barium hydroxide, but this is difficult to implement on a microscale and sample may be lost by coprecipitation with barium sulfate. Sample losses may be minimized by using a water-immiscible *N,N*-dioctylmethylamine and chloroform solution to neutralize the acid. This is preferably used after hydrolysis with sulfuric acid. However, hydrochloric acid can also be neutralized in this way.[14] Internal standards are usually added to the sample immediately prior to, or after, the hydrolysis step and carried through the remaining sample-handling steps.

C. Prederivatization Cleanup

Following the hydrolysis step, many researchers next perform derivatization without using any cleanup procedures. Solvent extraction or disposable hydrophobic columns, however, have been employed by some workers to remove lipids released by the hydrolysis reaction.[15,16] Cation exchange resins have also been used at this point to remove amino acids and other charged compounds.[17] Amino sugars are unfortunately separated from the neutral sugars by ion-exchange, and require separate analysis when this approach is adopted.

D. Reduction

There has been little controversy concerning the reduction step of the alditol acetate method. Most workers have added sodium borohydride in aqueous solution to reduce the aldose to the sugar alditol. This step can be accomplished in under 2 hr at room temperature or at 37°C; or more conveniently, overnight at 4°C. Blakeney, et al. suggested the addition of sodium borohydride in a dimethylsulfoxide (DMSO) solution, which is considerably more stable than an aqueous solution and facilitates storage and handling.[18] More simply, sodium borohydride can be dissolved in methylimidazole to form a stable solution, eliminating the additional DMSO solvent from the reaction mixture. However, as noted below, *N*-methylimidazole or DMSO can be difficult to remove from the sample and can be a source of extraneous peaks.

E. Acetylation

The next step in the alditol acetate method is the acetylation reaction. Prior to this, however, multiple evaporations with methanolacetic acid or, less commonly, methanol-hydrochloric acid, have usually been used to remove borate (as tetramethyl borate gas) to avoid inhibition of the acetylation. Borate may also be removed selectively by a resin containing covalently attached 1-deoxy-1-methylamino-D-glucitol.[19] Unfortunately, the samples must first be treated with a cation exchange resin to convert borate ions to boric acid. This procedure would be inappropriate for mixtures of neutral and amino sugars since the latter would be lost on the columns.

Pyridine or sodium acetate have been traditionally employed as catalysts for acetylation, but methylimidazole may be more convenient. The sample is thoroughly dried before proceeding to the acetylation step if pyridine or sodium acetate are used as catalysts, particularly if muramic acid or other amino sugars are to be analyzed. McGinnis[20] demonstrated that,

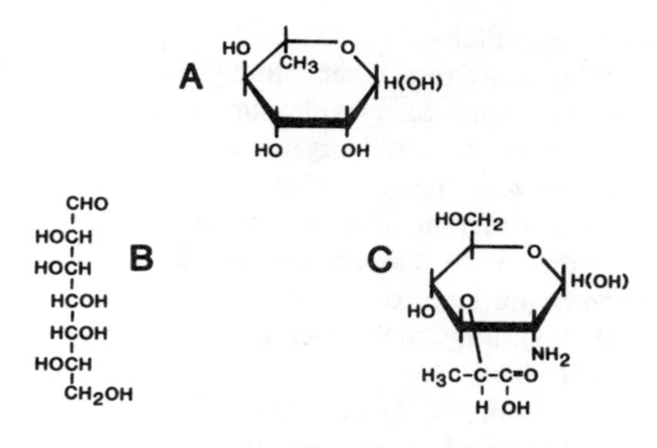

FIGURE 2. Structures of some important sugars A, rhamnose (fucose is one of its isomers); B, L-glycero-D-mannoheptose; C, muramic acid.

using methylimidazole as a catalyst, acetylation can proceed without removal of water or borate generated during the reduction step, thus allowing for rapid and simple derivatization. Blakeney, et al.[18,21] further developed this approach. Methylimidazole is an excellent catalyst for the acetylation of many neutral and amino sugars; however, it is not suitable for the analysis of amino sugars much as muramic acid.[22]

When pyridine, or its analogue, methylimidazole, is used as a catalyst for acetic anhydride acetylation, unfortunately browning of the sample may occur, producing interfering chromatographic peaks.[15] Tarry reaction products may also be deposited in the GC injection port. These products have been shown to result from the reaction between the organic base (pyridine or methylimidazole) and acetic anhydride since heating either reagent alone produces no browning. Another problem of using methylimidazole is reduced sensitivity for certain sugars such as rhamnose, fucose, and the heptoses (structures shown in Figure 2). Whiton, et al.[22] reported that acetylation for 4 hr at 37°C in the absence of water increased yields of these neutral sugars only slightly, although background peaks were reduced. The relative yields of rhamnose, fucose, and the heptoses were increased dramatically by raising the acetylation temperature to 115°C in the absence of water. Unfortunately, these more extreme reaction conditions caused the methylimidazole-acetic anhydride browning reaction to create a significant amount of tarry product. This is very difficult to remove from the sample and, subsequently, deposits in the GC injection port and column. The improved yields at extreme reaction conditions did not improve the analytical results sufficiently to compensate for this problem. In another application, the analysis of the neutral sugars in plant polysaccharides, *N*-methylimidazole catalyzed acetylation with short acetylation times, provided similar or slightly better yields of alditol acetates compared to sodium acetate catalyzed derivatization.[23] Amino sugars were not included in this work and mg or higher levels of each sugar component were present in the material analyses. Methylimidazole may have utility as a catalyst when sugars are in high abundance (above 10 μg/mg of starting material) and if the sample matrix is of limited complexity.

Acetylation using sodium acetate catalysis may be preferred in trace analysis work because of the low chromatographic background noise and excellent yields of all neutral and amino sugars, including muramic acid.[22] Sodium acetate does not react with acetic anhydride; the acetylating reagent and more extreme acetylation conditions may be used to give higher yields. Some authors report that additional sodium acetate should be added before acetylation to adequately catalyze the reaction. If sufficiently large amounts (several mg) of sodium borohydride are added in the reduction step, however, additional sodium acetate usually need not be added. Sodium borohydride decomposes to sodium borate in the presence of acetic acid, and after evaporation with methanol-acetic acid, sodium acetate is left behind. A further discussion of acetylation catalysts may be found in Chapter 6.

F. Postderivatization Cleanup Steps

Following the acetylation, many researchers simply dry the sample and inject aliquots into the GC without further cleanup. Sample cleanup at this stage, however, reduces contamination of the GC system by salts and organic acids or bases initially present in the samples or resulting from the derivatization. Cleanup at this point often prolongs column life and may clear the chromatogram of extraneous peaks. After evaporation of acetic anhydride, one simple method is to dissolve the sample in chloroform and extract with water. Residual water can be difficult to remove, and alternatively, the chloroform-water mixture may be passed through a hydrophilic cleanup column such as Chem-Elut (Analytichem, Inc., Harbor City, Calif.).

Trifluoracetyl derivatives of aldoses[4,24] and alditols[25] which can be used for high sensitivity electron capture or negative ionization mass spectrometry (MS) detection are unfortunately not amenable to these aqueous cleanup procedures. These derivatives are readily hydrolyzed on contact with water and, consequently, would not be seen on subsequent GC analysis.

Trace analysis of complex samples requires more extensive cleanup. Acetic acid produced by the addition of water to acetic anhydride and any basic compounds present may be removed by extracting the sample with chloroform. Acidic compounds may be removed by extraction with base (ammonium hydroxide) and applying the mixture to a hydrophilic cleanup column. The use of these columns is rapid and has been found to cause no appreciable sample loss while considerably reducing discoloration in derivatized biological samples and decreasing background peaks in the chromatogram.[22] These columns are normally disposed of after each use, but when samples are not too dirty, might be reused.

G. Analysis of Mixtures of Alditols, Ketoses, and Aldoses

The reduction of aldoses to alditols and their conversion to alditol acetates simplifies chromatograms by producing only one peak for each aldose. There may, however, be a significant loss of information incurred in the use of this reduction step for certain mixtures of sugars. Mixtures of aldoses and their corresponding alditols would yield only alditols, because they are not reduced further by sodium borohydride (e.g., glucose and glucitol both produce glucitol). Certain pairs of sugars will produce the same alditol (e.g., gulose and glucose). Aldoses and the related ketoses may also produce the same alditols on reduction and cannot be differentiated (e.g., glucose and fructose both produce glucitol). The presence of ketoses can confound the analysis by producing two sugar alcohols, each of which produces a separate chromatographic peak (e.g., fructose produces glucitol and mannitol).[26] Methods for the analysis of alditols in body fluids have usually featured ways to avoid interference from aldoses, especially glucose. In some work, aldoses have been removed before derivatization.[27] Other derivatization methods such as the aldononitrile acetate or *O*-methyloxime acetates produce different products for each of the three groups of sugars which can be resolved chromatographically.[5,7] If one is particularly interested in alditols and not aldoses or ketoses, samples may be acetylated without prior reduction.[28] Aldoses or ketoses can be differentiated from alditols using sodium borodeuteride in place of sodium borohydride for reduction. Newly formed alditols (produced from aldoses or ketoses on reduction) are deuterated, while preexisting alditols in the sample are not. After acetylation, one deuterium remains on the alditol (derived from aldoses or ketoses), allowing differentiation from the alditols originally present (which do not contain deuterium) by the mass/charge differences observed by MS.[29]

H. Analysis of Acidic Sugars

Acidic sugars are not amenable to analysis by GC or GC-MS without additional chemical manipulation, because adequate chromatography cannot be easily achieved for compounds containing a carboxyl group. The difficulty of matching the column polarity to the sample polarity may cause excessive band broadening or irreversible sorption on passage of these compounds through many types of GC columns. One way to conduct the GC analysis of acidic sugars is to derivatize the free carboxyl group. Samples may be evaporated to dryness in the presence of hydrochloric acid to produce lactones, which are subsequently trimethylsilylated.[30] Sugars are partially destroyed, however, in the presence of hydrochloric acid. There are several variations of the alditol acetate method which overcome this problem by converting the carboxyl group to an alcohol or ester prior to acetylation. The carboxyl group may be reacted with carbodiimide and reduced to an alditol, which is then converted to an alditol acetate.[31] Alternatively, aldonic acids have been dehydrated to lactones by simple evaporation and then reacted with alkylamines to produce *N*-alkylaldonamides.[32] This latter method is able to differentiate aldoses from aldonic acids because aldoses yield alditol acetates, while aldonic acids give alkylaldonamide acetates. Color was noted, however, to occur in acetylated samples, indicating the formation of significant side reaction products.

I. Analysis of Amino Sugars

Some researchers have reported difficulties in the GC analysis of amino sugar alditol acetates. However, following careful analytical practice and using appropriate GC columns, the alditol acetate procedure can be successfully employed for the analysis of amino sugars. It can also be used for the simultaneous analysis of mixtures of neutral and amino sugars. In order to avoid GC of alditol acetates of amino sugars, some researchers have converted amino sugars into neutral sugars using nitrous acid deamination prior to reduction and acetylation.[9] The use of ion-exchange resins to separate neutral from amino sugars followed by independent derivatization prior to GC analysis has also been proposed.[17] Some researchers, finding GC analysis of alditol acetates of amino sugars difficult, have recommended using an amino acid analyzer, which is based on liquid chromatography followed by post-column derivatization of the amino group.[17] Difficulties in the GC analysis of amino sugars are not primarily due to problems with the derivatization, but due to problems of long retention times, poor sensitivity, and chemical changes during the chromatographic analysis.[33] These chromatographic problems can be solved with deactivated columns of appropriate polarity as will be discussed more fully below.

The analysis of muramic acid, an amino sugar component found only in bacterial cell walls and not elsewhere in nature, presents slightly different problems. Muramic acid, 3-*O*-lactyl-D-glucosamine, contains lactic acid in an ether linkage on the number three carbon of the sugar ring and a free carboxyl group in this lactyl moiety (Figure 2). The structure of the alditol acetate derivative of muramic acid is unusual: under dehydrating conditions, the elimination of water between the amino and carboxyl group of muramicitol forms a C–N bond creating a lactam ring, thus avoiding the problems of a free carboxyl group.[22] Muramic acid has been shown to readily form a lactam ring in some other circumstances, such as in bacterial endospores, which are highly dehydrated where muramic acid occurs predominantly in this form.[34] The aldonitrile acetate of muramic acid is also known to be a lactam.[35] Under mild acetylation conditions, incomplete acetylation of muramicitol produces two separate chromatographic peaks; more rigorous acetylation conditions (100°C for 13 to 16 hr) are required to produce a single peak.

J. Selection of Internal Standards

An internal standard for GC of alditol acetates should be as similar as possible to the components of interest and should not naturally be present in the sample, nor should it coelute either with the components of interest or with background peaks. For the analysis of neutral sugars, any neutral sugar not present in the analytical sample could be employed; arabinose and xylose are often used for this reason rather than the more common sugars such as ribose or glucose. Inositol is sometimes used as an internal standard but, because it is already in the reduced form, does not control for losses in the reduction step. The internal standard could be added to the sample prior to the hydrolysis step or immediately after this step. The purpose of the internal standard is to correct for any physical or chemical losses that might occur in the amount of the sugar components during sample processing and chromatographic analysis. Adding the internal standard after the hydrolysis step does not properly allow for correction of any losses in sugar components that might have occurred during the hydrolysis. Adding the internal standard prior to the hydrolysis may cause variability due to differing rates of breakdown of the polymer sugar components compared to the added internal standard monomer. Ideally, an internal standard whose degradation matches the degradation of the polymer sugar components should be employed, but this information and a suitable compound may not be available and compromises are made in practice. Alternately, hydrolysis losses may be determined (Chapter 3).

Many workers use a single internal standard for measuring amounts of both neutral and amino sugars.[15] Dissimilarities in the structures of neutral and amino sugars, however, can lead to variability in the yields of derivatization reactions. Amino sugars also elute much later in the chromatogram than neutral sugars and thus may present different peak integration problems. Losses of amino sugars relative to neutral sugars during chromatography are also greater. These sources of variability will result in poor reproducibility if a single internal standard is used for such mixtures.[15] Most accurate quantitative analysis of amino sugars requires a second internal standard such as methylglucamine which is effective for quantitation and does not occur in any biological samples.

K. Gas Chromatographic (GC) Column Selection

The success of the alditol acetate method in many applications is often related to factors affecting the quality of the GC separation. Care must be taken in every aspect of the GC analysis; for example, trace levels of moisture or oxygen must be removed from the carrier gas using molecular sieve driers and oxygen scrubbers. Stability of derivatives may be influenced by time spent in the GC system and temperatures to which components are exposed. High carrier gas velocities and columns coated with thin layers of stationary phase can help to minimize the duration in passage through the column and, consequently, reduce analysis time. The use of helium as a carrier gas further decreases analysis time without any substantial loss in column efficiency.

The most significant improvement in GC separation of alditol acetates came with recent advances in capillary column technology.[36,37] Even capillary columns coated with nonpolar and nonselective stationary phases can sometimes perform creditably for many applications, whereas the same separation on a packed column may require a selective stationary phase. In comparison to capillary columns made of borosilicate glass, fused-silica columns provide decreased reactivity and greater column life.

Columns coated with so-called "bonded phases" are now the most popular type. In these, the stationary phase is cross-linked and immobilized by means of covalent bonding to the column wall. Cross-linking imparts greater temperature stability, resistance to bleed, and the columns may be washed with solvent after contamination with sample residue because the stationary phase is not readily extractable. Capillary-column GC differs from packed-column GC in the need to split the sample prior to introduction onto the column; capillary columns tolerate much less sample than packed columns and can easily be overloaded. The splitless injection technique introduced by Grob[38] in 1969 allows for more complete transfer of sample to the analytical column and is the preferred method for trace analysis of alditol acetates.[39]

Figure 3 illustrates a packed column separation of a mixture of neutral and amino sugar alditol acetates. The early eluting neutral sugar peaks are not completely resolved. The later eluting amino sugar peaks show severe band broadening possibly due to both the long column

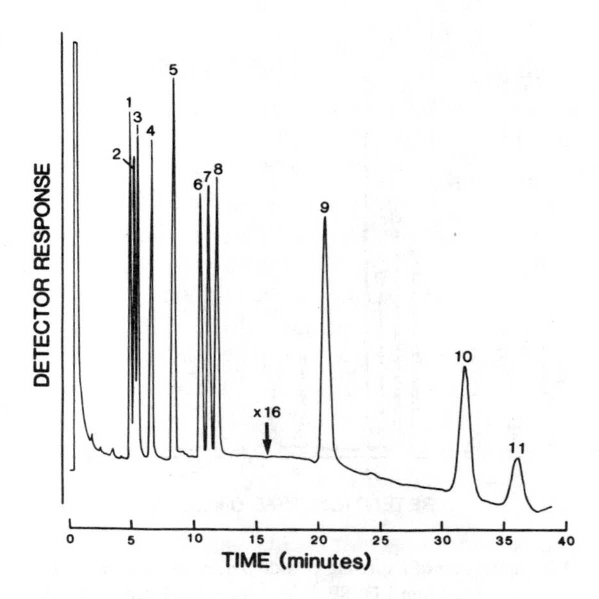

FIGURE 3. Separation of alditol acetates of a standard mixture of neutral and amino sugars separated on a 6 ft × 2 mm I.D. glass column packed with Supelcoport (100 to 200) mesh coated with 3% SP-2330. Other conditions include: initial oven temperature 200°C, programmed at 4°C/min to 245°C, helium gas flow rate 45 mℓ/min, on-column injection, flame ionization detection. Peak identification: 1, deoxyribose; 2, rhamnose; 3, fucose; 4, ribose; 5, xylose; 6, mannose; 7, galactose; 8, glucose; 9, muramic acid; 10, glucosamine; 11, galactosamine. (From Fox, A., Morgan, S. L., Hudson, J. R., Zhu, Z.-T., and Lau, P. Y., *J. Chromatogr.*, 256, 429, 1983. With permission.)

retention and adsorption in the GC system. On this packed column, galactosamine and mannosamine cannot be resolved at all. In comparison, Figure 4 illustrates a separation of a similar group of neutral and amino sugars on a fused-silica capillary column coated with SP-2330. Some additional sugars are present in this sample that are not present in the sample of Figure 3, including arabinose, inositol, several heptoses, methyl glucamine, and mannosamine. All components are baseline-resolved and the retention time of the last eluting peak is only 22 min. Of even greater interest is the baseline separation of mannosamine and galactosamine around 22 min; this separation has previously been difficult, particularly when neutral sugars must be resolved in the same chromatogram.[16,40]

Several stationary phases that have been used for GC analysis of alditol acetates are listed in order of increasing relative polarity in Table 2. The resolution of alditol acetates improves from more nonpolar and nonselective columns to more polar and selective columns. Figure 5 illustrates the separation of the same standard mixture that was separated in Figure 4, but on a fused-silica capillary column coated with OV-1701. In the OV-1701 separation, the alditol acetate of glucose (Peak 9) coelutes with that of inositol (Peak 10), the alditol acetate of muramic acid (Peak 13) is poorly resolved from that of glucosamine (Peak 15), and the alditol acetate of mannosamine (Peak 16) is not well separated from that of galactosamine (Peak 17). Hexoses are resolved on both columns, but the overall separation of the mixture of neutral and amino sugars is superior on SP-2330. Cross-linked OV-1701 columns are available and provide greater temperature stability and low bleed. Resolution is better on an SP-2330 column, but at present, cross-linked SP-2330 columns are not available.

Figures 3 to 5 illustrate the general elution order of alditol acetates: neutral pentoses and methyl pentoses elute first, followed by hexoses, then inositol and the heptoses, and finally

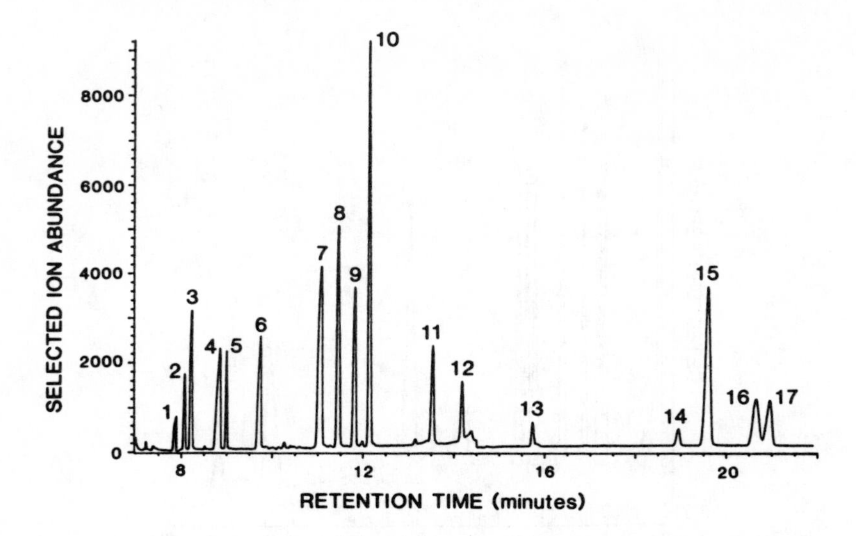

FIGURE 4. Chromatogram of a mixture of alditol acetates of neutral and amino sugars separated on a 30 m × 0.32 mm I.D. SP-2330 coated fused-silica column. Selected ion monitoring was employed with two groups of ions: first, from 6 to 14.5 min, m/z 159 for deoxyribose, 210.1 for inositol, m/z 303.1 for rhamnose and fucose, and 289.05 for other neutral sugars; secondly, from 14.5 to 22.0 min, m/z 318.0 for glucosamine, mannosamine, and galactosamine, m/z 327.25 for methylglucamine, and m/z 403.15 for muramic acid. Peak identification: 1, deoxyribose; 2, rhamnose; 3, fucose; 4, ribose; 5, arabinose (internal standard); 6, xylose; 7, mannose; 8, galactose; 9, glucose; 10, *myo*-inositol; 11, L-glycero-D-mannoheptose or D-glycero-L-mannoheptose; 12, D-glycero-D-mannoheptose; 13, muramic acid (pentaacetate); 14, methylglucamine (internal standard); 15, glucosamine; 16, mannosamine; 17, galactosamine.

the amino sugar hexoses. It is notable that methylglucamine (Peak 14) elutes last from the OV-1701 column and elutes before the amino sugar hexoses on the SP-2330 column.

Less polar columns such as SE-54 or OV-1 usually will not adequately resolve early eluting neutral sugars, although overlapping peaks may be deconvoluted by reconstructed ion profiles on a GC-MS instrument.[39] Separation of amino sugar alditol acetates has been shown to be poor on SP-2100 (similar to OV-1 except a fluid phase). Silar 10C (Chrompack) is a more polar phase and provides considerably better separation, but does not completely resolve galactosamine and mannosamine.[21] Columns coated with Carbowax® 20M can baseline separate these two amino alditol acetates, but on this phase, ribose and galactose are poorly resolved.[41] The two common amino sugar isomers, galactosamine and glucosamine, and the less common sugar, mannosamine, have been resolved on a chiral polysiloxane phase, although lengthy analysis times are required.[42]

Long analysis times for the separation of mixtures of neutral and amino sugars, even with capillary columns, have been a problem in some chromatographic systems.[9,42] A major difficulty in analyzing alditol acetates of amino sugars is poor sensitivity and long retention times on certain packed or capillary columns.[9] These problems also occur with other derivatives of amino sugars.[43-46] Aldononitrile acetates of amino sugars often do not produce detectable chromatographic peaks.[7] Selective decomposition of amino sugars or adsorption of amino sugars to columns and other parts of the chromatographic system clearly can occur.[33] Hudson et al.[16] performed a systematic investigation in which equal amounts of a mixture of a neutral sugar alditol acetate (glucitol hexaacetate) and an amino sugar alditol acetate (glucosaminitol hexaacetate) were separated on various capillary GC columns prepared with different treatments. Relative peak area ratios for the two alditol acetates varied from 0 to 0.9, depending on the column treatment; for example, amino sugars were found

Table 2
STATIONARY PHASES WHICH HAVE BEEN USED FOR GAS CHROMATOGRAPHY OF ALDITOL ACETATES

Phase	Chemical composition	Upper temp. (°C)	Relative polarity index (Squalane = 0%, OV-275 = 100%)
OV-1	Methyl silicone gum (100% methyl)	350	5.0
SE-30	Methyl silicone gum (100% methyl)	300	5.2
SP-2100	Methyl silicone fluid (100% methyl)	350	5.3
SE-54	Methyl silicone polymer (1% vinyl, 5% phenyl)	300	8.0
SE-52	Methyl silicone gum (5% phenyl, 95% methyl)	300	8.2
OV-1701	Methyl silicone polymer (7% cyanopropyl, 7% phenyl, 86% dimethylsiloxane)	250	18.0
OV-17	Methyl silicone polymer (50% methyl, 50% phenyl)	350	21.0
OV-225	Methyl silicone polymer (25% cyanopropyl, 25% phenyl, 50% methyl)	265	42.0
Carbowax® 20M	Polyethylene glycol ($HO-[-CH_2-CH_2-O-]_n-H$)	225	55.0
SP-2300	Methyl silicone fluid (68% cyanopropyl, 32% methyl)	275	57.0
SP-2330	Methyl silicone fluid (68% bis cyanopropyl, 32% dimethylsiloxane)	250	83.0
Silar 10C (CP Sil 88)	100% Cyanopropyl silicone	275	87.0
SP-2340	Methyl silicone fluid (75% cyanopropyl, 32% dimethylsiloxane)	250	87.2

Note: Phases are listed in order of increasing polarity.

Information supplied by Chrompack (Bridgewater, N.J.) and Supelco (Bellefonte, Pa.).

to completely adsorb to glass capillary columns in which the internal surface had been roughened by barium carbonate. Using glass liners in injection ports and deactivated glass columns (packed or capillary) with minimal adsorptive properties, the detection limits of alditol acetates are excellent with either flame ionization detection (FID) or GC-MS. Nevertheless, lower response factors for amino sugars compared to neutral sugars should be expected.

The conversion of aldoses to alditol acetates does not change the optical configuration of anomeric carbons, derivatized L- and D- sugar isomers should therefore be resolvable on an appropriate chiral phase. Such separations, currently not common, could be very useful for specific applications. D-Arabinitol, a product of certain fungi, is a chemical marker for detection of *Candida* in human serum. Normal serum may contain L-arabinitol which must

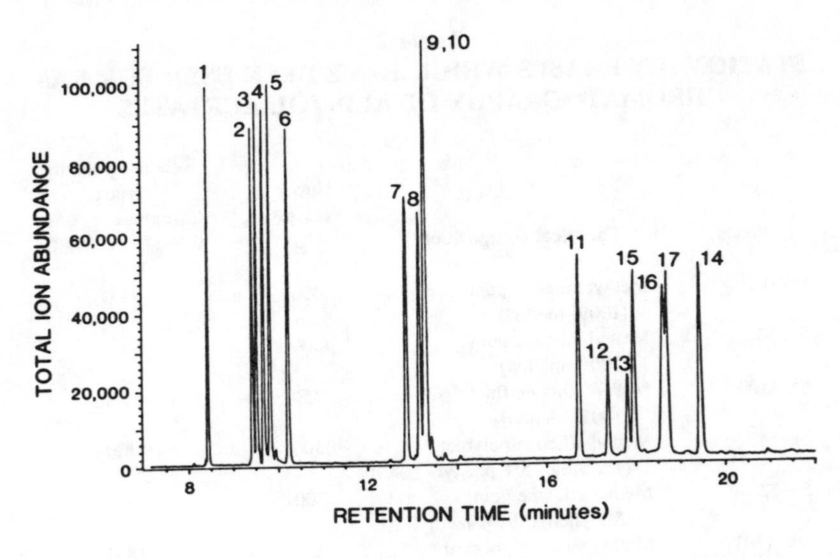

FIGURE 5. Chromatogram of a mixture of alditol acetates of neutral and amino sugars separated on a 25 m × 0.22 mm I.D. OV-1701 (BP10) fused-silica capillary column and detected by total ion monitoring. Peak identification as in Figure 4.

be resolved from the D-isomer produced by the yeast. Larsson has suggested that resolution of D- and L- arabinitol might improve the diagnosis of invasive candidiasis.[47] The D- and L- isomers of alditol trifluoroacetates (mannitol, fucitol, and arabinitol) have been separated on a column coated with a chiral phase.[48] Green et al. have described the use of a *N*-propionyl-L-valine *t*-butylamide chiral column for analysis of alditol acetates, but did not note whether D- and L-isomers of alditol acetates can be separated on this particular phase.[42] TMS enantiomers have also been resolved on chiral phase columns.[49] A more detailed discussion of chiral separations is presented in Chapter 12.

L. Detectors

Selective GC detectors aid the analysis of complex samples through elimination of extraneous background peaks derived from the sample matrix. Although detection of derivatized sugar mixtures down to a concentration of approximately 1 μg/mg of starting sample can be satisfactorily achieved using FID, selective detectors are often more sensitive.[15]

Whenham demonstrated the utility of the nitrogen/phosphorus detector (NPD) in the determination of amino sugars as their alditol acetates.[50] Pritchard et al. have also shown enhanced sensitivity using an electron capture detector (ECD) with trifluoroacetate derivatives.[4] The disadvantage of the NPD is that only amino sugars can be detected, while neutral sugars give no signal. The use of an ECD requires a derivatization reaction that introduces a halogen into the compound of interest. Although highly sensitive to such derivatives and more economical than a mass spectrometer, the ECD is notoriously unstable and subject to rapid contamination. A mass spectrometer in the selected ion monitoring (SIM) mode is a much more versatile, stable, and selective detector. Through appropriate selection of the monitored ions, all of the components of interest may be detected, while ignoring the background. GC-MS with SIM is not limited to a particular class of compounds as is the NPD and does not require specific chemical derivatives as does the ECD.

M. Mass Spectrometry (MS) of Alditol Acetate Derivatives

Alditol acetates do not generally give a molecular ion in MS with electron ionization (EI) at 70 eV and the spectra of isomers tend to be very similar. Structural analysis of carbohydrates by MS is well established, although the majority of applications have used derivatives other than alditol acetates.[51-53] The base peak in the EI spectra of alditol acetates is the acetylinium ion, $CH_3C\overset{+}{\equiv}O$ (m/z 43). Primarily fragments are produced by elimination of the acetoxyl group CH_3CO_2·, (m/z 59), acetic acid (m/z 60), ketene (m/z 42), or by cleavage of the alditol chain.

The 70 eV EI mass spectrum of glucitol hexaacetate, characteristic of hexose alditol acetates, is shown in Figure 6A. An interpretation of the 70 eV EI mass spectral fragmentation is given in Figure 7. Significant ions produced from glucitol hexaacetate (mol wt 434) include: m/z 375 (produced by loss of an acetoxyl group, m/z 59), m/z 361 and m/z 73 (produced by cleavage between C-1 and C-2 in the alditol chain), m/z 289 and m/z 145 (cleavage between C-2 and C-3 or C-4 and C-5 since glucitol acetate is symmetrical), and m/z 217 (cleavage between C-3 and C-4).

The 70 eV EI mass spectrum of glucosaminitol hexaacetate, characteristic of amino hexose derivatives, is shown in Figure 8A, and the interpretation of the mass spectral fragmentation is given in Figure 9. Fragmentation is strongly influenced by the presence of the acetamido group.[51] Significant ions produced from glucosaminitol hexaacetate (mol wt 433) include: m/z 375 (produced by loss of 59), m/z 360 (produced on cleavage between C-1 and C-2 in the alditol chain), m/z 144 (a nitrogen containing fragment derived from cleavage between C-2 and C-3), and m/z 145 (cleavage between C-4 and C-5). Other significant ions include: m/z 374 (loss of 59), m/z 318 (loss of 42 from 360), m/z 300 (loss of 60), m/z 259 (loss of 59 and 42), m/z 102 (loss of 42 from 144), and m/z 84 (loss of 60 from 144).

Glucosaminitol hexaacetate is asymmetrical (unlike glucitol hexaacetate), having an amide group on the number 2 carbon. Fragments which include this amide group are consequently one m/z unit less than those derived by cleavage from the opposite end of the molecule. Thus in the mass spectrum of glucosaminitol hexaacetate, the following pairs of ions are seen: m/z 102 and 103, 144 and 145, 216 and 217, 288 and 289, whereas glucitol hexaacetate shows only the second ion of each pair. This is illustrated by comparing the enlarged portions of the mass spectrum of glucitol hexaacetate (inset in Figure 6A) and glucosaminitol hexaacetate (inset in Figure 8A). In this example, glucosaminitol hexaacetate produces ions (m/z 144 and m/z 145, whereas glucitol hexaacetate produces only m/z 145.

For reference purposes, the mass spectra of other neutral sugar derivatives (rhamnose, ribose, deoxyribose, inositol, and L-glycero-D-α-mannoheptose) are shown in Figures 6B through F, respectively. Mass spectra of two other amino sugar derivatives (an aminodideoxyhexose and methylglucamine) are given in Figures 8B and C.

In the discussion of acetylation conditions, the formation of a lactam from muramic acid was described. The acetylation conditions affect the yield of muramic acid derivatives; under mild conditions, muramicitol produces two peaks, but in more extreme acetylation conditions, it produces a single peak. The mass spectra of these two muramic acid derivatives are shown in Figure 10. The spectrum of the earlier eluting derivative shows a strong ion at m/z 445 corresponding to the molecular ion of the pentaacetate of muramicitol lactam. The spectrum of the latter eluting derivative contains a molecular ion at m/z 403, corresponding to a tetraacetate.[22] Fox et al.[8,29] further confirmed the formation of a lactam derivative of muramic acid by observing ions of m/z 446 and 474 in methane chemical ionization mass spectra of muramic acid representing the (M + H) ion and the ($M + C_2H_5$) ions, respectively.

III. ANALYTICAL METHODOLOGY

The alditol acetate methodology as described above tends to be time consuming because it involves multiple manual operations and may produce rather dirty chromatograms if care is not taken with purity of reagents and choice of derivatization conditions. Optimal conditions

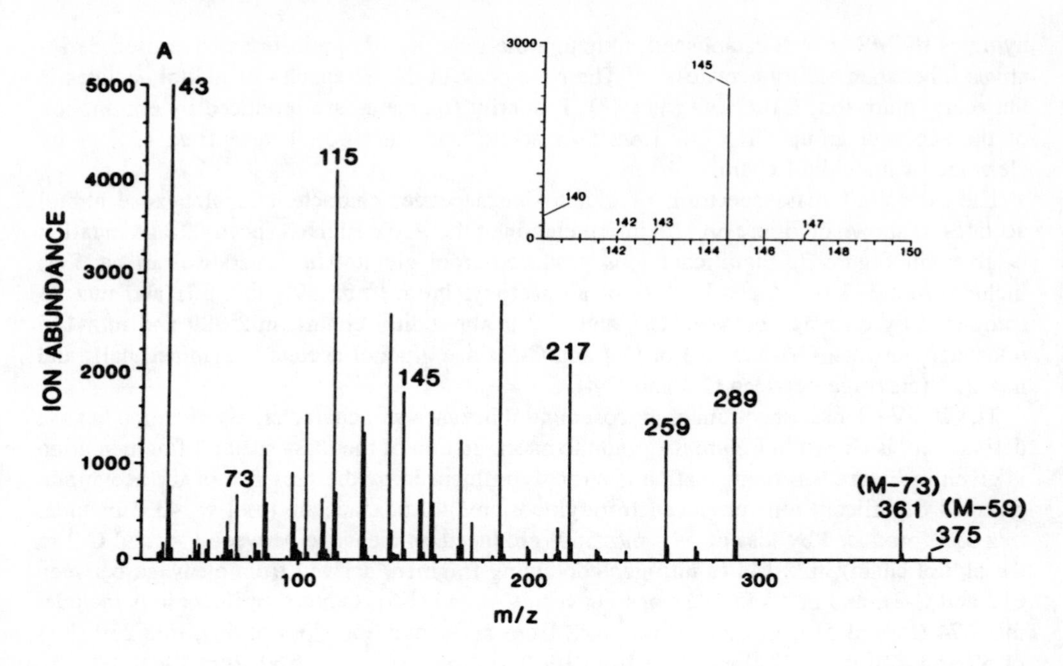

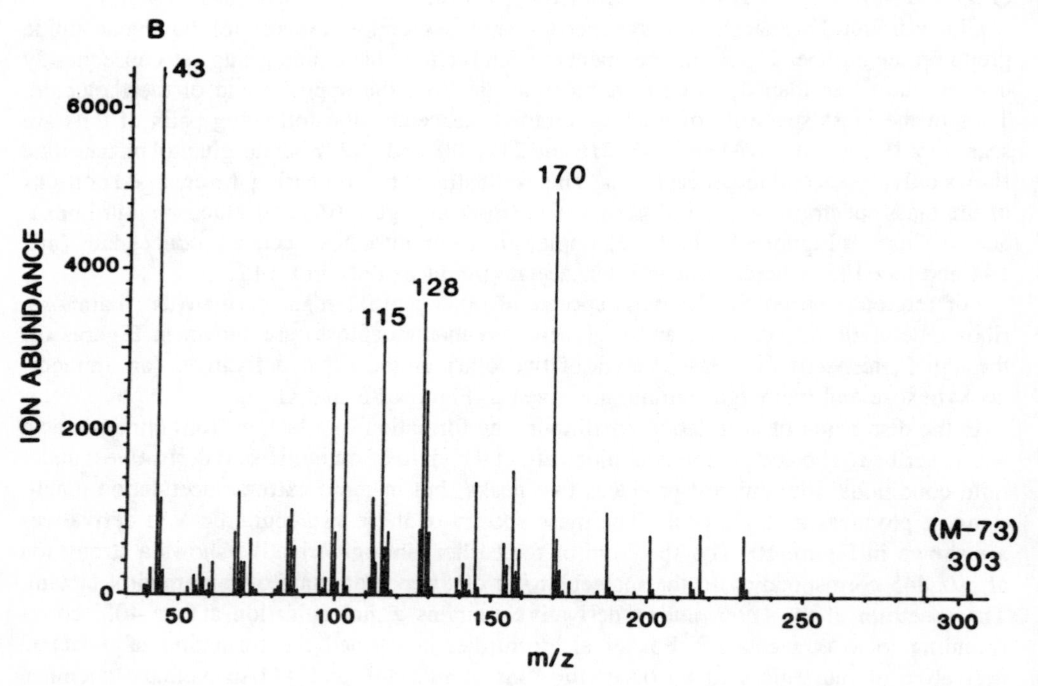

FIGURE 6. EI mass spectra (70 eV) of neutral sugar alditol acetates: A, glucose (mannose and galactose are similar). The enlarged portion, inset top right, shows ion m/z 145 (resulting from cleavage between C-2 and C-3 or C-4 and C-5). Note the absence of ion m/z 144 (compare with glucosamine in Figure 8A); B, rhamnose (fucose is similar); C, ribose (arabinose and xylose are similar); D, deoxyribose; E, inositol; F, L-glycero-D-mannoheptose (D-glycero-L-mannoheptose and D-glycero-D-mannoheptose are similar). Note in these spectra all peaks have been magnified such that the base peak m/z 43 is off-scale. The abundance of m/z 115 in all these neutral sugars is normally between 10 to 16% of ion m/z 43.

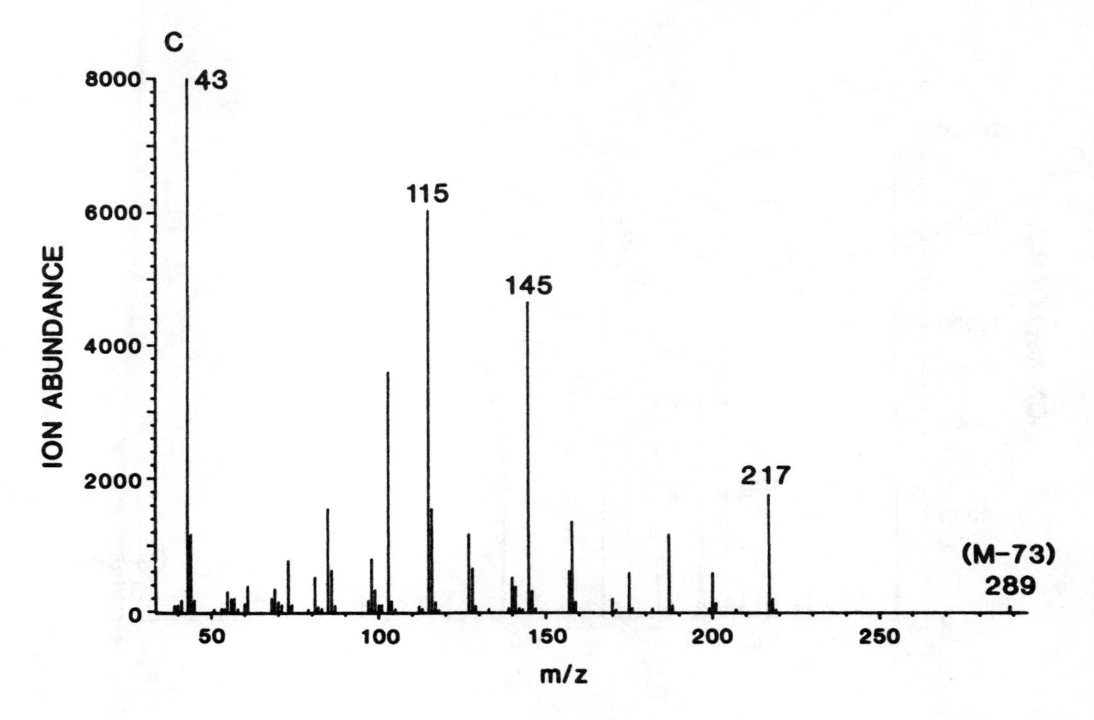

FIGURE 6C

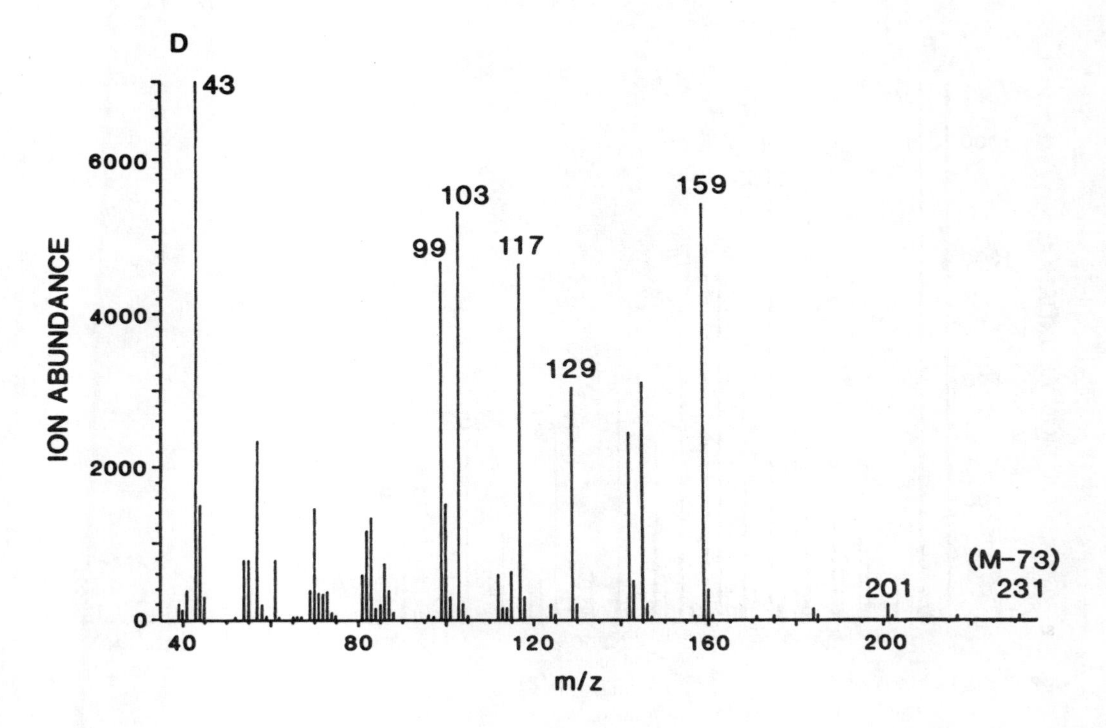

FIGURE 6D

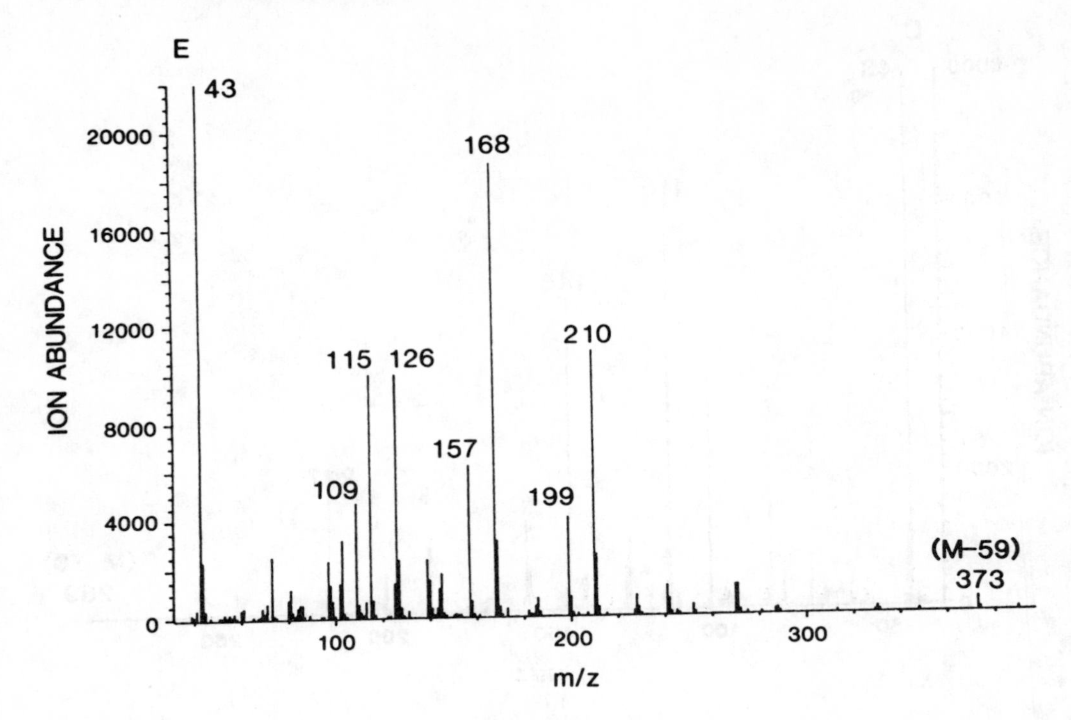

FIGURE 6E

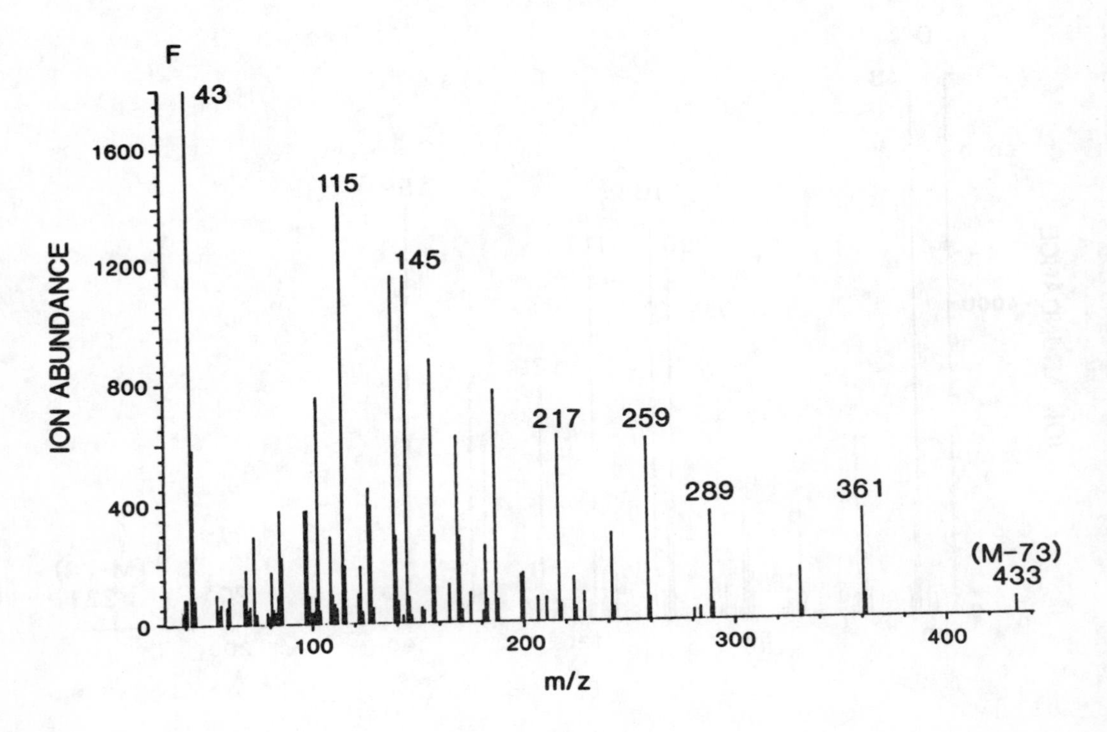

FIGURE 6F

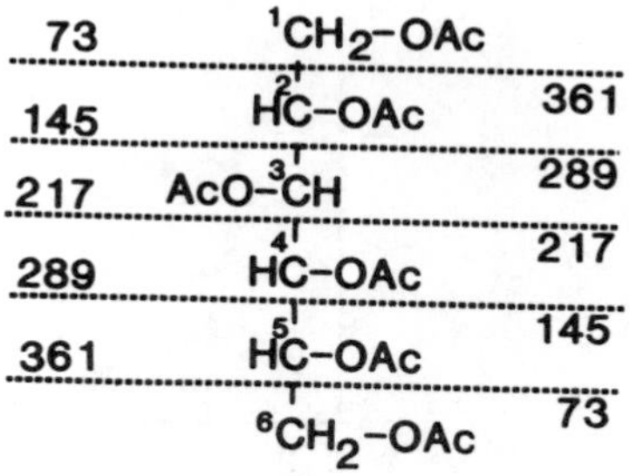

FIGURE 7. Interpretation of the 70 eV EI mass spectral fragmentation of glucitol hexaacetate as shown in Figure 6A.

for hydrolysis, reduction, and acetylation on large amounts of samples containing only common sugars have been previously investigated.[54] The method described below is designed for trace analysis of carbohydrates in complex material and overcomes many difficulties with small samples and manual processing.[8,15,16,22,39] Approximately 30 samples can be routinely analyzed in a simultaneous batch procedure consisting of the following five steps.

A. Prehydrolysis Sample Processing

Depending on the type of material to be analyzed, different sample preprocessing steps may be need to manipulate the sample into a form ready for hydrolysis and further derivatization. In tissue analysis, the sample is usually weighed, mechanically homogenized in phosphate buffer/detergent, and dialyzed. Dialysis of the sample prior to performing hydrolysis is useful for trace analysis of sugars occurring as monomeric constituents of bacterial cell wall macromolecules present in mammalian tissues. For other less complex samples, this step may not be necessary. Tissue homogenates in 0.05% Tween® 80 in phosphate buffered saline are dialyzed three times against distilled water. Low molecular weight free sugars such as glucose are removed from the sample by this dialysis procedure.

B. Hydrolysis and Neutralization

The pH of the aqueous samples is adjusted by adding sulfuric acid to reach a concentration of 2 *N* in a volume of 1 mℓ in a hydrolysis tube. These hydrolysis tubes are commercially available (Pierce, Rockford, Ill.) or can be made by sealing high vacuum stopcocks at one end of a hydrolysis tube (Chemglass, Inc., Vineland, N.J.). Figure 11A shows the custom-made apparatus permitting ten hydrolysis tubes to be alternately evacuated and flushed with nitrogen to remove oxygen. The evacuation apparatus consists of a nitrogen tank and vacuum pump, each connected to a copper-welded manifold and isolated by needle valves. Samples are frozen in the hydrolysis tubes prior to evacuation to reduce bumping, then unfrozen under vacuum and flushed several times with nitrogen. The PTFE valves on the hydrolysis tubes are closed under vacuum and hydrolysis performed in a Pierce Reacti-Therm heating module at 100°C for 3 hr. Figure 11B shows a heating module containing 18 hydrolysis tubes. After the tubes are cooled to room temperature, 2 μg of arabinose (as an internal standard for neutral sugars) and 2 μg methylglucamine (as an internal standard for amino sugars) in 200 μℓ water and 2.5 mℓ of 40% *N,N*-dioctylmethylamine in chloroform is added. The mixtures are vigorously mixed on a Vortex mixer and allowed to settle into two layers on standing. A 1 mℓ C-18 column (Analytichem) is prepared for each sample by rinsing with 2 mℓ methanol followed by 2 mℓ of distilled water. The upper, aqueous layer of each hydrolysis mixture is applied to a column. Each mixture is then pulled through the column by vacuum into a reaction vial, subsequently fitted with a screw cap and a PTFE silicone liner (Pierce) and the column is washed with 1.0 mℓ of distilled water. Figure 11C shows

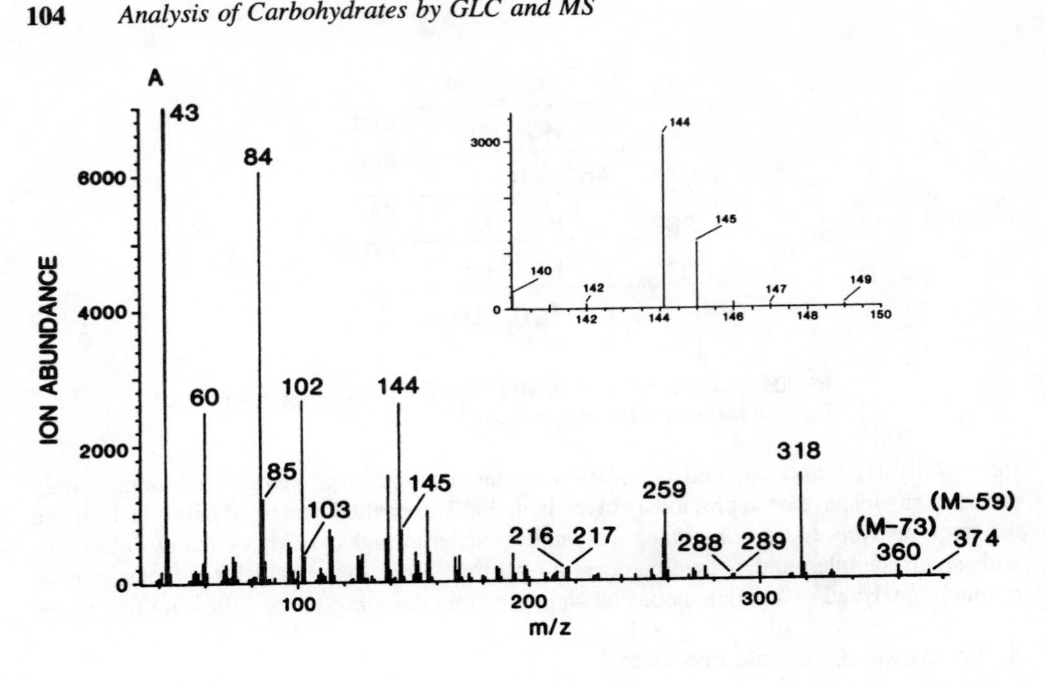

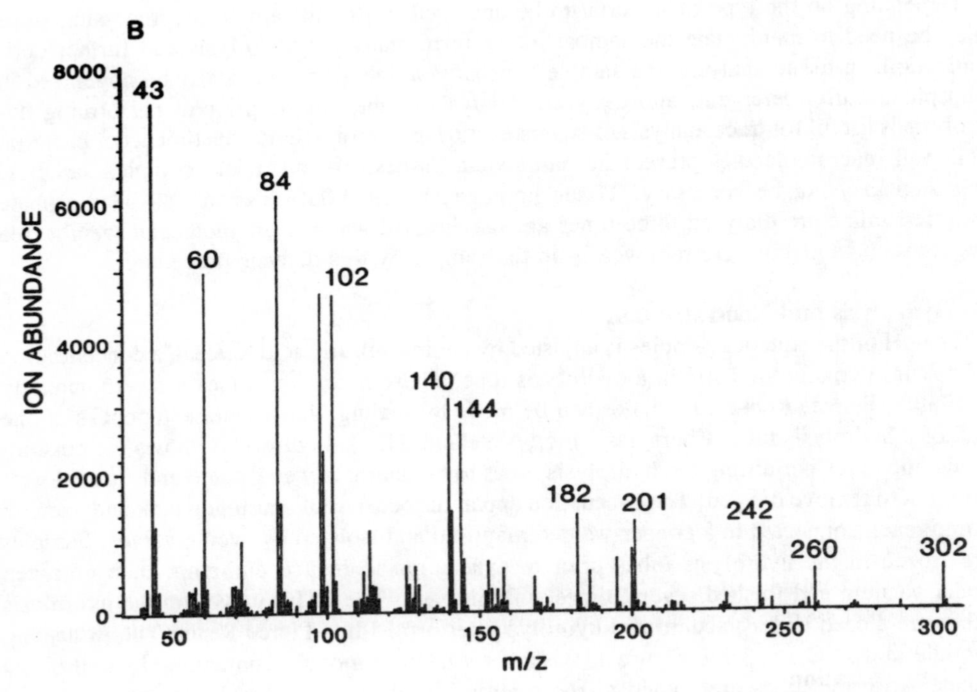

FIGURE 8. EI mass spectra (70 eV) of several amino sugars; A, glucosamine (galactosamine and mannosamine are similar). The enlarged portion of the spectrum, inset top right, shows an example of an ion which contains the amide group: m/z 144 (resulting from cleavage between C-2 and C-3) and the corresponding ion; m/z 145 (cleavage at the opposite end of the molecule, between C-4 and C-5). This ion does not contain the amide group (compare with Figure 6A); B, aminodideoxyhexose, X1 from *Fluoribacter (Legionella) bozemanae* and C, methylglucamine. Note in these spectra, all peaks have been magnified such that the base peak m/z 43 is off-scale. The abundance of ion m/z 84 (or m/z 86 in the case of methylglucamine) is approximately 50% of the base peak, m/z 43.

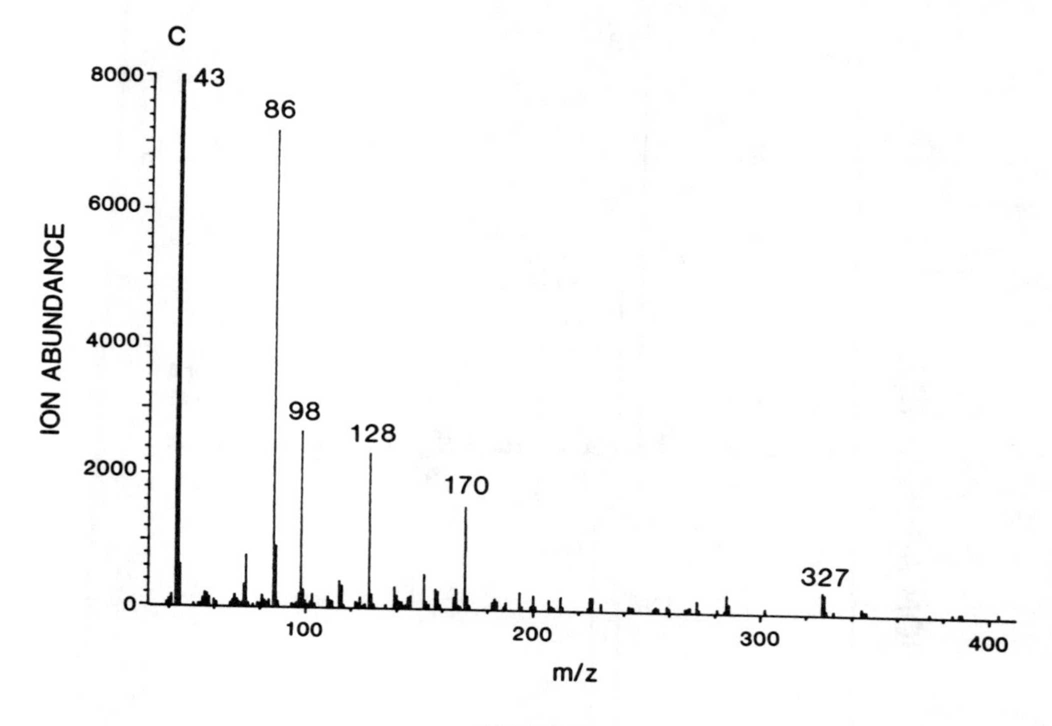

FIGURE 8C

	$^{1}CH_2$-OAc	
73		
144	^{2}CH-NH-Ac	360
		289
216	^{3}CH-OAc	
		217
288	^{4}CH-OAc	
		145
360	^{5}CH-OAc	
		73
	$^{6}CH_2$-OAc	

FIGURE 9. Interpretation of the 70 eV EI mass spectral fragmentation of glucosaminitol hexaacetate as shown in Figure 8A.

the Vac-Elut container opened to show reaction vials positioned to collect the column eluant. The Vac-Elut system (Analytichem) permits up to ten samples to be run simultaneously.

C. Derivatization

Added to the sample is 50 μℓ of sodium borohydride (100 mg/mℓ in water). The reduction is allowed to proceed overnight in a refrigerator or for 90 min at 37°C. Excess sodium borohydride is destroyed by adding 2 mℓ of acetic acid/methanol (1:200 v/v) to the sample which is then evaporated to dryness in a Vortex evaporator (Buchler, Fort Lee, N.J.) at 60°C under vacuum. Figure 11D shows the Vortex evaporator and solvent trap. Up to 36 samples can be evaporated to dryness simultaneously. The evaporator can be heated up to 80°C while shaking. The evaporation step is repeated four additional times to ensure complete removal of borate. The samples are allowed to dry for 3 hr after the last evaporation. After

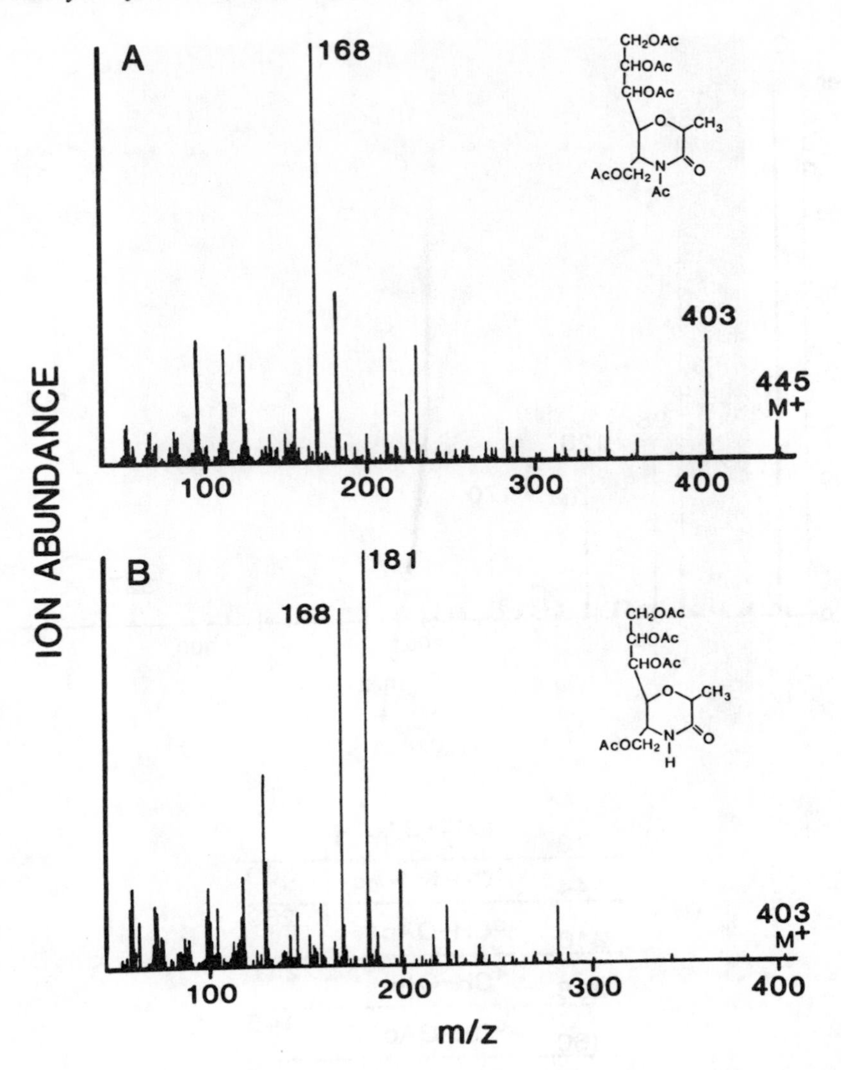

FIGURE 10. Electron-impact (EI) mass spectra (70 eV) of A, muramicitol pentaacetate and B, muramicitol tetraacetate. (From Whiton, R. S., Lau, P., Morgan, S. L., Gilbart, J., and Fox, A., *J. Chromatogr.*, 347, 109, 1985. With permission.)

cooling to room temperature, 300 μℓ of acetic anhydride is added to each vial and the sample heated at 100°C for 13 to 16 hr in the Reacti-Therm heating module.

D. Postderivatization Sample Cleanup

Samples are cooled in an ice bath, and 0.75 mℓ of water is added to each and left for 30 min in order to decompose excess acetic anhydride to acetic acid. Following the addition of 1 mℓ of chloroform, the sample is mixed by vortexing and the aqueous phase is removed and discarded. To the chloroform phase, 0.8 mℓ of cold ammonium hydroxide (80%, v/v) is added. This addition is carried out slowly in order to minimize the heat generated during neutralization. The mixture is poured onto a magnesium sulfate (Chem-Elut) column and eluted with 2 mℓ chloroform. Figure 11E shows ten columns being loaded to extract a batch of samples. Note the vials positioned beneath the columns to collect the eluant. The chloroform solution is evaporated to dryness under vacuum and redissolved in about 40 μℓ of chloroform before analysis.

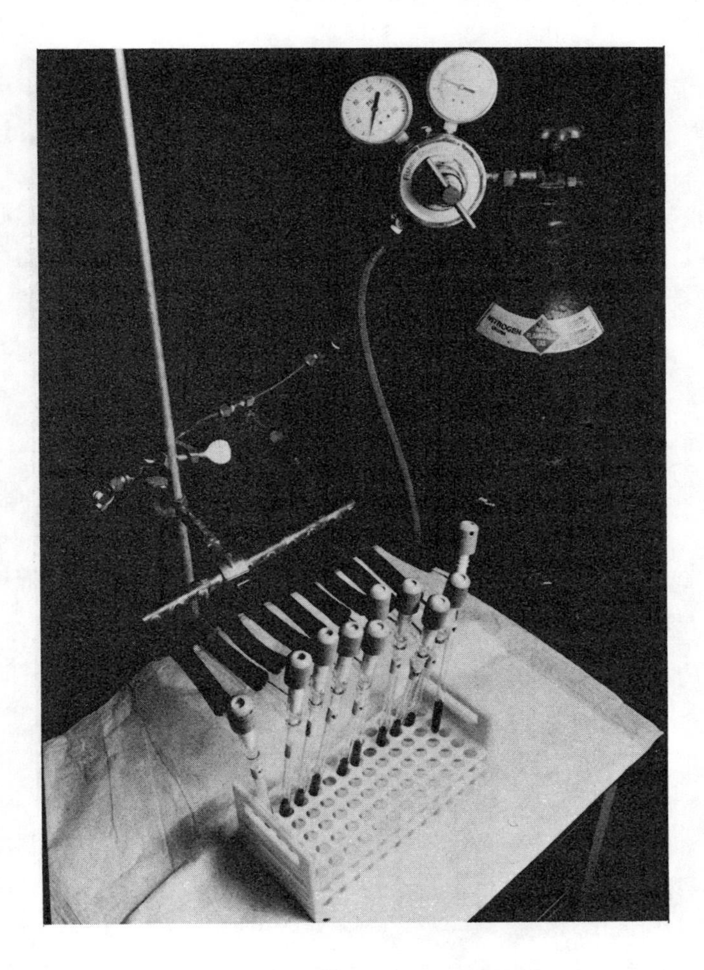

FIGURE 11. A, A custom-made apparatus permitting ten hydrolysis tubes to be alternately evacuated and flushed with nitrogen to remove oxygen; B, A heating module containing 18 hydrolysis tubes; C, The Vac-Elut container opened to show reaction vials positioned to collect eluant from a set of C-18 columns; D, The Vortex evaporator (with lid raised) connected via a solvent trap to a vacuum pump (latter not in picture); (Up to 36 samples can be vacuum-evaporated to dryness simultaneously. The evaporator can be heated up to 80° while shaking.); E, Ten Chem-Elut (magnesium sulfate) columns being loaded to extract a batch of samples. (Note the vials positioned beneath the columns to collect the eluant.)

E. GC-MS Analytical Conditions

GC-MS analyses are routinely performed using a 5970 Mass Selective Detector (Hewlett-Packard, Palo Alto, Calif.), a bench-top mass spectrometer, interfaced via a capillary direct inlet to a HP-5890 gas chromatograph equipped with a capillary column. However, other low-cost MS instruments including the Ion Trap (Finnigan, San Jose, California), as well as more complex quadropole or magnetic sector systems, are equally suitable. Fused-silica capillary columns, typically 25 m × 0.22 mm, and coated with either BP10 (an OV-1701 bonded phase from SGE, Austin, Tex.) or with SP-2330 (Supelco, Bellefonte, Pa.) are used for the analyses. Samples are injected (0.5 to 1.0 $\mu\ell$) in the splitless mode and after 0.75 min the split valve is reopened at a 1:20 split ratio with a column helium flow of 0.84 $m\ell$/min. The splitless liner is loosely packed with glass wool to avoid accumulation of

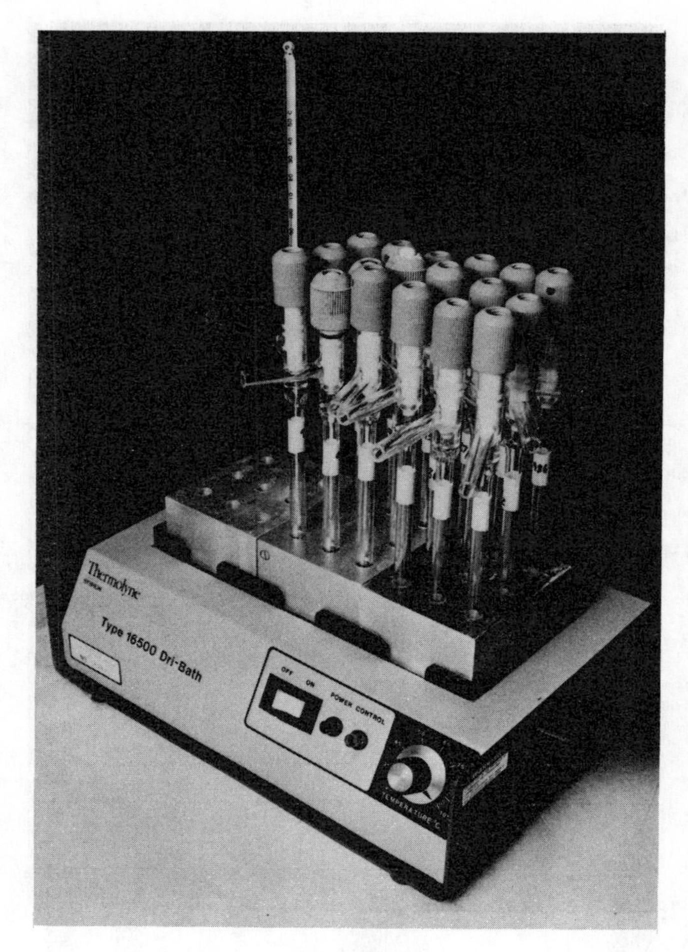

FIGURE 11B

nonvolatile material on the column head. GC injection is performed with the column oven held at 100°C, for 45 sec, programmed at 30°C/min to 230°C, at 4°C/min to 250°C and at 3°C/min to 265°C and then held for 4.5 min. The GC injector temperature is set at 250°C and the GC-MS interface is set to 270°C.

Mass spectral data may be accumulated in a total ion abundance mode with complete scanning of the mass spectra of eluting components, or in a selected ion mode that focuses for greater sensitivity on only a few of the prominent ions present in the sugars of interest. Variations in the selected ion monitoring parameters can be employed depending on the target carbohydrate and the purpose of the analysis. As an example, analysis of most common sugars can be carried out by monitoring three groups of ions at the appropriate retention times: first, ion at m/z 289.05 and m/z 303.1 for detection of pentoses/hexoses and methylpentoses, respectively; second, ions at m/z 318.0 and m/z 403.1 for detection of aminohexoses and muramic acid, respectively; third, ions at m/z 327.25 for detection of methylglucamine.

IV. APPLICATIONS

The alditol acetate procedure is widely used for quantitating or identifying the monomeric constituents of samples in which sugars are major components.[9,23] Common biological

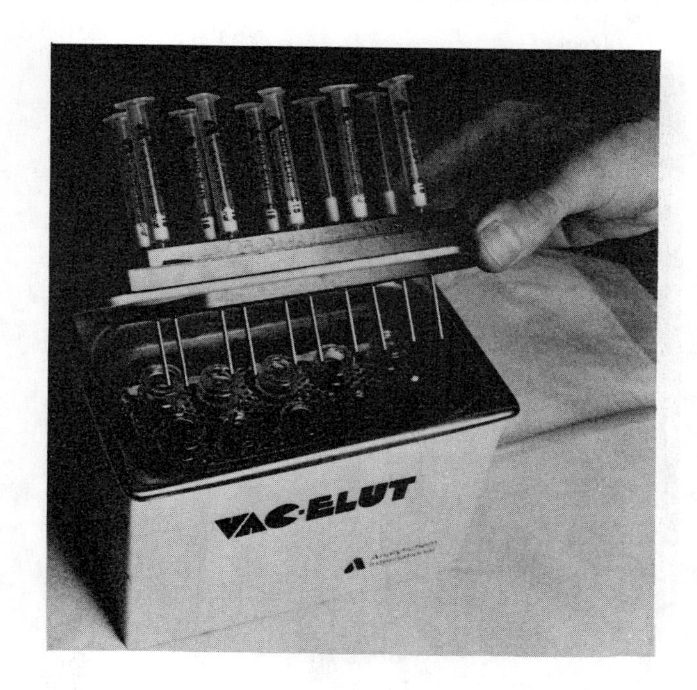

FIGURE 11C

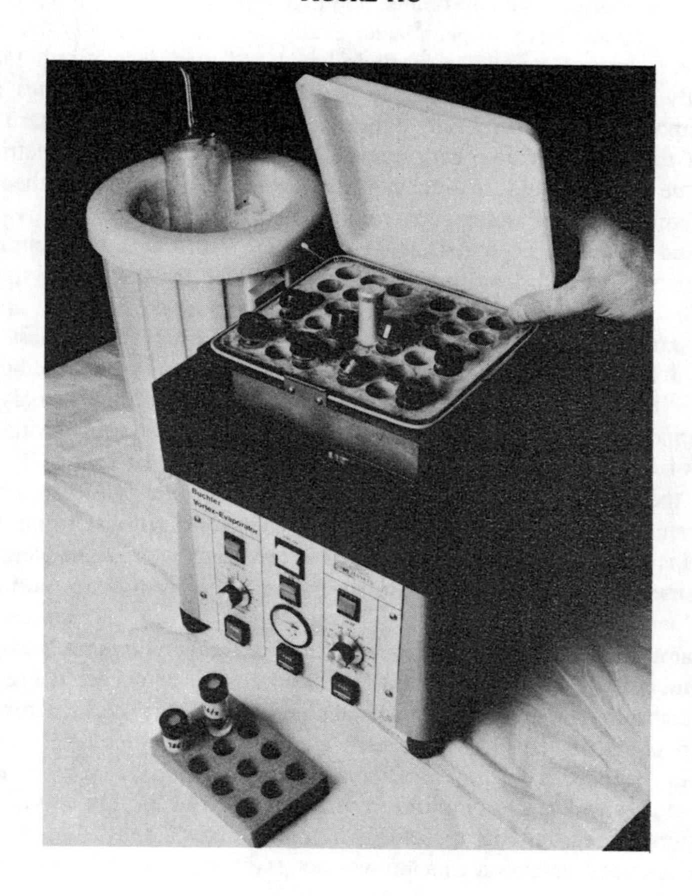

FIGURE 11D

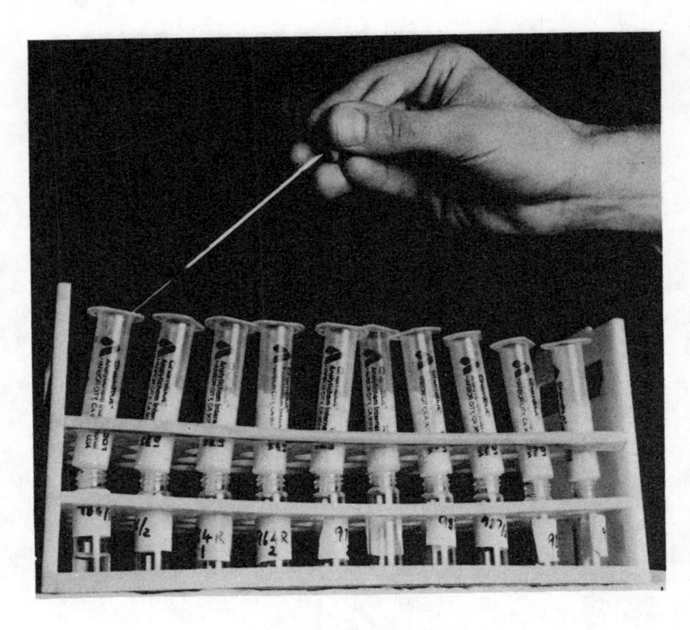

FIGURE 11E

compounds (polysaccharides, glycoproteins, and glycolipids) isolated from plant and animal tissues or microorganisms have all been successfully analyzed.[10] Usually the method starts with one to several mg of sample to be hydrolyzed and then derivatized. Problems may occur when only small amounts of sample are available (less than a mg) or when the monomeric component of interest is only a minor constituent (<1% of the starting material). The analysis of trace carbohydrate constituents in a complex biological matrix, such as a mammalian tissue or body fluid, is a considerably more difficult problem than the analysis of major sugar components of isolated macromolecules.[15,21,55] Greater selectivity in both the derivatization and cleanup procedures, as well as in the instrumental detection conditions, are usually required. An additional problem encountered in the trace analysis of sugars in complex samples is that side reactions between the component of interest and other constituents of the sample may occur resulting in decreased sensitivity.[39] The mass spectrometer may be operated in different modes depending on the requirement, including quantitative trace analysis or qualitative mass spectral identification. For definitive analysis of minor sugar components of a complex matrix where other sugars are present in high abundance, SIM GC-MS is essential.

To illustrate the practical use of the alditol acetate procedure, we include examples of alditol acetate analyses of increasing complexity. The modifications in the derivatization procedure and instrumental conditions described in the previous sections were essential for the successful use of the method in these examples taken from work performed in our laboratories.

The first example is a quantitative determination of the carbohydrate composition of a purified bacterial cell wall sample isolated from Group A streptococci.[15] Figure 12 shows a representative chromatogram of alditol acetates present in a hydrolysate from a sample of *Streptococcus pyogenes* cell walls. A borosilicate-glass capillary column coated with SP-2330 was custom-made for efficient separation and improved performance with amino sugars. The method started with 1 mg of lyophilized cell wall material. In this example, the sample preparation differed from the above protocol in that the acetic anhydride was removed by evaporation and postderivatization cleanup was not extensive. An FID was adequate for the

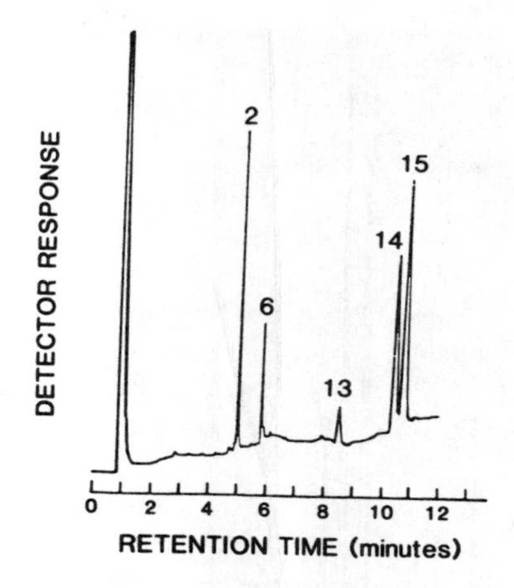

FIGURE 12. Chromatogram of alditol acetates derived from streptococcal cell walls separated on a 30 m × 0.32 mm I.D. SP2330 glass capillary column and detected with a flame ionization detector. Peak identification as in Figure 4. (From Fox, A., Morgan, S. L., Hudson, J. R., Zhu, Z.-T., and Lau, P. Y., *J. Chromatogr.*, 256, 429, 1983. With permission.)

detection of the major carbohydrate components. The cell walls were found to contain approximately 17.1% rhamnose, 4% muramic acid, and 16% glucosamine by quantitation with xylose and methylglucamine as internal standards for neutral and amino sugar components, respectively. In later work using GC-MS, glucose has also been shown to be present as a minor component of such cell wall preparations and is derived from contamination of the cell walls with membrane glycolipids. An FID is not selective enough to definitively demonstrate the low levels of glucose present in this sample.

A second example of the application of alditol acetate analysis is the analysis of a mammalian serum glycoprotein — transferrin. Here, the carbohydrate is present as a minor component in comparison to protein. A similar application was the carbohydrate analysis of ferritin, another mammalian glycoprotein.[55] In this work, the levels of component sugars found using FID were sufficiently low that the signal-to-noise ratio was of concern, although the background noise levels did not invalidate the analysis. Figure 13 illustrates a selected ion chromatogram of alditol acetates derived from human transferrin separated on an SP-2330 fused-silica capillary column. Human transferrin has been reported to contain six mannose, four galactose, eight glucosamine, and four neuraminic acid residues per molecule.[56] The major sugars found in the chromatogram of the hydrolyzed and derivatized commercial preparation were mannose, galactose, and glucosamine. Neuraminic acid is an acidic sugar and thus it cannot be detected using the alditol acetate method. Small amounts of fucose, glucose, and xylose were also found to be present in this sample; these minor components cannot be detected using an FID. Thus the MS provides qualitative information for identification of peaks, but also may be necessary for quantitation of minor components.

The third application of the alditol acetate method is a carbohydrate profiling study of *Legionella pneumophila*, the pathogen of legionnaires' disease, and some related organisms.[29,57] The taxonomy of the *Legionellaceae* had been studied previously by a variety of methods including genetic and biochemical data, ubiquinone patterns, fatty acid analysis, peptide profiles, and antigenic composition.[58,59] Our objective was to show that carbohydrate profiling can be a valuable diagnostic aid in reference microbiology laboratories for differentiating bacteria. Neutral and amino sugar content of whole bacteria was determined by the modified alditol acetate derivatization method described above in the cell wall analysis, employing GC with FID on a fused-silica capillary column coated with SP-2330. As an

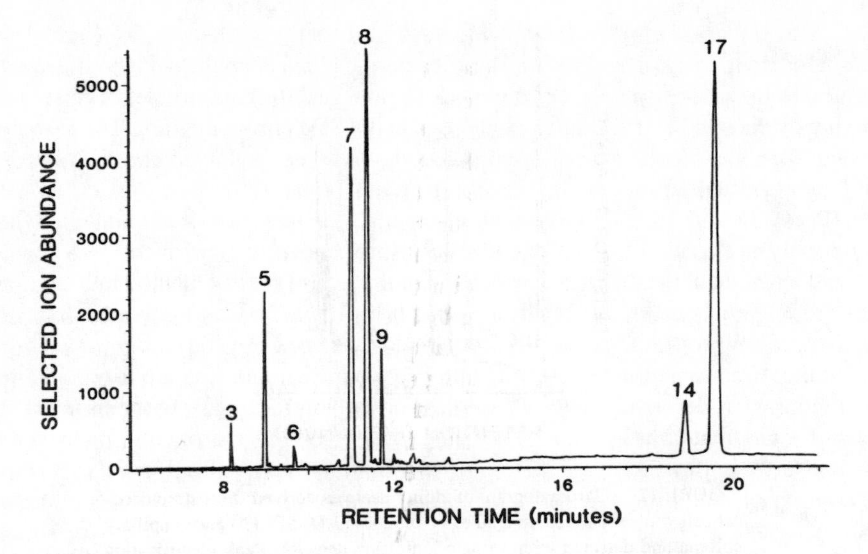

FIGURE 13. Selected ion chromatogram of alditol acetates derived from transferrin separated on a 30 m × 0.32 mm I.D. SP 2330 fused silica capillary column. Peak identification as in Figure 4.

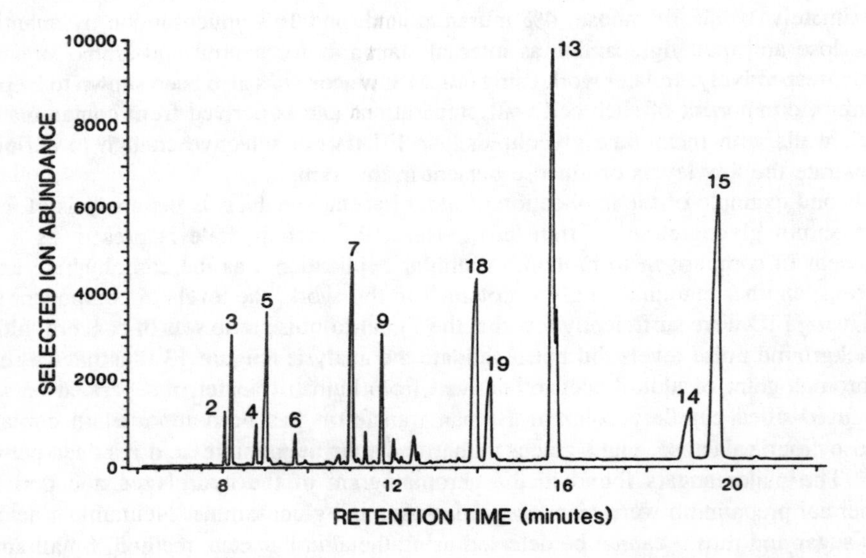

FIGURE 14. A selected ion chromatogram of alditol acetates of *Fluoribacteria (Legionella) bozemanae* cells separated on a 30 m × 0.32 mm I.D. SP-2330 capillary column. Peak identification and monitoring parameters as in Figure 4. An additional ion m/z 260 was monitored from 6 to 14.5 min to detect aminodideoxyhexoses (Peaks 18 and 19).

example of the quantitative results from this study, the amounts of sugars as percent sample dry weight for a hydrolysate of *Fluoribacteria (Legionella) bozemanae* (WIGA) were 0.3% rhamnose, 0.31% fucose, 0.31% ribose, 0.36% mannose, a trace level of glucose, two unusual aminodideoxyhexoses (X1 at 1.13% and X2 at 0.88%), 0.49% muramic acid, and 0.35% glucosamine. It should be realized that 0.3% sugar in such a sample represents only 3 μg of carbohydrate component in the original sample. Figure 14 presents a chromatogram for a hydrolysate of *Fluoribacteria (Legionella) bozemanae* performed more recently using

SIM GC-MS. In this chromatogram, arabinose and methylglucamine were employed as internal standards. In earlier work on these bacteria, glucose could only be detected at an unmeasurable trace level; however, this peak is now readily measurable. Xylose was also not previously detected by FID and is easily seen in the SIM chromatogram. The carbohydrate components are seen as sharp peaks well above the baseline. As noted above, this organism contains two unusual aminodideoxyhexoses that elute close to the heptoses on an SP-2330 column (Peaks 18 and 19).[29,57] Following our report, the presence of aminodedioxyhexoses in the lipopolysaccharide of legionella has been independently confirmed.[60]

The final example of the practical application of the alditol acetate method involves samples of tissue from animals in which polyarthritis had been elicited by the systemic administration of streptococcal cell wall fragments.[8,22,39,61] Samples of liver were dialyzed prior to hydrolysis and the alditol acetate methodology, including extensive sample cleanup using alkaline and acid extraction post-derivatization (as described in Section II above), was employed. Figure 15A shows a chromatogram of alditol acetates from a 100 mg sample of a liver taken from a rat 4 days after intraperitoneal injection of streptococcal cell walls. Figure 15B shows the chromatogram of alditol acetates from a normal control liver. Both samples contain large amounts of mannose, galactose, glucose, glucosamine, and lesser amounts of fucose and galactosamine. Deoxyribose was not detected in these samples since it is destroyed by acid hydrolysis. Mannosamine, not a common component of mammalian tissues, was also not detected. In the samples from animals injected with the bacterial cell walls, two additional sugars, muramic acid and rhamnose, were found. These sugars are remnants of the bacterial cell walls initially injected into these animals. Rhamnose and muramic acid, when analyzed as alditol acetates, can be used as chemical markers for bacterial cell walls in mammalian tissues. These sugars can be detected at levels as low as 1 ng/mg of tissue by SIM GC-MS. A number of other unique sugars found in bacteria and fungi may be of use for the detection of bacteria in body fluids and tissues, including heptoses and ketodeoxyoctonic acid as markers for Gram-negative bacteria[62] and arabinitol as a marker for *Candida* and other fungi.[28]

V. CONCLUSION

In this review we have outlined the major steps in the alditol acetate method for preparation of volatile derivatives of carbohydrates. These steps include release of the individual sugars by hydrolysis from complex polysaccharides, reduction of the monomer to an alditol, and then acetylation to produce a volatile derivative. The major factors influencing these steps have been described in some detail along with specific recommendations based on experiences reported in the literature. We have also described a variety of cleanup steps and precautions that may be employed to help prevent the introduction of extraneous contaminating peaks into the chromatogram. A detailed description of the alditol acetate method as implemented in our laboratories has been given, including specific recommendations for suitable hydrolysis and derivatization equipment. Finally, we have described some applications taken from our own work applying the alditol acetate method to the analysis of the carbohydrate composition of samples of purified bacterial cell walls and mammalian glycoproteins, to carbohydrate profiling of whole bacteria and, last, to trace analysis of carbohydrate markers for bacterial cellular constituents in mammalian tissues. The successful application of the alditol acetate method in these examples demonstrates the value of the method for a wide range of situations.

Although significant improvements in acetylation catalysts have recently occurred simplifying derivatization, some common versions of this method still require tedious evaporations to remove the borate before acetylation and the simplification of this step would be a great advance. Future research on alditol acetates might be profitably focused on simplifying and automating sample handling. Decreasing the time required for analysis, finding other

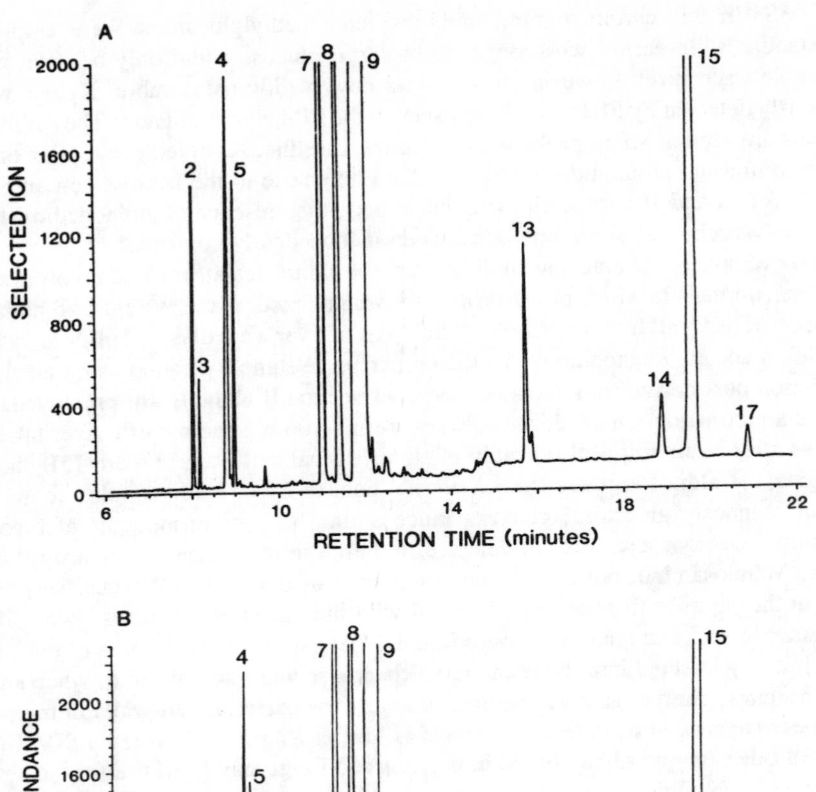

FIGURE 15. Selected ion monitoring chromatogram of alditol acetates prepared from hydrolysates of 100 mg (wet weight) of liver samples from A, a rat 4 days after intraperitoneal injection of streptococcal cell walls and B, a control rat injected with PBS. Analyses were performed on a 30 m × 0.32 mm I.D. SP-2330 fused-silica capillary column. Two groups of ions were monitored: first, from 6 to 14.5 min, m/z 303.1 for rhamnose and fucose and 289.05 for other neutral sugars; secondly, from 14.5 to 22.0 min, m/z 318.0 for glucosamine, mannosamine, and galactosamine, m/z 327.25 for methylglucamine and m/z 403.15 for muramic acid. Peak identification as in Figure 4.

improved catalysts for acetylation, and further increasing the sensitivity and selectivity of the analytical method would be extremely advantageous.

ACKNOWLEDGMENTS

This work was supported by grants from the U.S. Army Research Office, from the National Institutes of Health (NIH Grant No. EY04715), from the American Heart Association, and from the National Science Foundation Biological Instrumentation Program. The work of Joseph R. Hudson, Pauline Lau, Larry Eudy, Michael D. Walla, and Robert S. Whiton is gratefully acknowledged.

REFERENCES

1. **Gunner, S. W., Jones, J. K. N., and Perry, M. B.,** Analysis of sugar mixtures by gas-liquid partition chromatography, *Chem. Ind. (London),* 255, 1961.
2. **Sawardeker, J. S., Sloneker, J. H., and Jeanes, A.,** Quantitative determination of monosaccharides as their alditol acetates, *Anal. Chem.,* 37, 1602, 1965.
3. **Sweeley, C. C., Bentley, R., Makita, M., and Wells, W. W.,** Gas-liquid chromatography of trimethylsilyl derivatives of sugars and related substances, *J. Am. Chem. Soc.,* 85, 2497, 1963.
4. **Pritchard, D. G. and Niedermeier, W.,** Sensitive gas chromatographic determination of the monosaccharide composition of glycoproteins using electron capture detection, *J. Chromatogr.,* 152, 487, 1978.
5. **Varma, R. and Varma, R. S.,** Simultaneous determination of neutral sugars and hexosamines in glycoproteins and acid mucopolysaccharides (glycosaminoglycans) by gas-liquid chromatography, *J. Chromatogr.,* 128, 45, 1976.
6. **Chen, C. C. and McGinnis, G. D.,** The use of 1-methylimidazole as a solvent and catalyst for the preparation of aldononitrile acetates of aldoses, *Carbohydr. Res.,* 90, 127, 1981.
7. **Mawhinney, T. P., Feather, M. S., Barbero, G. J., and Martinez, J. R.,** The rapid, quantitative determination of neutral sugars (as aldononitrile acetates) and amino sugars as (*O*-methyloxime acetates) in glycoproteins by gas-liquid chromatography, *Anal. Biochem.,* 101, 112, 1980.
8. **Fox, A., Schwab, J. H., and Cochran, T.,** Muramic acid detection in mammalian tissues by gas-liquid chromatography-mass spectrometry, *Infect. Immun.,* 29, 526, 1980.
9. **Niedermeier, W.,** Gas chromatography of neutral and amino sugars in glycoproteins, *Anal. Biochem.,* 40, 465, 1971.
10. **Kakehi, A. and Honda, S.,** Profiling of carbohydrates, glycoproteins and glycolipids, *J. Chromatogr.,* 379, 27, 1986.
11. **Robards, K. and Whitelaw, M.,** Chromatography of monosaccharides and disaccharides, *J. Chromatogr.,* 373, 81, 1986.
12. **Furneaux, R. H.,** Separatory and analytical methods, *Carb. Chem.,* 16, 249, 1985.
13. **Dutton, G. G. S.,** Applications of gas liquid-chromatography to carbohydrates, *Adv. Carbohydr. Chem. Biochem.,* 28, 11, 1973.
14. **Hough, L., Jones, J. V. S., and Wusteman, P.,** On the automated analysis of neutral monosaccharides in glycoproteins and polysaccharides, *Carbohydr. Res.,* 21, 9, 1972.
15. **Fox, A., Morgan, S. L., Hudson, J. R., Zhu, Z-T., and Lau, P. Y.,** Capillary gas chromatographic analysis of alditol acetates of neutral and amino sugars in bacterial cell walls, *J. Chromatogr.,* 256, 429, 1983.
16. **Hudson, J. R., Morgan, S. L., and Fox, A.,** High-resolution glass capillary columns for the gas chromatographic analysis of alditol acetates of neutral and amino sugars, *J. High Res. Chromatogr. Chromatogr. Commun.,* 5, 285, 1982.
17. **Kim, J. H., Shome, B., Liao, T.-H., and Pierce, J. G.,** Analysis of neutral sugars by gas-liquid chromatography of alditol acetates: application to thyrotropic hormone and other glycoproteins, *Anal. Biochem.,* 20, 258, 1967.
18. **Blakeney, A. B., Harris, P. J., Henry, R. J., and Stone, B. A.,** A simple and rapid preparation of alditol acetates for monosaccharide analysis, *Carbohydr. Res.,* 113, 291, 1983.
19. **Hicks, K. B., Simpson, G. L., and Bradbury, A. G. W.,** Removal of boric acid and related compounds from solutions of carbohydrates with a boron-selective resin (IRA-743), *Carbohydr. Res.,* 147, 39, 1986.
20. **McGinnis, G. D.,** Preparation of aldononitrile acetates using *N*-methylimidazole as catalyst and solvent, *Carbohydr. Res.,* 108, 284, 1982.
21. **Henry, R. J., Blakeney, A. B., Harris, P. J., and Stone, B. A.,** Detection of neutral and aminosugars from glycoproteins and polysaccharides as their alditol acetates, *J. Chromatogr.,* 256, 419, 1983.
22. **Whiton, R. S., Lau, P., Morgan, S. L., Gilbart, J., and Fox, A.,** Modifications in the alditol acetate method for analysis of muramic acid and other neutral and amino sugars by capillary gas chromatography-mass spectrometry with selected ion monitoring, *J. Chromatogr.,* 347, 109, 1985.

23. **Englyst, H. N. and Cummings, J. H.**, Simplified method for the measurement of total non-starch polysaccharides by gas-liquid chromatography of constituent sugars as alditol acetates, *Analyst (London)*, 109, 937, 1984.
24. **Bryn, K. and Jantzen, E.**, Analysis of lipopolysaccharides by methanolysis, trifluoroacetylation, and gas chromatography on a fused silica capillary column, *J. Chromatogr.*, 240, 405, 1982.
25. **Matsui, M., Okada, M., Imanari, T., and Tamura, Z.**, Gas chromatography of trifluoroacetyl derivatives of alditols and trimethylsilyl derivatives of aldonolactones, *Chem. Pharm. Bull.*, 16, 1383, 1968.
26. **Cowie, G. L. and Hedges, J. L.**, Determination of neutral sugars in plankton, sediments and wood by capillary gas chromatography of equilibrated isomeric mixtures, *Anal. Chem.*, 56, 497, 1984.
27. **Jakobs, C., Warner, T. G., Sweetman, L., and Nyhan, W. L.**, Stable isotope dilution analysis of galactitol in amniotic fluid: an accurate approach to prenatal diagnosis of galactosemia, *Pediatr. Res.*, 18, 714, 1984.
28. **Roboz, J., Suzuki, R., and Holland, J. F.**, Quantification of arabinitol in serum by selected ion monitoring as a diagnostic technique in invasive candidiasis, *J. Clin. Microbiol.*, 12, 594, 1980.
29. **Walla, M. D., Lau, P. Y., Morgan, S. L., Fox, A., and Brown, A.**, Capillary gas chromatography-mass spectrometry of carbohydrate components of legionellae and other bacteria, *J. Chromatogr.*, 288, 399, 1984.
30. **Morrison, I. M. and Perry, M. B.**, The analysis of neutral glycoses in biological materials by gas-liquid partition chromatography, *Can. J. Biochem.*, 44, 1115, 1966.
31. **Taylor, R. L. and Conrad, H. E.**, Stoichiometric depolymerization of polyuronides and glycosaminoglycuronans to monosaccharides following reduction of their carbodiimide activated carboxyl group, *Biochemistry*, 11, 1383, 1972.
32. **Lehrfeld, J.**, GLC determination of aldonic acids as acetylated aldonamides, *Carbohydr. Res.*, 135, 179, 1985.
33. **Bishop, C. T., Cooper, F. P., and Murray, R. K.**, Reactions of carbohydrate derivatives during gas-liquid chromatography, *Can. J. Chem.*, 41, 2245, 1963.
34. **Warth, A. D. and Strominger, J. L.**, Structure of the peptidoglycan from spores of *Bacillus subtilis*, *Biochemistry*, 11, 1389, 1972.
35. **Findlay, R. H., Moriarty, D. J. W., and White, D. C.**, Improved method of determining muramic acid from environmental samples, *Geomicrobiol. J.*, 3, 135, 1983.
36. **Jennings, W.**, *Gas Chromatography with Glass Capillary Columns*, 2nd ed., Academic Press, New York, 1980.
37. **Kaiser, M. A. and Klee, M. S.**, Current status of high resolution column technology for gas chromatography, *J. Chromatogr. Sci.*, 24, 369, 1986.
38. **Grob, K. and Grob, G.**, Splitless injection on capillary columns, I. The basic technique; steroid analysis as an example, *J. Chromatogr. Sci.*, 7, 584, 1969.
39. **Gilbart, J., Fox, A., Whiton, R. S., and Morgan, S. L.**, Rhamnose and muramic acid: chemical markers for bacterial cell walls in mammalian tissues, *J. Microbiol. Meth.*, 5, 271, 1986.
40. **Kiho, T., Ukai, S., and Hara, C.**, Simultaneous determination of the alditol acetate derivatives of amino and neutral sugars by gas-liquid chromatography, *J. Chromatogr.*, 369, 415, 1986.
41. **Oshima, R., Kumanotani, J., and Watanabe, C.**, Fused silica capillary gas chromatographic separation of alditol acetates of neutral and aminosugars, *J. Chromatogr.*, 250, 90, 1982.
42. **Green, R. C., Doctor, V. M., Holzer, G., and Oro, J.**, Separation of neutral and amino sugars by capillary gas chromatography, *J. Chromatogr.*, 207, 268, 1981.
43. **Griggs, L. J., Post, A., White, E. R., Finkelstein, J. A., Moeckel, W. E., Holden, K. G., Zarembo, J. E., and Weisbach, J. E.**, Identification and quantitation of alditol acetates of neutral and amino sugars from mucins by automated gas-liquid chromatography, *Anal. Biochem.*, 43, 369, 1971.
44. **Richey, J. M., Richey, H. G., and Schraer, R.**, Quantitative analysis of carbohydrates using gas-liquid chromatography, *Anal. Biochem.*, 9, 272, 1964.
45. **Oates, M. D. G. and Shrager, J.**, The determination of sugars and aminosugars in the hydrolysates of mucopolysaccharides by gas-liquid chromatography, *J. Chromatogr.*, 28, 232, 1967.
46. **Zanetta, J. P., Breckenridge, W. C., and Vincendon, G.**, Analysis of monosaccharides by gas-liquid chromatography of the *O*-methyl glycosides as trifluoroacetate derivatives, *J. Chromatogr.*, 69, 291, 1972.
47. **Larsson, L.**, Gas chromatography and mass spectrometry, in *Automation in Clinical Microbiology*, Jorgensen, J., Ed., CRC Press, Boca Raton, Fla., 1987, 153.
48. **König, W. A. and Benecke, I.**, Enantiomer separation of polyols and amines by enantiomer selective gas chromatography, *J. Chromatogr.*, 269, 19, 1983.
49. **Leavitt, A. L. and Sherman, W. R.**, Direct gas chromatographic resolution of DL-myo-inositol 1-phosphate and other sugar enantiomers as simple derivatives on a chiral capillary column, *Carb. Res.*, 103, 203, 1982.
50. **Whenham, R. J.**, Sensitive assay for amino sugars using capillary gas chromatography with nitrogen selective detection, *J. Chromatogr.*, 303, 380, 1984.

51. **Longren, J. and Svensson, S.,** Mass spectrometry in structural analysis, *Adv. Carbohydr. Chem. Biochem.*, 29, 41, 1974.
52. **Beynon, J. H., Saunders, R. A., and Williams, A. E.,** *The Mass Spectra of Organic Molecules*, Elsevier, Amsterdam, 1968.
53. **Biemann, K., DeJongh, D. C., and Scnoes, H. K.,** Application of mass spectrometry to structure problems. XIII. Acetates of pentoses and hexoses, *J. Am. Chem. Soc.*, 85, 1763, 1963.
54. **Torello, L. A., Yates, A. J., and Thompson, D. K.,** Critical study of the alditol acetate method for quantitating small quantities of hexoses and hexosamines in gangliosides, *J. Chromatogr.*, 202, 195, 1980.
55. **Collawn, J. F., Lau, P. Y., Morgan, S. L., Fox, A., and Fish, W. W.,** A chemical and physical comparison of ferritin subunit species fractionated by high performance liquid chromatography, *Arch. Biochim. Biophys.*, 233, 260, 1984.
56. **Dorland, L., Haverkamp, J., Schut, B. L., Vliegenhart, J. F. G., Spik, G., Strecker, G., Fournet, B., and Montreuil, J.,** The structure of the asialo-carbohydrate units of human serotransferrin as proven by 360 MHz proton magnetic resonance spectroscopy, *FEBS Lett.*, 77, 15, 1977.
57. **Fox, A., Lau, P., Brown, A., Morgan, S. L., Zhu, Z.-T., Lema, M., and Walla, M. D.,** Capillary gas chromatographic analysis of carbohydrates of *Legionella pneumophila* and other members of the Legionellaceae, *J. Clin. Micro.*, 19, 326, 1984.
58. **Brenner, D. J., Steigerwalt, A. G., Gorman, G. W., Wilkinson, H. W., Bibb, W. F., Hackel, M., Tyndall, R. L., Campbell, J., Feeley, J. C., Thacker, W. L., Skaliy, P., Martin, W. T., Brake, B. J., Fields, B. S., McEachern, H. V., and Corcoran, L. K.,** Ten new species of *Legionella*, *Int. J. Syst. Bacteriol.*, 35, 50, 1985.
59. **Campbell, J., Bibb, W. F., Lambert, M. A., Eng, S., Steigerwalt, A. G., Allard, J., Moss, C. W., and Brenner, J.,** *Legionella sainthelensi:* a new species of *legionella* isolated from water near Mt. St. Helens, *Appl. Environ. Microbiol.*, 47, 369, 1984.
60. **Otten, S., Iyer, S., Johnson, W., and Montgomery, R.,** Serospecific antigens of *Legionella pneumophila*, *J. Bacteriol.*, 167, 893, 1986.
61. **Gilbart, J. and Fox, A.,** Elimination of group A streptococcal cell walls from mammalian tissues, *Infect. Immun.*, 55, 1526, 1987.
62. **Fox, A. and Morgan, S. L.,** The chemotaxonomic characterization of microorganisms by capillary gas chromatography and gas chromatography-mass spectrometry, in *Instrumental Methods for Rapid Microbiological Analysis*, Nelson, W. H., Ed., Verlag Chemie, Deerfield Beach, Fla., 1985, 135.

Chapter 6

ANALYSIS OF MONOSACCHARIDES AS PER-*O*-ACETYLATED ALDONONITRILE (PAAN) DERIVATIVES BY GAS-LIQUID CHROMATOGRAPHY (GLC)

Gary D. McGinnis and Christopher J. Biermann

TABLE OF CONTENTS

I. INTRODUCTION

Gas chromatographic (GC) analysis of carbohydrates has been performed employing several types of derivatives. The trimethylsilyl (TMS) ethers of carbohydrates have been widely used in the past because of their ease of formation and their relatively high volatility. The increased volatility is particularly important for the analysis of oligosaccharides. TMS derivatives of carbohydrates have several disadvantages. Water reacts with the silylating reagents and hydrolyzes the silylated products; consequently, the sample to be derivatized must be reasonably dry, less than about 10% water. Moreover, individual sugars give two or more peaks each, due to the various ring and anomeric forms of each sugar in solution. Quantitative analysis of samples containing only a very few simple sugars is very good, but accurate analysis of complex mixtures is much more difficult because of the overlap of some of the peaks. Accurate quantitation depends on obtaining a constant ratio between anomeric forms of a particular carbohydrate; unfortunately, the ratio is actually dependent on the composition of the sample. One approach taken by some investigators is to convert the sugars into oximes before TMS. When the *syn* and *anti* isomers are not resolved, each sugar shows only one peak. Nevertheless, this derivative still inherits the other disadvantages of TMS ethers.

The alditol acetate, aldononitrile acetate (per-*O*-acetylated aldononitrile, [PAAN]), and, more recently, *O*-methyloxime acetate derivatives of sugars are becoming more widely used. The major advantage of the alditol acetate method is that the reaction gives only one product for each saccharide. With aldononitrile acetates and *O*-methyloxime acetates, more than one product is obtained with certain carbohydrates (e.g., glucose and mannose). The reaction of the oxime with acetic anhydride gives a mixture of the nitrile and glycosylamine. Fortunately, the amount of glycosylamine produced is not dependent on the reaction conditions or on the ratio of anomeric forms in solution and does not interfere with the analysis, providing the GC detector is calibrated for the different carbohydrates after conversion into their PAAN derivatives.

With the *O*-methyloxime acetates, introduction of capillary columns to carbohydrate analysis allows separation of complex mixture of carbohydrates in an effective fashion, with each carbohydrate giving two peaks, in a reproducible ratio, corresponding to the *syn* and *anti* forms.

Acetylated derivatives are generally employed for monosaccharides only, since derivatives of oligosaccharides are not sufficiently volatile for analysis by gas-liquid chromatography (GLC) with polar packings. Exceptions to this are the study of methylated oligosaccharides and the analysis of some di- and trisaccharides.[1,2]

Neutral and amino sugars may be analyzed by this method, though the latter group with a bit more difficulty. With the introduction of acetylation catalysts more powerful than pyridine,[3-6] the formation of the various acetylated derivatives has become much easier and faster. With pyridine as the solvent, the formation of alditol acetates is time consuming because of the necessity of removing boric acid, formed from the reduction of sodium borohydride, before acetylation.[7] However, when 1-methylimidazole is used as the solvent, borate removal is not necessary, and acetylation can be done in <10 min.[3,4]

Due to recent advances in the formation and separation of acetylated derivatives, the choice of which derivative to use for a particular sample is not always clear. In many cases, several methods may be suitable for a particular analysis. For samples where only the neutral monosaccharides in aqueous solutions are of concern, the method developed in our laboratory is particularly simple and fast.[3,4] When sulfuric acid is used in the hydrolysis step, which is very common for hydrolysis of wood and other samples, its removal is not necessary prior to derivatization, which is not true for other methods of derivatization. It is quite likely that other acids such as trifluoroacetic acid and hydrochloric acid would not have to be

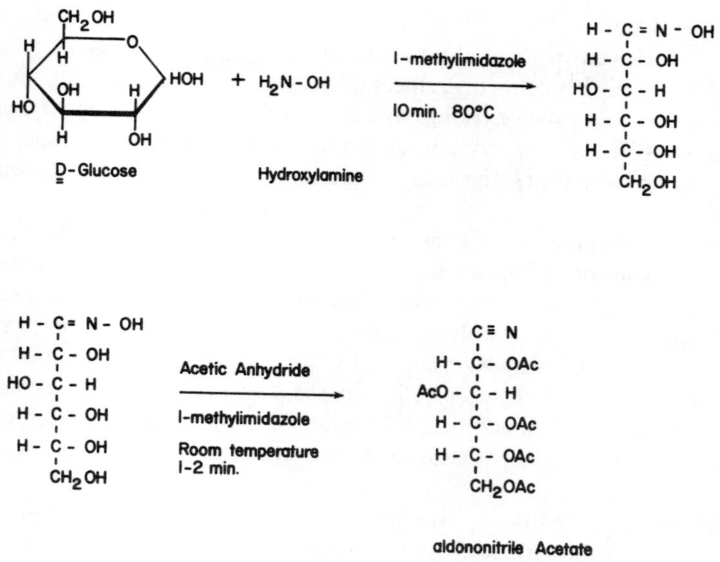

FIGURE 1. General reaction scheme for the formation of peracetylated aldononitriles.

removed prior to derivatization, but this hypothesis has not been tested and reported in the literature.

II. PREPARATION OF THE PAAN DERIVATIVES

Various methods available for the formation of peracetylated aldononitriles share a common scheme shown in Figure 1 for D-glucose. In the first reaction, the aldehyde is converted to the corresponding oxime by reacting the sugars with hydroxylamine hydrochloride with a suitable catalyst/solvent such as pyridine, 1-methylimidazole, or, occasionally, sodium acetate. The presence of water does not interfere with the reaction. In the second step, excess acetic anhydride is added so that all available hydroxyl groups and water are acetylated. The oxime is probably also acetylated during dehydration to form the nitrile. The catalyst in the first step is usually used as the catalyst in the second step. It can be noted that the formation of *O*-methyl oxime acetate derivatives is analogous, except that *O*-methylhydroxyamine hydrochloride is employed in the first step. In this case, though, the methoxy group is a poor leaving group, so that the integrity of the oxime remains in the second step.

Although the derivatization of neutral aldoses is straightforward, derivatization of amino sugars is much more difficult. When 1-methylimidazole is used as the catalyst, there is some degradation of the 2-amino-2-deoxyaldoses during derivatization;[3,4] when 1-dimethylamino-2-propanol was used, the derivatization of these amino sugars was not reproducible.[8] Pyridine is not a suitable catalyst for the formation of aldononitrile acetates from the amino sugars either,[9-12] although the amino sugars are sometimes deaminated with sodium nitrite before their analysis.[11,12] Deamination is sometimes used prior to the formation of alditol acetates or prior to methanolysis as well. The use of 4-(dimethylamino)pyridine (DMAP), a very powerful acylation catalyst, has recently been successfully used to form the aldononitrile acetate derivatives from these hexosamines and neutral sugars simultaneously.[13] Alditols, if present, are converted to the corresponding alditol acetate derivatives simultaneously.

Methods are given here only for convenience. Some details may be missing; consequently, before a method is actually employed for sample analysis, the original work should be consulted.

A. Pyridine Catalyst

Lance and Jones[9] were perhaps the first to prepare PAAN derivatives of sugars. They prepared PAAN derivatives of methyl ethers of xylose. Their method calls for 1 to 3 mg of sugar dissolved in 5 drops of pyridine to be treated with 1 to 3 mg of hydroxylamine hydrochloride at 90°C for 1 hr. Acetic anhydride is then added (15 drops) with heating continued for an additional hr. The samples are then injected into the GC without further sample cleanup.

Most methods employing pyridine as the only catalyst do not vary much from this procedure; however, solvent tailing on the GC can be reduced and GC column life can be increased by using a sample cleanup procedure after derivatization. The method used by Varma et al.[10] calls for 1 mg hydrolyzed guar gum sample (or other sample or standard) in 5 drops pyridine and 1 mg of hydroxylamine hydrochloride to be heated at 90°C for 30 min in a sealed ampule. Acetic anhydride (15 drops) is then added to the cooled ampule, with the resealed ampule heated an additional 30 min. The residue is evaporated at 40°C under reduced pressure. The residue is dissolved in 0.1 mℓ of dry chloroform for injection into the GC.

Typical deamination conditions, used before derivatization, from Varma, et al.[11] are as follows: hydrolyzed residue (2 to 5 mg) is taken up in 0.03 to 0.04 mℓ of concentrated hydrochloric acid and diluted with 1 mℓ of distilled water in a 25 mℓ evaporating flask. Sodium nitrite (2 to 3 mg) is added to the cooled flask (0 to 5°C), and the flask stoppered and stirred occasionally for 10 to 15 min. The stopper is removed and the flask is placed in boiling water for 2 to 3 min. Distilled water (10 mℓ) is added to the cooled solution, and its pH adjusted to 5 to 6 with a trace of Bio-Rad® AG 1-X2(HCO_3^-) resin, and the resin removed by filtration. The filtrate and washings are evaporated to dryness by rotoevaporation at ambient temperature or freeze-drying.

B. 1-Methylimidazole Catalyst

A slight modification of the published method[3,4] is very useful for sugar hydrolyzates containing up to 2 *N* sulfuric acid. Hydrolyzate or sugar standard solution (0.2 mℓ, 1 to 20 mg carbohydrate) containing an internal standard is added to a 6.3 mℓ vial (vinyl-lined screw cap) along with 0.4 mℓ of catalyst/solvent stock solution (1 g of hydroxylamine hydrochloride in 20 mℓ of 1-methylimidazole) and mixed. The capped vial is heated at 80°C for 10 min (5 min for analysis of amino sugars), cooled to room temperature, and 1 mℓ of acetic anhydride is added carefully as the reaction is quick and exothermic. After 5 min chloroform (1 mℓ) is added and the solution washed twice with 2 mℓ of water; the water layer is conveniently removed by a pipette. Anhydrous sodium sulfate (~ 0.40 g) is then added.

Preliminary experiments using *O*-methyloxime hydrochloride, instead of hydroxylamine hydrochloride, with this method did not give the *O*-methyloxime derivatives in a reproducible fashion, and was not as suitable as the method of Mawhinney[8] or Neeser and Schweizer.[5]

C. 1-Dimethylamino-2-Propanol Catalyst

Mawhinney, et al.[8] prepared PAAN acetates of neutral sugars using 1-dimethylamino-2-propanol as a catalyst. Neutral sugars (0 to 5 mg in no more than 0.04 mℓ water) are oximated using 0.2 mℓ of reagent (0.6 g hydroxylamine hydrochloride in 2.0 mℓ methanol and 5.47 mℓ pyridine to which 0.53 mℓ of 1-dimethylamino-2-propanol is added). The solution is heated at 70°C for 5 min, cooled to room temperature, and dried with a stream of dry air (10 min). Pyridine-acetic anhydride solution (1.0 mℓ; 1:3, v/v) is added, the contents mixed, heated to 70°C for 25 min, cooled to room temperature, the concentrated to a syrup using a stream of dry air. The contents are dissolved in 1.0 mℓ of chloroform, washed with 1.0 mℓ of 1.0 *N* hydrochloric acid (to remove pyridine), and washed three times with 1.0 mℓ of distilled water.

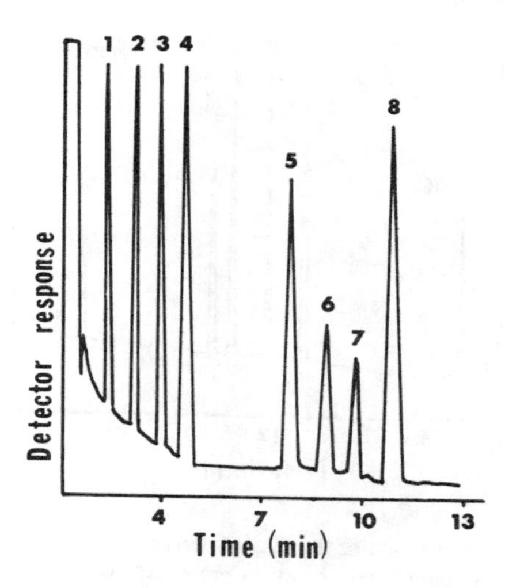

FIGURE 2. Separation of the PAAN derivatives of neutral sugars on a packed column with 1% diethylene glycol adipate (190°C 2 min, 3°C/min to 210°C, hold). 1, rhamnose (2.41 min); 2, fucose/ribose (3.22); 3, arabinose (4.04); 4, xylose (4.78); 5, mannose, (7.95); 6, glucose (9.09); 7, galactose (9.96); and 8, myoinositol (internal standard) (11.18).

D. 4-(Dimethylamino)pyridine (DMAP) Catalyst

Guerrant and Moss[13] were able to form PAAN derivatives of neutral and amino sugars simultaneously using DMAP as a catalyst. Hydroxylamine reagent (0.3 ± 0.05 mℓ; 32 mg/mℓ of hydroxylamine hydrochloride and 40 mg/mℓ of DMAP in 4:1 v/v pyridine-methanol) is added to a dry sample of 0.15 mg of each standard carbohydrate in a 13 × 100 mm Teflon®-lined, screw-capped culture tube. The sample is sonicated for 1 min, heated at 75°C for 25 min, and cooled. Then 1.0 mℓ of acetic anhydride is added, and the tube is closed and sonicated for 1 min and reheated for 15 min. The tube is cooled and 2.0 mℓ of 1,2-dichloroethane is added. Excess derivatization reagents are removed by two extractions with 1.0 mℓ of 1 *N* HCl followed by three extractions with 1.0 mℓ of water. Extractions were performed as quickly as possible to minimize hydrolysis. The 1,2-dichloroethane is then evaporated using dry nitrogen and heating to 50°C, and the sample reconstituted in 0.4 mℓ of ethyl acetate-hexane (1:1, v/v).

III. ANALYSIS OF PAAN DERIVATIVES BY GAS-LIQUID CHROMATOGRAPHY (GLC)

Of course the method of derivatization has little bearing on the separation of the derivatized monosaccharides. Early separation of PAAN derivatives such as Varma et al.,[10] while state of the art for the time, are not impressive in terms of modern packed columns or, better yet, capillary columns. However, these earlier studies are useful for obtaining retention times of derivatized sugars relative to each other. Second, in most cases, the number of carbohydrates obtained from natural sources are rather limited and the resolution obtained on a capillary column is not critical. For most analyses of some of the common neutral carbohydrates, a packed column works very well, as shown in Figure 2. Wide-bore glass-capillary columns may add to the separation (Figure 3).

Seymour, et al.[14-16] studied the GC and mass spectrometry of a number of PAAN derivatives. One study looked at separations using GLC columns packed with OV-17 and neopentyl glycol succinate.[14] The other studies are concerned with methodology for structural analysis.[15,16] PAAN derivatives have been used to analyze component neutral monosaccharides of gums[17] and foods.[18]

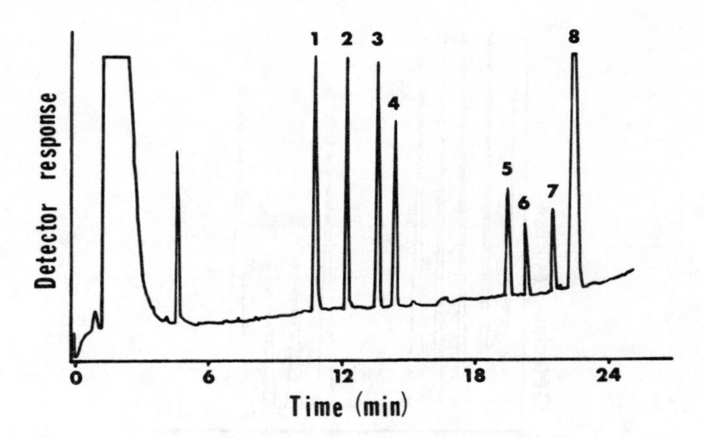

FIGURE 3. Separation of the PAAN derivatives on a 60-m-wide-bore glass capillary column with bonded Carbowax® phase (H_2 carrier gas at 65 cm/sec, 210°C 2 min, 2°C/min to 250°C, hold). 1, rhamnose (10.90 min); 2, fucose/ribose (12.41); 3, arabinose (13.69); 4, xylose (14.59); 5, mannose, (19.40); 6, glucose (20.30); 7, galactose (21.48); and 8, glucitol (internal standard) (22.43).

For more elaborate separations, capillary columns are required. Guerrant and Moss,[13] one of the few sources for capillary column separation of PAAN derivatives, separated 28 carbohydrates, (neutral and amino, including neuraminic acids and hydrolyzates from whole bacteria) and alditols as PAAN and alditol acetate derivatives on a fused-silica capillary column (0.2 mm × 50 m) coated with OV-1. Future research with other column coatings will extend the range of separations possible by GLC of the PAAN derivatives.

IV. MASS SPECTROMETRY (MS) OF PAAN DERIVATIVES

Much of the work on mass spectrometry (MS) of PAAN derivatives has been performed by Seymour and co-workers[14-16] and by Li and co-workers.[18] MS studies using both electron-impact and chemical ionization MS have been done using a variety of PAAN derivatives. By using both ionization techniques, Seymour was able to determine not only the molecular weight, but the number of aldehyde and alcohol groups and their relative position. Li and co-workers used chemical ionization MS to identify methylated carbohydrates obtained from laminaran, a polysaccharide found in seaweed. The PAAN derivatives have also been used by Varma and co-workers to identify arabinose in brain hyaluronic acid.

PAAN derivatives have the presence of the nitrile to distinguish the (former) reducing end of the sugar in MS. Consequently the PAAN derivatives of sugars are very useful for MS studies of carbohydrates. The nitrile resists cleavage, thus relatively large fragments arise from the nitrile end-group.

REFERENCES

1. **Kärkkäinen, J.,** Structural analysis of trisaccharides as permethylated trisaccharide alditols by gas-liquid chromatography-mass spectrometry, *Carbohydr. Res.*, 17, 11, 1971.
2. **Schwind, H., Scharbert, F., Schmidt, R., and Katterman, R.,** Gas chromatographic determination of di- and trisaccharides, *J. Clin. Chem. Biochem.*, 16, 145, 1978.
3. **McGinnis, G. D.,** Preparation of aldononitrile acetates using *N*-methylimidazole as catalyst and solvent, *Carbohydr. Res.*, 108, 284, 1982.

4. **Chi, C. C. and McGinnis, G. D.,** The use of 1-methylimidazole as a solvent and catalyst for the preparation of aldononitrile acetates of aldoses, *Carbohydr. Res.*, 90, 127, 1981.
5. **Neeser, J.-R. and Schweizer, T. F.,** A quantitative determination by capillary gas-liquid chromatography of neutral and amino sugars (as *O*-methyl oxime acetates), and a study of hydrolytic conditions for glycoproteins and polysaccharides in order to increase sugar recoveries, *Anal. Biochem.*, 142, 58, 1984.
6. **Henry, R. J., Blakeney, A. B., Harris, P. J., and Stone, B. A.,** Detection of neutral and aminosugars from glycoproteins and polysaccharides as their alditol acetates, *J. Chromatogr.*, 256, 419, 1983.
7. **Albersheim, P., Nevins, D. J., English, P. D., and Karr, A.,** A method for the analysis of sugars in plant cell-wall polysaccharides by gas-liquid chromatography, *Carbohydr. Res.*, 5, 340, 1967.
8. **Mawhinney, T. P., Feather, M. S., Barbero, G. J., and Martinez, J. R.,** The rapid, quantitative determination of neutral sugars (as aldononitrile acetates) and amino sugars (as *O*-methyloxime acetates) in glycoproteins by gas-liquid chromatography, *Anal. Biochem.*, 101, 112, 1980.
9. **Lance, D. G. and Jones, J. K. N.,** Gas chromatography of derivatives of the methyl ethers of D-xylose, *Can. J. Chem.*, 45, 1995, 1967.
10. **Varma, R., Varma, R. S., and Wardi, A. H.,** Separation of aldononitrile acetates of neutral sugars by gas-liquid chromatography and its application to polysaccharides, *J. Chromatogr.*, 77, 222, 1973.
11. **Varma, R. S., Varma, R., Allen, W. S., and Wardi, A. H.,** A gas chromatographic method for determination of hexosamines in glycoproteins and acid mucopolysaccharides, *J. Chromatogr.*, 93, 221, 1974.
12. **Turner, S. H. and Cherniak, R.,** Total characterization of polysaccharides by gas-liquid chromatography, *Carbohydr. Res.*, 95, 137, 1981.
13. **Guerrant, G. O. and Moss, C. W.,** Determination of monosaccharides as aldononitrile, *O*-methyloxime, alditol, and cyclitol acetate derivatives by gas chromatography, *Anal. Chem.*, 56, 633, 1984.
14. **Seymour, F. R., Chen, E. C. M., and Bishop, S. H.,** Identification of aldoses by use of their peracetylated aldononitrile derivatives: a G.L.C.-M.S. approach, *Carbohydr. Res.*, 73, 19, 1979.
15. **Seymour, F. R., Slodki, M. E., Plattner, R. D., and Jeanes, A.,** Six unusual dextrans: methylation structural analysis by combined G.L.C.-M.S. of per-*O*-acetyl-aldononitriles, *Carbohydr. Res.*, 53, 153, 1977.
16. **Seymour, F. R., Plattner, R. D., and Slodki, M. E.,** Gas-liquid chromatography-mass spectrometry of methylated and deuteriomethylated per-*O*-acetyl-aldononitriles from D-mannose, *Carbohydr. Res.*, 44, 181, 1975.
17. **Lawrence, J. F. and Iyengar, J. R.,** Gas chromatographic determination of polysaccharide gums in foods after hydrolysis and derivatization, *J. Chromatogr.*, 350, 237, 1985.
18. **Li, B. W. and Stewart, K. K.,** Quantitative analysis of simple carbohydrates in foods, in Proc. 9th Materials Res. Symp., National Bureau of Standards, Gaithersburg, Md., 1979, 271.

Chapter 7

CARBOHYDRATE TRIFLUOROACETATES

Peter Englmaier

TABLE OF CONTENTS

I. INTRODUCTION

Carbohydrates and related substances need a marked increase in volatility to make them suitable for gas chromatography (GC). Various derivatizations are applicable for this purpose. All of them involve the hydroxyl groups to avoid hydrogen bonds, the major reason for the poor volatility of carbohydrates. Besides the commonly used trimethylsilyl (TMS) ethers, an acylation, meaning the esterification with a carboxylic acid, mostly acetic acid and trifluoroacetic acid meet this requirement.

While the use of carbohydrate acetates for GC has been well known since 1961,[1,2] trifluoroacetylation of sugars was first reported in 1966 by Vilkas et al.[3] and polyols in 1969 by Shapira,[4] documenting a high sensitivity and good resolution of these derivatives.

Compared with TMS, the acetates, especially the trifluoroacetates, offer some special features that make them interesting for the carbohydrate analyst. First, trifluoroacetates are one of the most volatile carbohydrate derivative groups used in GC.[5] Table 1 compares them with acetates and TMS, exemplified with a mono-, di-, and tetrasaccharide, a polyol, and a cyclitol.

Furthermore, acylesters may be separated on highly polar columns, resulting in a much better resolution than usually achieved with TMS. Of special interest is the complete separation of most plant cyclitols, polyols, and monosaccharides as their trifluoroacetates previously reported.[6] Figure 1 gives a presentation of the retention behavior of these substances as trifluoroacetates and TMS, documenting the superior separation of trifluoroacetyl esters.

Besides this, acylesters are chemically and thermally much more stable than TMS, which slightly degrade in the reaction solution. This fact was reported already by Kim et al.[7] and became rather important in an automated analysis device.[8] The decomposition rate of trifluoroacetates and TMS is compared in Table 2; the increase in reproducibility of the quantitative determination by use of trifluoroacetyl esters is documented in Table 3.

Despite these facts, the use of trifluoroacetates was rather unpopular due to their difficult preparation procedure. The introduction of catalysts such as 1-methylimidazole[9] and the use of acyl donors like bis-acylamides[10] were important steps to significantly facilitate the esterification, and today a trifluoroacetylation is as easy to handle as a TMS.

II. DERIVATIZATION PROCEDURES

A carboxylic acid itself does not react spontaneously with hydroxyls. A reactive acid derivative in a proper solvent is needed to achieve a fast and quantitative esterification under mild conditions to avoid side reactions, especially a hydrolyzation of oligosaccharides.

Three different types of trifluoroacetylating reagents are in use for GC applications: acid anhydrides (trifluoroacetic anhydride), bis-acylamides (*N*-methyl-bis-trifluoroacetamide [MBTFA] and bis-trifluoroacetamide [BTFA]), and acylimidazoles (trifluoroacetylimidazole [TFAI]), and all of them are suited to carbohydrate derivatization.

Since these reagents are highly reactive, impurities or corrosive materials must be carefully eliminated during the derivatization. Only suitable glass vials with tight caps (crimped aluminum seals or Teflon®-tightened screw caps) are useful. Syringes for volumetric operations must be of glass-stainless steel-Teflon® and of the highest accuracy available, especially if used in the microrange. The recommended size of vials and amounts of reagents are the lowest ones to be handled accurately. They may be suitably increased if desired. Carefully clean all glassware, use acetone and/or dichloromethane. Store all the reagents in their shipping containers in a cool, dark place, or, best of all, in a desiccator to ensure complete dryness. Heating procedures should be carried out in a metal block thermostat.

Since carbohydrates and related polyols are only slightly soluble in the derivatization reagents, an additional solvent is required, mostly pyridine, which also acts as a catalyst.[11] The sample should be fully dissolved when adding the reagent.[8]

Table 1
RETENTION BEHAVIOR OF CARBOHYDRATES AND RELATED SUBSTANCES AS TRIMETHYLSILYL-, ACETYL-, AND TRIFLUOROACETYL DERIVATIVES UNDER IDENTICAL CHROMATOGRAPHIC CONDITIONS

Substance	Trimethylsilylates		Acetates		Trifluoroacetates	
	RT	T	RT	T	RT	T
Mannitol	10.5	205	6.8	168	5.7	157
Myo-Inositol	12.2	222	8.2	182	7.1	171
Glucose (first peak)	10.3	202	6.6	166	5.0	150
Sucrose	20.1	300 (isotherm)	17.4	274	14.8	248
Stachyose	32.0	300 (isotherm)	21.8	300 (isotherm)	18.4	284

Note: 6 ft × 2 mm packed glass column, 3% Dexsil® 300 on Chromosorb® W-HP 100 to 120 mesh as a nonpolar partition medium, 10°C/min linear temperature program from 100 to 300°C and hold of the upper limit, carrier gas N_2 at 20 mℓ/min. The values are the mean of 4 individual tests. RT is the retention time in min; T°C, the actual temperature at elution.

A. Acid Anhydrides

The use of acid anhydrides is the standard method for esterification in routine work. This procedure is commonly applied to the synthesis of alditol acetates, but also mono- and oligosaccharides may be acetylated.[2,12] Most of the trifluoroacetylation methods also use this derivatization procedure.[3,4,13-19]

Although acylations with acid anhydrides may be used without any catalysis,[14,18,19,20] they are effectively forced by a nucleophilic catalyst, such as pyridine[11,17,21] (successfully usable as a solvent), 4-(dimethylamino)pyridine,[22] or 1-methylimidazole.[9,23-25] Since these reagents also work as proton acceptors, the free acid resulting from the acylation is neutralized, and side reactions like a hydrolysis of oligosaccharides are suppressed. Sodium acetate[25,26] acts in a similar manner. Acidic catalysis, as reported by Oades,[21] is of minor importance in the acylation routine.

The acid anhydride is usually introduced in 30-fold excess to the dry sample, together with an equal amount of dry pyridine. The reaction conditions depend on the sample composition; 30 min at room temperature or 10 min at 75°C may ensure quantitative yield.

It should be noted that borate ions interfere with a pyridine-catalyzed acylation.[23] This fact may become important when monosaccharides are reduced to their alditols by means of sodium borohydride. The preparation of acylated alditols is often preferred, because each monosaccharide produces one single sharp peak and thus leads to a simple peak identification. Since the evaporation of borate as its methyl ester is not in all cases quantitative,[23] only an anion exchange step[41] results in the complete removal of borate ions. Alternatively, 1-methylimidazole is used as a catalyst,[9,23,24] since it works in the presence of borates and can be effectively used with trifluoroacetylating reagents. Otherwise a 10 to 12 hr refluxing at 100°C may be necessary to achieve a quantitative acylation.[20,25]

The acid anhydride should be removed prior to gas-liquid chromatographic analysis because of its excessive tailing. Usually it is evaporated, but an extraction of the derivatives (e.g., with dichloromethane) can also be effectively used.[23] In the latter case, water must be added in excess to decompose the remaining acid anhydride.

Although complicated by an evaporation or extraction step, the following procedure is applicable to the trifluoroacetylation of monosaccharides, alditols, and cyclitols and obtains good results. By the use of acetic anhydride, it yields acetates without any further modification. The method is not recommended for oligosaccharides, because hydrolysis of monosaccharides may occur and so it does not work quantitatively (see Figure 2).

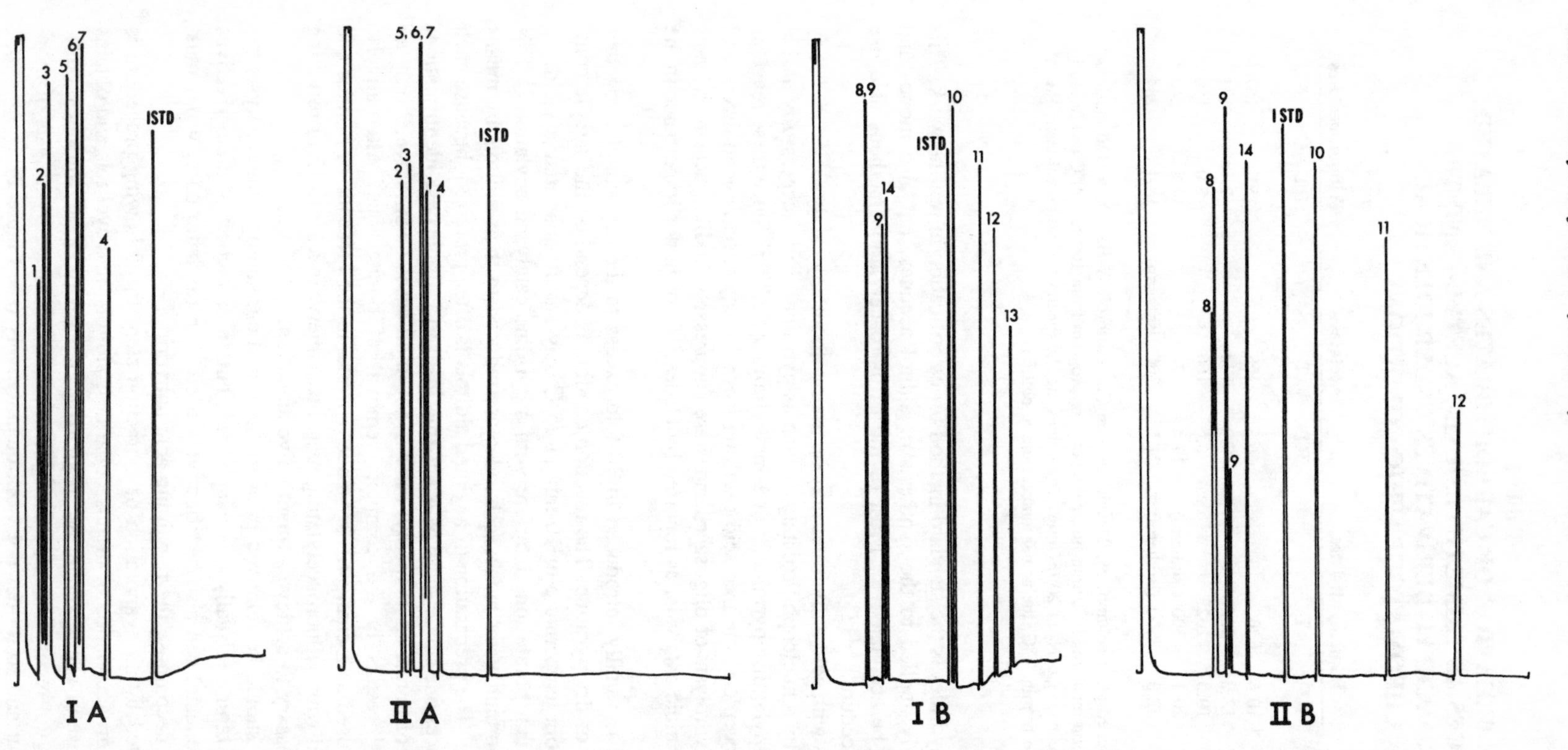

FIGURE 1. Test mixtures of cyclitols, sugar alcohols, mono-, and oligo-saccharides as trifluoroacetates (I) and as TMS (II). Sample preparation and chromatography:

I. Trifluroroacetylation with *N* -methyl-bis(trifluoroacetamide), MBTFA in pyridine. Column: Dexil®410, 3% on Chromosorb®W-HP, 100 to 120 mesh, 6 ft x 2 mm glass column. Temperature range: 100 to 310°C, heating rate stepwise increased from 3.5°C/min to 6°C/min to 25°C/min. Detector: flame ionization. Run time: 25 min.

II. Treatment with hydroxylammoniumchloride in pyridine (converts mono- and oligosaccharides into their oximes) followed by silylation with *N,O*-bis(trimethylsilylacetamide), BSA + 5% trimethylchlorosilane, TMCS as a catalyst. Column: Dexsil®300, loading as described above. Temperature range: 160 to 350°C, heating rate increased from 10°C/min to 30°C/min. Detector: flame ionization. Run time: 30 min.

The components are:

A. Cyclitols and polyols: 1, L-chiro-inositol; 2, D-pinitol; 3, L-quebrachitol; 4, scyllo-inositol; 5, D-mannitol; 6, D-glucitol; 7, galactitol; 14, *myo*-inositol.

B. Mono- and oligosaccharides: 8, D-fructose; 9, D-glucose; 10, sucrose; 11, 6-kestose-type-fructotetraose; 13, kestose-type-fructopentaose.

As trifluoroacetate, glucose (9) form peaks; as oxime-TMS, fructose (8) and glucose (9) do so. ISTD is the internal standard, phenyl-β-D-glucopyranoside®.

Table 2
STABILITY OF CARBOHYDRATES AND RELATED SUBSTANCES AS TRIMETHYLSILYL- OR TRIFLUOROACETYL DERIVATIVES

Substance	Silylated (30 min to 9 hr)	Trifluoroacetylated (15 min to 72 hr)
Mannitol	$C = 0.991 + 3.872.10^{-5}t$	$C = 1.000 + 0.928.10^{-6}t$
Myo-Inositol	$C = 0.988 + 3.623.10^{-5}t$	$C = 1.008 - 0.136.10^{-6}t$
Glucose (first peak)	$C = 0.992 + 3.896.10^{-5}t$	$C = 0.997 + 4.328.10^{-6}t$
Sucrose	$C = 0.998 + 1.088.10^{-5}t$	$C = 0.994 + 5.126.10^{-6}t$
Stachyose	$C = 1.015 - 3.626.10^{-5}t$	$C = 0.999 + 1.042.10^{-6}t$

Note: The values of C are proportional to the peak areas. C is set 1.000 at t = 30 min (silylated) and 15 min (trifluoroacetylated). The table shows the equation of the linear regression graph C vs. t. For each graph, n = 64; individual determinations were carried out in the period mentioned above.

Table 3
REPRODUCIBILITY OF THE QUANTITATIVE DETERMINATION OF CARBOHYDRATES EITHER AS TRIMETHYLSILYLATES OR AS TRIFLUOROACETATES USING PHENYL-β-D-GLUCOPYRANOSIDE AS AN INTERNAL STANDARD (ISTD)

Substance	Silylated (30 min to 9 hr)	Trifluoroacetylated (15 min to 72 hr)
Mannitol	$Q = 0.986 + 3.286.10^{-5}t$	$Q = 1.001 - 0.631.10^{-6}t$
Myo-Inositol	$Q = 0.994 + 1.800.10^{-5}t$	$Q = 1.012 - 1.411.10^{-6}t$
Glucose (first peak)	$Q = 0.997 + 2.272.10^{-5}t$	$Q = 0.999 + 2.389.10^{-6}t$
Sucrose	$Q = 0.999 + 0.378.10^{-5}t$	$Q = 0.995 + 4.804.10^{-6}t$
Stachyose	$Q = 1.004 - 2.414.10^{-5}t$	$Q = 1.006 - 0.662.10^{-6}t$

Note: The values of Q are the quotient area x(t)/area ISTD (t). Q is set 1.000 at t = 30 min (silylated) and 15 min (trifluoroacetylated). Presentation of the results as in Table 2.

The amounts of reagents are suited to samples with a maximum total carbohydrate content of 2 mg. Use 1-mℓ vials with tight caps. A reduction of monosaccharides to their alditols with sodium borohydride may be done prior to the following derivatization.

1. Ensure complete dryness (store the samples in an evacuated desiccator over freshly regenerated silicagel).
2. Add 40 μℓ of pyridine (containing 0.2 mg/sample of the internal standard substance, e.g., phenyl-β-glucopyranoside, if desired) and dissolve by heating to 75°C[8] for 20 min. If borate ions are present, insert additional 10 μℓ of 1-methylimidazole together with the pyridine.
3. Add 80 μℓ of trifluoroacetic anhydride, shake vigorously, cap the vial, and heat to 75°C for 10 min.
4. a. Add 50 μℓ of distilled water to decompose excess anhydride and allow to cool down for 10 min. Then add 100 μℓ of dichloromethane and shake vigorously. After complete separation of the phases, remove the dichloromethane layer for GC. Or,
   b. Evaporate the liquids in a nitrogen stream or in a vacuum, redissolve in 100 μℓ of pyridine (or preferably dichloromethane because of its higher volatility), and heat at 75°C for 20 min.

The sample is now ready for GC.

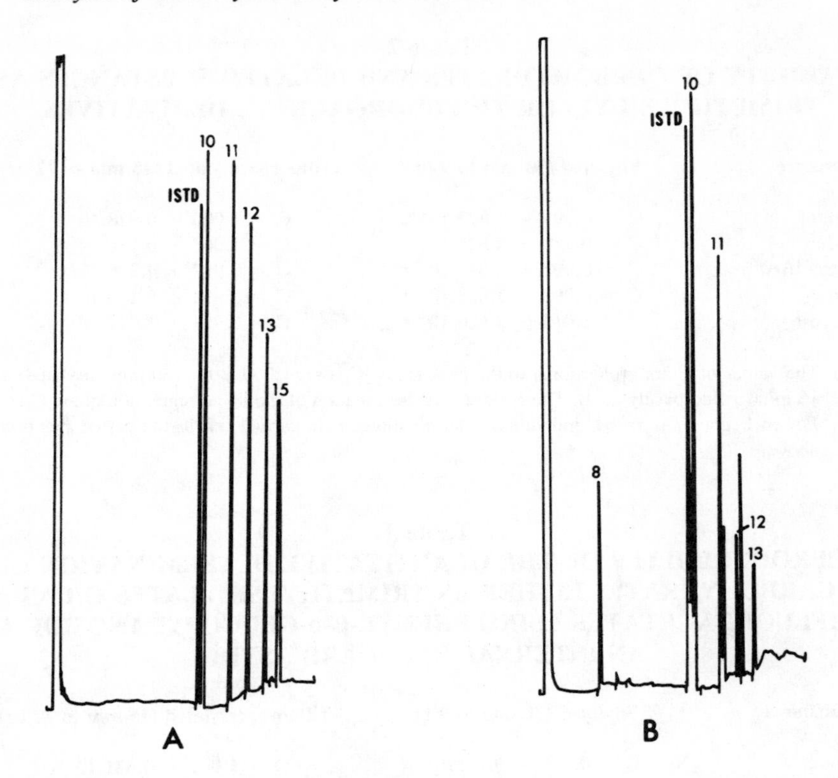

FIGURE 2. Oligosaccharide trifluoroacetylation with *N*-methyl-bis(trifluoroacetamide), MBTFA (A) and trifluoroacetic anhydride (B). The latter causes hydrolyzations as documented by the occurrence of free monosaccharides. The samples contain equal amounts of each individual component. Chromatography and peak identification as in Figure 1; 15, kestose-type fructohexaose.

B. Bis-Acylamides

Since their introduction by Donike,[10] these acylating reagents became a useful tool for derivatization of hydroxyl-, amino-, and sulfhydryl groups for GC applications. They are simple to use and their high volatility prevents interference with the derivatives to be analyzed, so a decomposition and removal, like with acid anhydrides, is unnecessary. They react under basic conditions and, therefore, pyridine can be successfully used as a solvent. Acid amides, as a reaction product, are as volatile as the reagent as well as being inert, contrary to the free acids resulting from an anhydride derivatization. This makes these reagents suitable for oligosaccharide derivatization, an important fact in carbohydrate analysis because of the superior volatility of trifluoroacetates compared with TMS. A derivatization and separation of oligosaccharides is possible up to the hexasaccharides[8] (Figure 2).

While BTFA is of moderate reactivity[10] and, therefore, suited only for amines, MBTFA, a highly reactive reagent, reacts spontaneously with hydroxyls, and a trifluoroacetylation of carbohydrates is easily achieved,[5,27,28] especially in a homogenous medium.[8] If the dry sample is dissolved in the reagent, prolonged reaction times or elevated temperature are necessary for a quantitative derivatization.[8,27,28]

MBTFA works well in the presence of borate ions, and so it is also suited for trifluoroacetylation in samples previously reduced with sodium borohydride. To achieve a rapid derivatization, 1-methylimidazole, as described in Step 2 of the acid anhydride method, may be added.

Since an evaporation or extraction is not needed, the following method is as easy to handle as TMS procedures and can be done in microvials with a volume of 200 μℓ or below. The maximum carbohydrate content is 1 mg per sample. Take special care to store the reagent in a cool and dry place. Check the quality of each individual charge before use.

1. Ensure complete dryness as mentioned in the acid anhydride method.
2. Add 20 μℓ of pyridine containing 0.1 mg of an internal standard and dissolve as described in Step 1 above.
3. Add 40 μℓ of MBTFA, shake vigorously, cap the vial, and heat to 75°C for 10 min.

If the sample is clear, it may be directly injected. Ketoses in the sample cause a yellow or reddish color. Turbidity indicates a decomposed reagent, a wet sample, or a too high carbohydrate content. Never inject a turbid sample.

C. Acylimidazoles

These acylating reagents usually work without an additional catalyst due to the nucleophilic properties of the imidazole group. Compared with MBTFA, they are less reactive and need a heating period of 1 hr at 75°C for complete derivatization. With heptafluorobutyrates, Leavitt and Sherman[29] suggest a similar treatment. Some cyclitols (e.g., scyllo-inositol) and monosaccharides are not quantitatively acylated even after overnight heating to 75°C. This is in agreement with findings of Sullivan and Schewe.[27]

Due to their low acylation power, acylimidazoles are the reagent of choice for selective acylations of reactive groups. BTFA and MBTFA are also suited for such special applications. For example, they can be successfully used for acylation of amino groups in the presence of hydroxyls, if the latter are protected by a TMS group. Through their use, some exchange reactions of TMS to trifluoroacetyl groups for analytical purpose became possible. Together with a silylating agent, this works with amines[10,30,31] as well as with highly reactive hydroxyls found in a steroid skeleton.[32,33]

This method is also applicable to the selective *N*-trifluoroacetylation of amino sugars. Following TMS, it allows the identification and quantification of all biologically interesting amino sugars such as glucosamine and galactosamine and *N*-acetylated derivatives that occur in bacterial cell wall constituents, glycolipids, and other biopolymers. As *N*-acetyl amino sugars remain acetylated, they are distinguishable from the free aldosamines, the latter becoming trifluoroacetylated during the derivatization.

The derivatization procedure is simple: The dry sample is dissolved in pyridine and trimethylsilylated (useful reagents are BSA plus TMCS or TMSI, which produce a clear, nonacidic reaction solution). After the reaction is complete, either TFAI, MBTFA, or BTFA is added, followed by 15 min heating at 75°C. Further information is available in References 10 and 31 to 33.

The per-*O*-trimethylsilyl-*N*-trifluoroacetyl derivatives are superior to TMS and aminoalditol acetates because of their better stability. Furthermore, it is not necessary to remove the derivatization reagents before injection.

An acetolysis of polysaccharides as described by Nilsson and Zopf[34] also produces *N*-trifluoroacetylated amino sugars. If the polysaccharide is previously permethylated, the remaining *N*-trifluoroacetyl-*O*-methylethers can be directly separated on a low-polarity column.

III. CHIRAL DERIVATIZATIONS

Diastereomeric carbohydrate derivatives are needed for chiral separations on commonly used nonchiral columns. Compared with a direct resolution of enantiomers on chiral columns, individually prepared partition liquids are not required and the diastereomeric derivatives are well separated on most of the columns recommended in Section IV.B.

Glycosides,[35] dithioacetals,[22] and oximes can be applied to chiral resolutions. All of them work with trifluoroacetylation as well as with TMS, but the first is preferable due to the higher volatility of the derivatives, which allows the use of chiral groups with relatively high molecular weight. Schweer presents some suitable oximation procedures[15,16,36] involving a derivatization in an aqueous medium forced by sodium acetate, but the oximation can be successfully achieved in a pyridinous solution, too. For trifluoroacetylation, a complete removal of the oximation reagents is required; TMS, however, can be done in the presence of a pyridinous oximation assay.

IV. GAS CHROMATOGRAPHY (GC)

Due to the high volatility of acylated carbohydrates, even complex mixtures easily separate in short run times at moderate temperatures. However, the choice of a proper column and reproducible GC conditions are of special importance to obtain good separations.

A. Column Selection

Acyl derivatives, especially trifluoroacetates, need a moderate or high polar partition medium for good resolution. The following phases were found to be suitable for peracylated carbohydrates.

1. Ethylene glycol succinate polyester and nitrile silicone (ECNSS-M), a copolymer of ethyleneglycol succinate and a methyl cyanoethylsilicone, is suitable for both alditol acetates and trifluoroacetates, but has too low a temperature limit (210°C) for other applications, and needs special care to minimize bleeding effects.
2. Cyanoalkylsilicones, such as OV 225, OV 275, Silar 10C, SP 2330, or SP 2340 are usually stable up to 275°C and, therefore, well suited for samples up to disaccharides and for alditols.
3. Dexsil 410®, a polycarborane-methylcyanoethylsilicone[6,8] is the only high-temperature-resistant cyanoalkylsilicone for general use. It is thermally stable up to 360°C.

The separation properties of phase mixtures prepared from nonpolar and polar components usually do not meet today's requirements. High-polar polyester phases have only a poor thermal stability. Trifluoropropylsilicones, such as OV 210 as sometimes used,[14,27] are slightly destroyed when trifluoroacetylating mixtures are directly injected, and so they are generally not suited for these derivatives. Figure 3 shows this degradation with QF 1.

Analyses are usually done on packed columns with 3% loading on a suitable diatomite support such as Chromosorb® W-HP, Chromosorb® 750 or Gas Chrom Q.® Use only narrow-bore (2 mm inner diameter), carefully deactivated glass columns.

The introduction of capillary columns increases the column efficiency drastically. They offer a much higher resolution in the same analysis time or, alternatively, shorten the run time at an equal separation power as with packed techniques. All of the cyanopropyl silicones recommended in Section IV.A.2 can successfully be used in capillary columns,[16,20,23,24] but bonded phases such as crosslinked OV 225 are of superior stability and lower bleed. Their temperature limit (250°C or higher) meets the requirements of routine analyses up to disaccharides. Wide-bore capillaries with thick film coating (0.25 μm), together with a conventional on-column injector, are as easy to operate as packed columns and, therefore, the best choice for general applications.

If a higher thermal stability is required (e.g., for oligosaccharide analyses) use a packed Dexsil® 410 column. For Dexsil® phases, packed columns (usually 3% on Chromosorb® W-HP 100-120 mesh or similar[6,8] are required due to the difficulties in forming a uniform and reproducible capillary coating with these partition media.

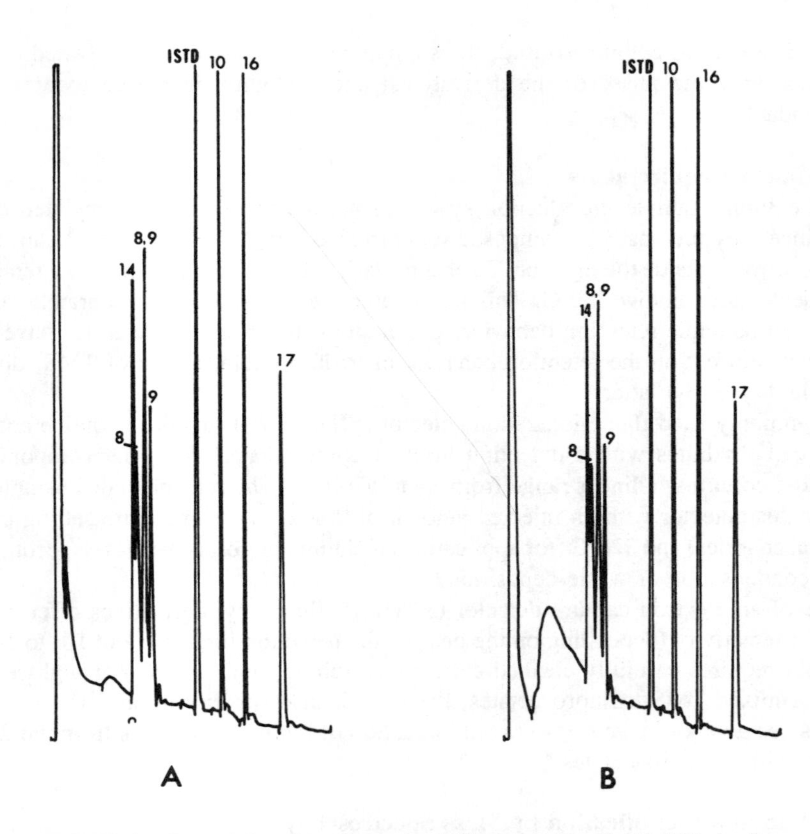

FIGURE 3. Degradation of a trifluoropropylsilicone partition liquid (QF1, loading as described in Figure 1) by use of a MBTFA reaction mixture. (A) freshly prepared column, (B) the same column after 25 test runs. Chromatographic conditions: temperature range 80 to 210°C, stepwise increased heating rate as described with Dexsil® 410 in Figure 1. Peak identification as in Figure 1. Additional components are 16, raffinose and 17, stachyose. Fructose (8) and glucose (9) produce two peaks.

Take special care to keep the injector end of the column clean. Packed columns should be regularly maintained by replacing the glass wool plug. With capillary columns, an uncoated, deactivated piece of capillary tubing as a precolumn will do this job.

For alditol trifluoroacetate analysis, an oven temperature program from 100 to 200°C and a heating rate of 6 to 10°C/min will be sufficient for all phases recommended above. If samples containing disaccharides are to be separated, increase the upper temperature limit to 250°C and use a two-step heating rate up to 15°C/min. For oligosaccharide analysis, together with alditols and cyclitols on Dexsil® 410, select a temperature range from 100 to 310°C with a stepwise increased heating rate[8] from 3.5°C/min to 25°C/min. For carrier gas suggestions see Chapter 1.

B. Chiral Applications

After alkaline treatment of cyanopropylsilicones, the remaining carboxyls can be coupled with a chiral amine (commonly a derivative of L-valine) resulting in a chiral partition liquid.

As an example, a trademark for this is Chirasil-Val®. Fluoroacylated carbohydrates are well suited to analysis on these phases as demonstrated by König et al., with trifluoroacetates[18,19,37,38] and Leavitt and Sherman with heptafluorobutyrates.[29] The upper temperature limit is only slightly lower than that of the cyanopropylsilicone itself and sufficient for alditol and monosaccharide analysis. Chiral phases work best in capillary columns. If a packed column is used, the support must be carefully deactivated to prevent interference with the phase or the derivatives, and a slightly increased loading (5%) is recommendable.

C. Instrument Requirements

An on-column, sample introduction system is needed for analysis of acylated carbohydrates, since they are easily decomposed on corrosive surfaces. Take special care to avoid hot metal surfaces inside the injector. To ensure flash vaporization, do not set the temperature of the injector area below 250°C. This temperature setting should be accurately stabilized to obtain reproducible retention behavior. Oscillations in system temperatures have a much more obvious effect on the retention behavior of trifluoroacetates than of TMS, due to the lower volatility of the latter.[6]

The commonly used flame-ionization detector (FID) is well suited for routine analysis of acylated carbohydrates with a detection limit of approximately 5 ng per component and, with packed columns, a linear range from 10 to 5700 ng. The recommended sample size is suited for this detector with an injected amount of 0.8 $\mu\ell$. A temperature setting of 275°C (up to disaccharides) and 320°C (for applications including oligosaccharides) is recommended to avoid condensation or waste deposition.[6]

By use of an electron capture detector (ECD)[31,39] fluoroacyl derivatives offer a marked increase in sensitivity (depending on the design, the detection limit is about 10^2 to 10^3 times lower), but requires carefully cleaned carrier gas, tubing, and valves. For high-sensitivity analysis of mixed TMS-trifluoroacetates, the ECD is also superior to an FID.

Always use a vapor exhauster to avoid inhalation of highly toxic gases from the detector, especially with trifluoroacetates.[8]

D. Detection and Identification by Mass Spectroscopy

Mass spectroscopy is an important tool for the carbohydrate analyst, especially due to the possibilities of identification of ring isomers and anomers. As both TMS and acylation do not involve isomerization steps,[17,40] they are well suited for these applications. While the TMS have a rather complex fragmentation pattern, acetates and trifluoroacetates offer very simple mass spectra, characterized by elimination of acyl groups. As trifluoroacetyl esters are advantageous, due to the presence of molecular ions, they are often used for problems involving structural investigations.[17,34,40] Table 4 shows a fragmentation pattern of aldohexoses (after König et al.,[17,40] modified). The detection limit for the leading fragment ion in a mass selective detector is similar to an FID.

V. SUGGESTIONS FOR METHOD SELECTION

Since a general method for derivatization and GC of carbohydrates does not exist, the choice of a suitable procedure depends on the user's requirements, mainly on the sample composition and the components to be analyzed.

Although trifluoroacetylation became rather easy to handle during the last 10 years, it is only a method suited for special applications. Compared to TMS, the derivatization and the chromatographic conditions need a much more accurate observation to obtain reproducible results. So TMS is preferable for general use, especially for routines with biological material and for the beginner in carbohydrate analysis.

The application of trifluoroacetates may be divided into the following three fields.

A. Analyses of Alditols, Cyclitols, and Monosaccharides

One of the main applications of trifluoroacetates includes separations of these components in natural samples, because most of the interesting cyclitols and polyols are baseline separated

Table 4
FRAGMENTATION PATTERN OF TRIFLUOROACETYLATED ALDOHEXOSES

PYRANOSIDES		FURANOSIDES	
Fragment ion	m/e	m/e	Fragment ion
M - F	641		
M - TFA	563		
M - TFAO	547	547	M - FAO
M - $TFAOCH_2$	**533**		
547 - TFAOH	433	433	547 - TFAOH
M-2TFAOH	432	432	M-2TFAOH
M - TFAOH - $TFAOCH_2$	419		
432 - F	413	413	432 - F
		407	M - TFAO-CH_2-CH=O^+TFA
M - TFAOH - TFAOCHO	**404**		
M - TFAOH - $TFAOCH_2CHO$	390		
		379	407 - CO
432 - TFAO	319	319	432 - TFAO
432 - $TFAOCH_2$	305	305	432 - $TFAOCH_2$
		293	407 - TFAOH
432 - TFAOCHO	290		
404 - $TFAOCH_2$	277		
TFAO-CH=CH-CH=O^+TFA	265	265	TFAO-CH=CH-CH=O^+TFA
		253	TFAO-CH_2-CH=O^+TFA

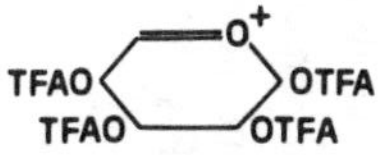

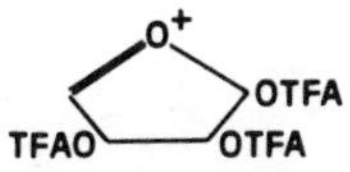

Note: m/e for the molecular ion M is 660. TFA = trifluoroacetyl, all other characters are the element symbols. For experimental details see References 17 and 40. The most important fragment ions are in boldface type.

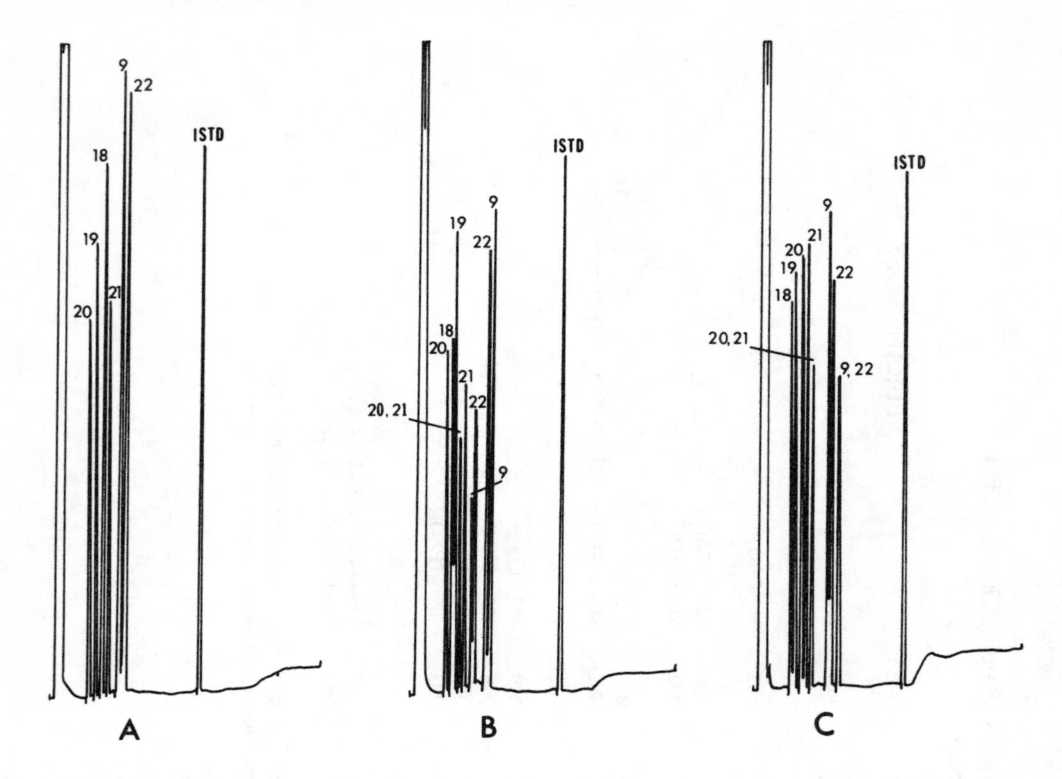

FIGURE 4. Identification of some aldoses as (A) their alditols, (B) in their free form, and (C) as oximes yielded by the reaction with hydroxylammoniumchloride in pyridine after trifluoroacetylation. Chromatographic conditions as in Figure 1. The peaks are numbered as follows: 9, D-glucose; 18, D-arabinose; 19, D-ribose; 20, L-rhamnose; 21, D-xylose; and 22, D-galactose. 9, 20, 21, and 22 appear as two peaks as oximes, as well as in their free form.

from each other and from the monosaccharides,[6] and this is not possible with TMS. For selective monosaccharide monitoring use an oximation[15] or mercaptalation[22] procedure, although they are complicated by the need for complete removal of the excess reagents prior to an acylation.

Derivatization is usually achieved by adding MBTFA to a pyridinous solution of the carbohydrates. If monosaccharides are only of low concentration, TFAI will also work, but it needs a relatively long heating period for a quantitative reaction. The trifluoroacetic anhydride procedure is also applicable.

For quantitative estimation of aldoses, the synthesis of trifluoroacetylated alditols can replace the commonly used alditol acetate procedures, resulting in a higher volatility of the trifluoroacetates compared with acetates (Table 1 and Figures 1 and 4). The method is especially suited to polysaccharide hydrolyzates. For derivatization, the acid anhydride method is usually preferred, but MBTFA and TFAI are also applicable.

Generally, the standard trifluoroacetylation using trifluoroacetic anhydride can be replaced by other trifluoroacetyl donors, which need no further evaporation or extraction prior to chromatography. The disadvantage of these reagents (MBTFA, TFAI) are their high costs and problems with obtaining a sufficient purity.

B. Oligosaccharide Estimation

Together with a high-temperature resistant phase, trifluoroacetates offer a separation and quantitation of oligosaccharides up to the hexasaccharides (Figure 5), a field usually covered by high-performance liquid chromatography (size exclusion or amine-modified columns).

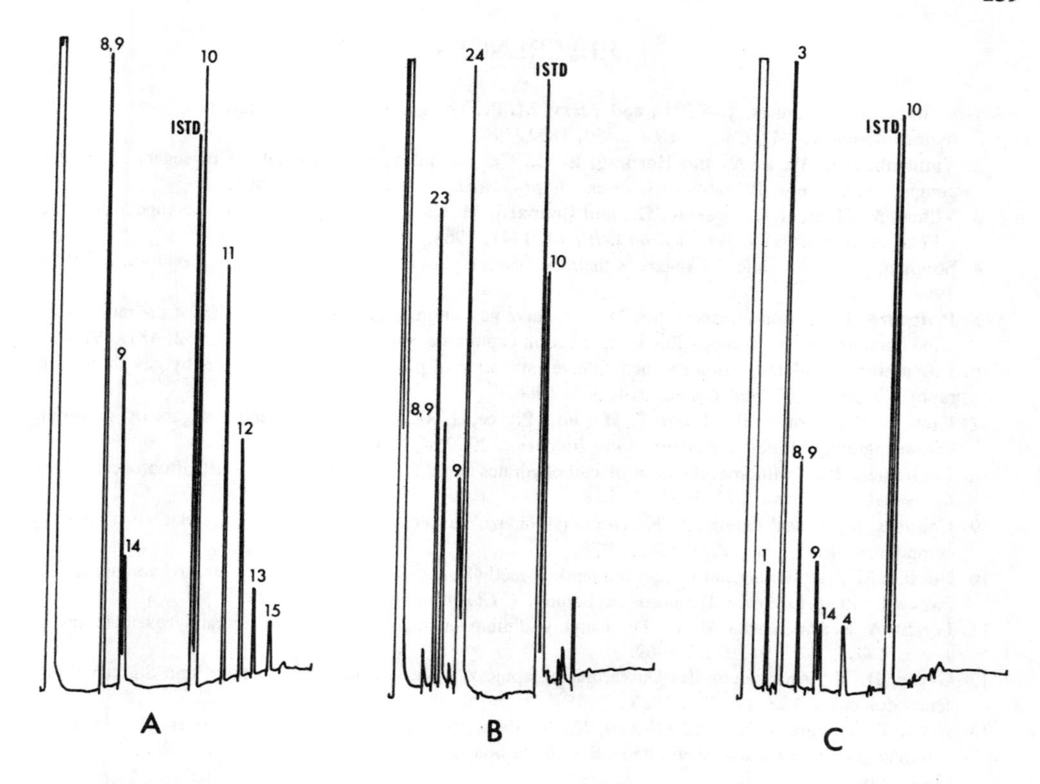

FIGURE 5. Some natural carbohydrate samples separated after trifluoroacetylation with MBTFA: (A) *Bromus erectus*, leaf bases; (B) *Tanacetum corymbosum*, flower discs; and (C) *Artemisia gmelinii*, leaves. Peak identification as in Figures 1 and 2, additional components are 23, L-leucanthemitol and 24, L-viburnitol.

MBTFA is the only reagent applicable for this purpose, because the acid anhydride procedure causes a partial hydrolyzation, especially with oligofructans. The monitoring of alditols and cyclitols, together with oligosaccharides, is also possible (Figure 5).

C. Amino Sugar Derivatizations

Since the stability of *N*-trifluoroacetylated or pertrifluoroacetylated aldosamines is superior to that of TMS, these derivatives are specially requested, for example, in structural studies with polysaccharides. The pertrifluoroacetates of aldosamines or alditolamines (after a reduction step) are achieved with MBTFA. For selective *N*-trifluoroacetylations, MBTFA and TFAI are applicable, but these are procedures for researchers interested in method development and not yet suited to routine analysis and to beginners. For suggestions see Section II.C.

If carefully carried out, the given procedures will work well in the application fields mentioned above. Due to the superior stability of the trifluoroacetates, the samples are suited to automatic chromatography devices, resulting in an increased sample output and a much better reproducibility than achieved with manual injection techniques.

REFERENCES

1. **Gunner, S. W., Jones, J. K. N., and Perry, M. P.,** The gas-liquid partition chromatography of carbohydrate derivatives. I., *Can. J. Chem.*, 39, 1892, 1961.
2. **Vandenheuval, W. J. A. and Horning, E. C.,** Gas chromatographic separations of sugars and related compounds as acetyl derivatives, *Biochem. Biophys. Res. Commun.*, 4, 399, 1961.
3. **Vilkas, M., Hiu, I. J., Boussac, G., and Bonnard, M. C.,** Chromatographie en phase vapeur de sucres à l'état trifluoroacetates, *Tetrahedron Lett.*, 14, 1441, 1966.
4. **Shapira, J.,** Identification of sugars as their trifluoroacetyl polyol derivatives, *Nature (London)*, 222, 792, 1969.
5. **Pritchard, D. G. and Niedermeier, W.,** Sensitive gas chromatographic determination of the monosaccharide composition of glycoproteins using electron capture detection, *J. Chromatogr.*, 152, 487, 1978.
6. **Englmaier, P.,** Identification and quantitative estimation of plant cyclitols and polyols by gas chromatography, *Fresenius Z. Anal. Chem.*, 324, 338, 1986.
7. **Kim, J. H., Shome, B., Liao, T. H., and Pierce, J. G.,** Analysis of neutral sugars by gas-liquid chromatography of alditol acetates, *Anal. Biochem.*, 20, 258, 1967.
8. **Englmaier, P.,** Trifluoroacetylation of carbohydrates for g.l.c., using *N*-methyl-bis(trifluoroacetamide), *Carbohydr. Res.*, 144, 177, 1985.
9. **Connors, K. A. and Pandit, N. K.,** *N*-methylimidazole as a catalyst for analytical acetylations of hydroxy compounds, *Anal. Chem.*, 50, 1542, 1978.
10. **Donike, M.,** Acylierung mit bis(acylamiden): *N*-methyl-bis(trifluoroacetamid) und bis (trifluoroacetamid), zwei neue Reagenzien zur Trifluoroacetylierung, *J. Chromatogr.*, 78, 273, 1973.
11. **Fersht, A. R. and Jencks, W. P.,** The acetylpyridinium ion intermediate in pyridine-catalyzed acyl transfer, *J. Am. Chem. Soc.*, 91, 2125, 1969.
12. **Geyer, H. V.,** Anwendung der Gaschromatographie zur Untersuchung von Zuckern und Stärkehydrolysenprodukten, *Stärke*, 17, 307, 1965.
13. **Ueno, T., Kurihara, N., and Minoru, N.,** Synthetic studies of carbohydrate antibiotics. III. Gas-liquid chromatography of trifluoroacetyl derivatives of carbohydrates including cyclitols, *Agric. Biol. Chem.*, 31, 1189, 1967.
14. **Zanetta, J. P., Breckenridge, W. C., and Vincendon, G.,** Analysis of monosaccharides by gas-liquid chromatography of the O-methyl glycosides as trifluoroacetate derivatives, *J. Chromatogr.*, 69, 291, 1972.
15. **Decker, P. and Schweer, H.,** Gas-liquid chromatography on OV 225 of tetroses and aldopentoses as their *O*-methoxime and *O*-butoxime pertrifluoroacetyl derivatives and of C_3– C_6 alditol pertrifluoroacetates, *J. Chromatogr.*, 236, 369, 1982.
16. **Schweer, H.,** Gas chromatographic separation of carbohydrate enantiomers as (−)menthyloxime pertrifluoroacetates on silicone OV 225, *J. Chromatogr.*, 243, 149, 1982.
17. **König, W. A.,** Gaschromatographie und Massenspektrometrie von Kohlehydraten, *Z. Naturforsch.*, 290, 1, 1974.
18. **König, W. A., Benecke, I., and Sievers, S.,** New results in the gas chromatographic separation of enantiomers of hydroxy acids and carbohydrates, *J. Chromatogr.*, 217, 71, 1981.
19. **König, W. A. and Benecke, I.,** Enantiomer separation of polyols and amines by enantioselective gas chromatography, *J. Chromatogr.*, 269, 19, 1983.
20. **Fox, A., Morgan, S. L., Hudson, J. R., Zhu, Z. T., and Lau, P. Y.,** Capillary gas chromatographic analysis of alditol acetates of neutral and amino sugars in bacterial cell walls, *J. Chromatogr.*, 256, 429, 1983.
21. **Oades, J. M.,** Gas-liquid chromatography of alditol acetates and its application to the analysis of sugars in complex hydrolysates, *J. Chromatogr.*, 28, 246, 1967.
22. **Little, M. R.,** Separation, by g.l.c., of enantiomeric sugars as diastereoisomeric dithioacetals, *Carbohydr. Res.*, 105, 1, 1982.
23. **Blakeney, A. B., Harris, P. J., Henry, R. J., and Stone, B. A.,** A simple and rapid preparation of alditol acetates for monosaccharide analysis, *Carbohydr. Res.*, 113, 291, 1983.
24. **Henry, R. J., Blakeney, A. B., Harris, P. J., and Stone, B. A.,** Detection of neutral and aminosugars from glycoproteins and polysaccharides as their alditol acetates, *J. Chromatogr.*, 256, 419, 1983.
25. **Whiton, R. S., Lau, P., Morgan, S. L., Gilbart, J., and Fox, A.,** Modifications in the alditol acetate method for analysis of muramic acid and other neutral and amino sugars by capillary gas chromatography-mass spectrometry with selected ion monitoring, *J. Chromatogr.*, 347, 109, 1985.
26. **Albersheim, P., Nevins, D. J., English, P. D., and Karr, A.,** A method for the analysis of sugars in plant cell-wall polysaccharides by gas-liquid chromatography, *Carbohydr. Res.*, 5, 340, 1967.
27. **Sullivan, J. E. and Schewe, L. R.,** Preparation and gas chromatography of highly volatile trifluoroacetylated carbohydrates using *N*-methyl-bis(trifluoroacetamide), *J. Chromatogr. Sci.*, 15, 196, 1977.
28. **Selosse, E. J. M. and Reilly, P. J.,** Capillary column gas chromatography of trifluoroacetyl trisaccharides, *J. Chromatogr.*, 238, 253, 1985.

29. **Leavitt, A. L. and Sherman, W. R.,** Direct gas chromatographic resolution of DL-myo-inositol-1-phosphate and other sugar enantiomers as simple derivatives on a chiral capillary column, *Carbohydr. Res.*, 103, 203, 1982.
30. **Schwedt, G. and Bussemas, H. H.,** Schnelle Derivatisierung von 3-Methoxytyramin, Normetanephrin und Metanephrin zu Trimethylsilyltrifluoroacetylderivaten für die gaschromato-graphische Analyse, *J. Chromatogr.*, 106, 440, 1975.
31. **Horning, M. G., Moss, A. M., Boucher, E. A., and Horning, E. C.,** The GLC separation of hydroxyl-substituted amines of biological importance including the catecholamines. Preparation of derivatives for electron capture detection, *Anal. Lett.*, 1, 311, 1968.
32. **Ikekawa, N., Hattori, F., Rubio-Lightbourn, J., Miyazaki, H., Ishibashi, M., and Mori, C.,** Gas chromatographic separation of phytoecdysones, *J. Chromatogr. Sci.*, 10, 233, 1972.
33. **Miyazaki, H., Ishibashi, M., Mori, C., and Ikekawa, N.,** Gas phase microanalysis of zooecdysones, *Anal. Chem.*, 45, 1164, 1973.
34. **Nilsson, B. and Zopf, D.,** Gas chromatography and mass spectrometry of hexosamine-containing oligosaccharide alditols as their permethylated, *N*-trifluoroacetyl derivatives, *Methods Enzymol.*, 83, 46, 1982.
35. **Leontein, K., Lindberg, B., and Lönngren, J.,** Assignment of absolute configuration of sugars by g.l.c. of their acetylated glycosides formed from chiral alcohols, *Carbohydr. Res.*, 62, 359, 1978.
36. **Schweer, H.,** Gas chromatographic separation of enantiomeric sugars as diastereomeric trifluoroacetylated (−)bornyloximes, *J. Chromatogr.*, 259, 164, 1983.
37. **König, W. A., Benecke, I., and Bretting, H.,** Gaschromatographische Trennung enantiomerer Kohlenhydrate an einer neuen chiralen stationären Phase, *Angew. Chem.*, 93, 688, 1981.
38. **Benecke, I., Schmidt, E., and König, W. A.,** Gas chromatographic resolution of carbohydrate enantiomers. A new chiral phase for pentoses, *J. High Res. Chrom.*, 4, 553, 1981.
39. **Eklund, G., Josefsson, B., and Roos, C.,** Gas-liquid chromatography of monosaccharides at the picogram level using glass capillary columns, trifluoroacetyl derivatization and electron-capture detection, *J. Chromatogr.*, 142, 575, 1977.
40. **König, W. A., Bauer, H., Voelter, W., and Bayer, E.,** Gas-chromatographie und Massenspektrometrie trifluoroacetylierter Kohlenhydrate, *Chem. Ber.*, 106, 1905, 1973.
41. **Hicks, K. B., Simpson, G. L., and Bradbury, A. G. W.,** Removal of boric acid and related compounds from solutions of carbohydrates with a boron-selective resin (IRA-743), *Carbohydr. Res.*, 147, 39, 1986.

Chapter 8

ANALYSIS OF CARBOHYDRATES AS *O*-ALKYLOXIME DERIVATIVES BY GAS-LIQUID CHROMATOGRAPHY (GLC)

Jean-Richard Neeser and Thomas F. Schweizer

TABLE OF CONTENTS

INTRODUCTION

Until recently, the most popular and widely used sugar derivatives for gas-liquid chromatography (GLC) and gas chromatography-mass spectrometry (GC-MS) purposes were undoubtedly the trimethylsilyl (TMS) ethers. Sweeley et al.[1] were the first to develop a convenient method to prepare these volatile and thermally stable derivatives. Bourne et al.[2] had earlier described the derivatization of glycosides with trifluoroacetic anhydride, and Gunner et al.[3] had proposed the separation of carbohydrates as acetyl derivatives, both types of ester being easily formed from saccharides. In this last study,[3] it was even found that the various anomeric forms of sugars in solution led to different acetylated compounds, giving separate peaks by GLC. This finding has been widely confirmed using the TMS derivatization procedure, which usually leads to the formation of two, three, or four peaks from each monosaccharide. Such peak multiplicity obviously results in complicated chromatograms and difficulties in the analysis of complex carbohydrate mixtures.

To simplify the chromatograms, some investigators have proposed reduction of the carbonyl group with sodium borohydride to give the corresponding sugar alcohol.[4] This conversion of aldose monosaccharides into alditols has found wide use, since it effectively suppresses the incidence of more than one peak per sugar on the resulting chromatogram. However, this reduction can lead to a loss of information since different ketoses and aldoses can be converted into the same alditol; for example, fructose yields two C_2-epimers, mannitol, and sorbitol, but the latter can also arise from glucose.[5] Thus, the original reducing sugar cannot necessarily be unambiguously identified.

An alternative approach to carbonyl reduction is the conversion of this group in reducing sugars to an oxime, prior to further derivatization of the free hydroxyl functions for GLC analysis.[1] Conversion of reducing sugars into oximes can give only two products (the *syn* and *anti* isomers) and, hence, a maximum of two peaks per carbohydrate. In addition, this reaction avoids the characteristic drawback of reduction into sugar alcohols, as each native sugar gives different products by oximation. Accordingly, derivatization into oxime appears as an ideal compromise, eliminating the problem of too many derivatives coming from a single sugar and, in consequence, the problem of peak multiplicity, while at the same time avoiding the information loss produced by chemical reduction of the carbohydrate-carbonyl group.

Sugar oximes have been prepared using either nonsubstituted hydroxylamine, or various *O*-alkylhydroxylamine reagents. Figure 1 shows the five main derivative families which have been produced via a preliminary oximation. Combined oximation/TMS by using hydroxylamine hydrochloride for the first step leads to per-TMS non-*O*-substituted sugar oximes, compounds which are discussed in Chapter 4 of this book. Further, combined oximation/acetylation, again by the use of hydroxylamine hydrochloride for the preliminary carbonyl-conversion, leads to the formation of a nitrile by an elimination reaction subsequent to acetylation, and, consequently, to the aldononitrile acetates (see Chapter 6). The present chapter deals with the three derivatization combinations involving conversion of carbohydrates into *O*-alkyloximes, followed by either trimethylsilylation acetylation, or trifluoroacetylation (Figure 1). In particular, derivatization procedures, GLC analysis, and MS-fragmentation patterns are reviewed. The reason for preparing *O*-alkyloximes from carbohydrates 15 years ago was certainly that the introduction of the *O*-methyloxime group was well known to increase the volatility of carbonyl-containing compounds. At that time, however, it was not suspected that the useful properties of these derivatives would lead to the important developments we now know.

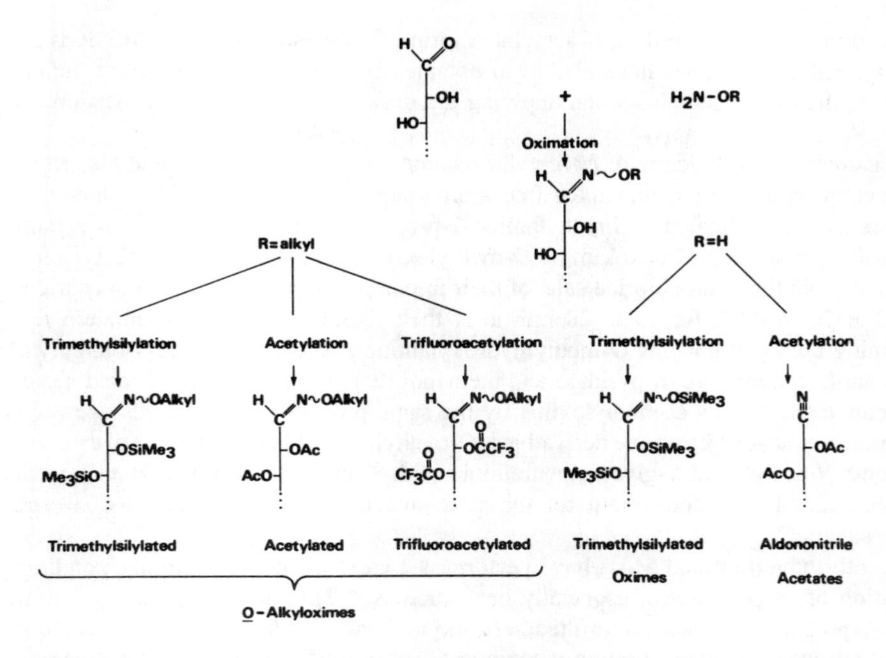

FIGURE 1. The main derivative families prepared for GLC via a preliminary oximation.

II. DERIVATIZATION PROCEDURES AND GAS-LIQUID CHROMATOGRAPHY (GLC) ANALYSIS

A. The *O*-Alkyloximation Reaction

The oximation reaction leading to the conversion of a sugar carbonyl group to an *O*-alkyloxime function is analogous to the derivatization to give simple non-*O*-substituted oximes, since the reactivity of *O*-alkylhydroxylamine hydrochlorides is similar to that of the corresponding nonsubstituted hydroxylamine salt. In most cases, neutral aldoses, ketoses, and dicarbonyl sugars readily react with about 2 *M*-equivalents of *O*-alkylhydroxylamine salt, when pyridine is used as both the solvent and the catalyst.[6-14] Some authors, however, especially those working with small amounts of carbohydrates from biological samples, have successfully carried out this derivatization step in the presence of a large excess of the oximation reagent.[15-18] Usually, the reaction is performed at 70 to 80°C for 1 or 2 hr.[6,7,14-17,19,20] Alternatively, the reaction has been conducted at 70°C for 10 min[18] or at room temperature for 1 to 2 hr.[8,10-12] In fact, no systematic study has ever been undertaken in order to determine the optimal conditions for the derivatization of neutral sugars to give *O*-alkyloximes. By performing the reaction with *O*-methyloxime-hydroxylamine in very concentrated solutions of pyridine-d_5 (for ^{13}C-NMR measurements), Funcke and Von Sonntag have shown that the oxime synthesis from various aldoses, deoxyaldoses, and ketoses is complete within a few minutes at room temperature.[9]

O-alkyloximations have also been successfully performed in water, by reacting the starting carbohydrates with an *O*-alkylhydroxylammonium salt in the presence of sodium acetate at 60 to 80°C for 1 hr.[21-29] Thus, neutral sugars including tetroses, pentoses, and hexoses, as well as oxidation products like tetruloses, pentuloses, and formose sugars, were conveniently converted into various *O*-alkyloximes with water as solvent.

Compared to the oximation of neutral sugar carbonyl groups, more difficulties have been encountered in derivatizing amino sugars, and it was noted that such *O*-methyloximes were not obtained when starting from the corresponding free amino sugars.[30] The per-trimethylsilylated *O*-methyloxime (MO-TMS) derivatives of these hexosamines could, however, be prepared in pyridine from their *N*-acetylated forms.[30] Interestingly, MO-TMS derivatives of *N*-acetylated hexosamines have also been obtained in mixtures resulting from the derivatization of biological samples when applying the usual *O*-alkyloximation procedure in pyridine.[16,17]

To increase the efficiency of *N*-acetylhexosamine *O*-alkyloximation, and also to develop a procedure equally convenient for free amino sugar derivatization, Mawhinney and coworkers have investigated 1-dimethylamino-2-propanol as a catalyst.[31,32] They reported the successful preparation of hexosamine *O*-methyloximes, either from the *N*-acetylated amino sugars or from the hydrochloride salts of their free amino forms. The monosaccharides were heated at 70 to 80°C for 15 to 20 min after their dissolution in an "oximation reagent" containing both components *O*-methylhydroxylamine hydrochloride and 1-dimethylamino-2-propanol, in a mixture of pyridine and methanol (2:1, v/v).[31,32] Muramic acid has equally been converted into its *O*-methyloxime by the same procedure,[33] which also permitted the preparation of acyclic ketoxime derivatives (*O*-methyl- and *O*-benzyloximes) of esters derived from both *N*-acetyl- and *N*-glycolylneuraminic acids.[34] In our laboratories, the above method has been found very convenient for the simultaneous *O*-methyloximation of neutral and amino sugars.[35]

Recently, Guerrant and Moss have performed a careful study to optimize conditions for oximation of carbohydrates, especially hexosamines.[36] They found that using 1-dimethylamino-2-propanol as a catalyst resulted in *O*-alkyloximation of both neutral and amino sugars, but did not give complete reaction when non-*O*-substituted hydroxylamine was used as the oximation reagent. They also found that with a more efficient catalyst, namely 4-(dimethylamino)-pyridine, neutral and amino sugars could be simultaneously oximated with all hydroxylamine hydrochlorides (*O*-alkylated and non-*O*-substituted).[36]

In the course of their pioneering studies on MO-TMS derivatives of sugars, Laine and Sweeley had already observed that usually both the *syn* and *anti* forms of these oximes were formed during synthesis from a single starting carbohydrate.[6,7] The ^{13}C-NMR study further performed by Funcke and Von Sonntag[9] established that *O*-methyloximation of an aldose gave principally the *syn* isomer, probably because of a steric effect due to the C_2-hydroxyl group. The fact that 2-deoxysugars and ketoses yielded *syn* and *anti* forms in equal proportion by the same synthetic procedure, supports such an hypothesis.[9]

In summary, *O*-alkyloximation is a very convenient procedure, chemically easy to carry out, and effective for converting reducing sugars to two isomeric acyclic oximes which have been further derivatized to form three different types of volatile compounds suitable for carbohydrate determination by GLC or combined GC-MS.

B. The per-Trimethylsilylated *O*-Methyloximes (MO-TMS) and *O*-Benzyloximes (BO-TMS)

The usefulness of *O*-methyloximation combined with TMS was first demonstrated by Laine and Sweeley.[6,7] In most cases, the TMS reaction is carried out directly after completion of the preliminary oximation step, without intermediary workup. For this purpose, TMS reagents are simply added to the reaction mixture in pyridine. Apparently, all methods currently used for carbohydrate TMS are compatible with sugar oximes. Among these, BSTFA[6,7,15-17] and hexamethyldisilazene (HMDS) with trimethylchlorosilane (TMCS) as a catalyst[10-13,30] are the preferred reagents to prepare MO-TMS. For all TMS reagents, the reaction is generally effected by heating at 80°C for 10 to 15 min. As for the oximation step, some investigators have, however, obtained satisfactory results at room temperature, especially when using BSTFA in pyridine, in the presence of TMCS as a catalyst.[8,9]

The first studies devoted to GC-MS properties of MO-TMS sugar derivatives, dealt with monosaccharide standards containing three to seven carbons.[6,7] Among those were aldoses, deoxyaldoses, and ketoses. Since both *syn* and *anti* forms of *O*-methyloximes were synthesized, MO-TMS sugars usually exhibited a major peak by GLC, with a small accompanying peak or shoulder. In cases where a single peak was observed, it was assumed that geometric isomers were unresolved, but could be separated by changing liquid phases.

At the time of the work of Laine and Sweeley, however, the limited investigation of different liquid phases had not yet led to the definition of appropriate GLC conditions for separating closely related MO-TMS sugars. In later studies, the GC-MS behavior of MO-TMS xylose, glucose, mannose, and fructose was analyzed using various stationary phases (OV-1, OV-17, QF-1) on packed columns, and compared with that of simple TMS-oximes derived from various monosaccharides.[8] Oximation/TMS has since proved useful for serum glucose derivatization and subsequent determination.[15] Since diulose sugars contain two carbonyl groups, their TMS derivatives can be too nonvolatile for convenient determination by GLC, and here *O*-methyloximation prior to TMS appears to offer the best solution for subsequent GLC analysis.[10,11]

Orme et al.[30] prepared the MO-TMS derivatives of three naturally occurring *N*-acetylated hexosamines (the 2-acetamido-2-deoxyaldoses: GlcNAc, GalNAc, and ManNAc). Individual analysis on OV-225 revealed in each case a typical feature characterized by a major peak, along with a more polar minor peak or shoulder.[30] Finally, the MO-TMS of the methyl and ethyl esters of both *N*-acetyl- and *N*-glycolylneuraminic acids have been studied by GC-MS, and their behavior has been compared with that of the *O*-benzylketoximes of the same *N*-acylneuraminic methyl esters.[34] With only one exception, all these esterified acyclic ketoximes, analyzed as MO-TMS derivatives on SE-52, produced a single symmetrical peak, reflecting that their *syn* and *anti* forms were not resolved.[34]

The introduction of capillary columns offered the possibility of resolving very complex sugar mixtures. Biological samples containing more than 20 monosaccharides, such as pentoses, deoxyhexoses, hexoses (aldo- and keto-sugars), and hexosamines have been derivatized into MO-TMS prior to separation on open tubular glass-capillary columns coated with SE-30.[16] In such experiments, MO-TMS have been seen to be stable for months, in contrast to simple TMS oximes. The fact that ketoses can be analyzed together with aldoses was another argument for MO-TMS; good separations of fructose, galactose, and glucose were obtained on such capillary columns.[16] In contrast, galactose and mannose could not be resolved on SE-30, but rather on the more-polar SP-2250.[17] Fused-silica columns gave complete resolution of all peaks arising from seven derivatives, whereas glass-capillary columns were not as efficient. Improvement of resolution for MO-TMS sugars by capillary GLC allowed testing of the separation of several tens of monosaccharides together.[17] Experimental mixtures of deoxy-, keto-, and dialdo-sugars were derivatized as either MO-TMS[12] or BO-TMS[13] and subsequently resolved successfully. In this case, it was claimed that BO-TMS, rather than MO-TMS, resulted in adequate resolution of alduloses and diuloses obtained from irradiated fructose, and also permit removal of unreacted material by solvent partitioning, due to their lower polarity.[13] Finally, a mixture of eight disaccharides has been separated on a capillary column coated with OV-101.[14] Each reducing disaccharide, except palatinose, gave rise to a large and a small peak. The GLC columns and conditions used for MO-TMS and BO-TMS analysis are summarized in Table 1.

C. The *O*-Methyloxime Acetates (MOA)

As an alternative to TMS, sugar hydroxyl groups can be acetylated after *O*-methyloximation. Usually, when pyridine is used both as solvent and catalyst for *O*-methyloximation (see Section II.A), acetylation is simply carried out by adding acetic anhydride to the mixture and heating again.[18,20] When a more efficient *O*-methyloximation catalyst is used (for example, 1-dimethylamino-2-propanol), pyridine/methanol is the preferred solvent for this initial reaction as discussed in Section II.A. In this case, the solvent mixture is usually evaporated using a stream of dry air or N_2, prior to the acetylation with pyridine/acetic

Table 1
STATIONARY PHASES, PACKED, AND CAPILLARY COLUMNS USED FOR GLC ANALYSIS OF *O*-ALKYLOXIME CARBOHYDRATE DERIVATIVES

Derivatives	Columns	Stationary phase	Ref.
MO-TMS	2 m × 2 mm	0.5% OV-1	8
	1.2 m × 3 mm	2% OV-1	45
	80 m, capillary	OV-101	9
	45 m × 0.25 mm	OV-101	14
	50 m × 0.3 mm	OV-101	19
	1.3 m × 2 mm	3% GE SE-30	6
	2 m × 2 mm	3% SE-30	7
	1.5 m × 3 mm	SE-30	7
	2 m × 3 mm	3% GE SE-30	10
	2.5 m × 2 mm	3% SE-30	15
	25 m × 0.28 mm	SE-30	16
	35 m × 0.3 mm	SE-30	19
	50 m × 0.2 mm	SP-2100	17
	2 m × 4 mm	3% GE SE-52	11
	1.8 m × 0.3 mm	1.5% GC SE-52	34
	123 m, capillary	Dexsil®	9
	123 m × 0.25 mm	Dexsil®	12
	2 m × 2 mm	3% OV-17	7
	2 m × 2 mm	0.5% OV-17	8
	1.8 m × 3 mm	3% OV-17	45
	2 m × 2 mm	3% DC QF-1	8
	30 m × 0.25 mm	SP-2250	17
	1.8 m × 0.3 mm	3% SP-2250	34
	1.9 m × 3 mm	3% OV-225	30
	2 m × 2 mm	1% GE XE-60	8
	2.7 m × 4 mm	3% ECNSS-M	18
BO-TMS	50 m × 0.5 mm	SE-30	13
AO-TFA	1.8 m × 3 mm	OV-101	20
	2 m × 3 mm	OV-225	23
	50 m, capillary	OV-225	21—29
MOA	50 m × 0.2 mm	OV-1	36[a]
	10 m × 0.2 mm	OV-101	33
	11 m × 0.2 mm	SE-54	33
	1.8 m × 3 mm	OV-225	20, 32
	25 m × 0.3 mm	Carbowax® 20 *M*	35, 37[a]
	1.8 m × 3 mm	DEGA	32, 43[a]
	1.8 m × 3 mm	1% ECNSS-M	32

Note: Classified with respect to liquid phases (by increasing Mc-Reynolds' constants) for each derivative family.

[a] MOA determined simultaneously with alditol acetates.

anhydride.[31-33,35] Alternatively, acetic anhydride has just been added to the pyridine/methanol mixture after oximation in the presence of 4-(dimethylamino)-pyridine, to prepare either per-acetylated aldononitrile or per-acetylated *O*-methyloxime derivatives.[36]

In any case, the sugar oximes obtained after the first derivatization step need not be isolated prior to the second step, which is readily achieved by heating the solution at 70 to 80°C for 10 to 60 min. MOA are then easily isolated for GLC analysis by evaporating pyridine and acetic anhydride, followed by partitioning the components of the residual mixtures between $CHCl_3$ or CH_2Cl_2 and water. In order to remove pyridine and excess reagents, the organic layer is preferably washed with a HCl solution and with several portions of water. A subsequent drying of this organic layer generally leads to samples ready for injection.

From the experimental procedures described above, it immediately appears that MOA are extremely easy to prepare. The cleanup procedures are clearly less cumbersome and time consuming than those involved in the preparation of alditol acetates by older methods. Such MOA were first introduced by Murphy and Pennock for GLC determinations of glucose, fructose, and galactose in biological fluids.[18] In the same paper, these authors reported the retention times of a variety of neutral hexoses and pentoses derivatized as MOA. Usually, acetylation of hydroxyl groups is limited to monosaccharides, because of the relatively low volatility of such esters, when derived from larger saccharides. Nevertheless, Schwind et al.[20] separated MOA of sucrose, maltose, and lactose on a packed OV-225 column at 260°C in less than 80 min, more than 2 hr being necessary to elute these disaccharides as alditol acetates in the same GLC conditions. Even if neither of these two derivatives can be recommended in practice for disaccharide separations, the above comparison points to another advantage of MOA over their alditol analogues, namely their higher volatility.

So far, no particular advantage of MOA over aldononitrile acetates has been apparent. In fact, these latter derivatives have proven useful for neutral sugar GLC analysis, giving only one peak per carbohydrate, and being as easy to prepare as their MOA analogues (see Chapter 6). However, aldononitrile acetates of amino sugars have been reported to exhibit erratic chromatographic properties,[32] whereas MOA of GlcNAc and GalNAc gave one symmetrical peak each, both being base-line separated on a number of polar phases coating-packed columns.[31,32] Again, because of their higher volatility compared to alditol acetates, MOA have appeared ideally suitable for the determination of amino sugars by GLC. Consequently, Mawhinney et al. have proposed to quantitate amino sugars from glycoproteins as their MOA, and neutral sugars as their aldononitrile acetates, after performing two separate hydrolyses of the native protein-linked oligosaccharides.[32]

As in the case of MO-TMS, the introduction of capillary columns has led here to an increase in the resolution efficiency for separating complex mixtures of MOA. First, muramic acid has been added to the group of amino sugars, conveniently measurable as MOA, even in the presence of GlcNAc.[33] In our laboratories, it has been shown that a fused-silica capillary column, coated with a medium polarity liquid phase, gives the complete baseline separation of *syn* and *anti* isomers of MOA derived from GlcNAc (two peaks), GalNAc (one peak), ManNAc (two peaks), and muramic acid (two peaks), simultaneously with a fully resolved mixture of MOA *syn* and *anti* isomers of eight neutral sugars.[35] Accordingly, MOA derivatives for the simultaneous GLC determination of neutral and amino sugars, combined with an almost nondestructive and quantitative hydrolysis procedure for glycoproteins and polysaccharides, appeared as an easy and clean method for carbohydrate determination from such complex biological molecules.[35] In another study, Guerrant and Moss have successfully added large monosaccharides like heptuloses, *N*-acetylneuraminic acid, and 2-keto-3-deoxyoctonate to their carbohydrate mixture to be resolved as MOA or aldononitrile acetates.[36] By using an apolar liquid phase, they have separated up to 28 carbohydrates as complex mixtures of MOA or aldononitrile acetate derivatives, together with alditol and cyclitol acetates.[36] From our side, we have demonstrated that sialic acids can be cleaved from their oligosaccharide backbones and converted to 2-amino-2-deoxymannose derivatives by the simultaneous action of both a neuraminidase and a neuraminic acid aldolase, and this, prior to acid hydrolysis of the remaining sugar chains from glycoproteins, and to monosaccharide derivatization into MOA for GLC analysis.[37] By contrast with the extreme acid sensitivity of sialic acids, their corresponding 2-amino-2-deoxymannose derivatives are stable towards acids, and the above methodology thus offers the unique advantage of a "one pot" method for all carbohydrate components from glycoproteins.[37]

Since an important difference exists between retention times of MOA and alditol acetates,

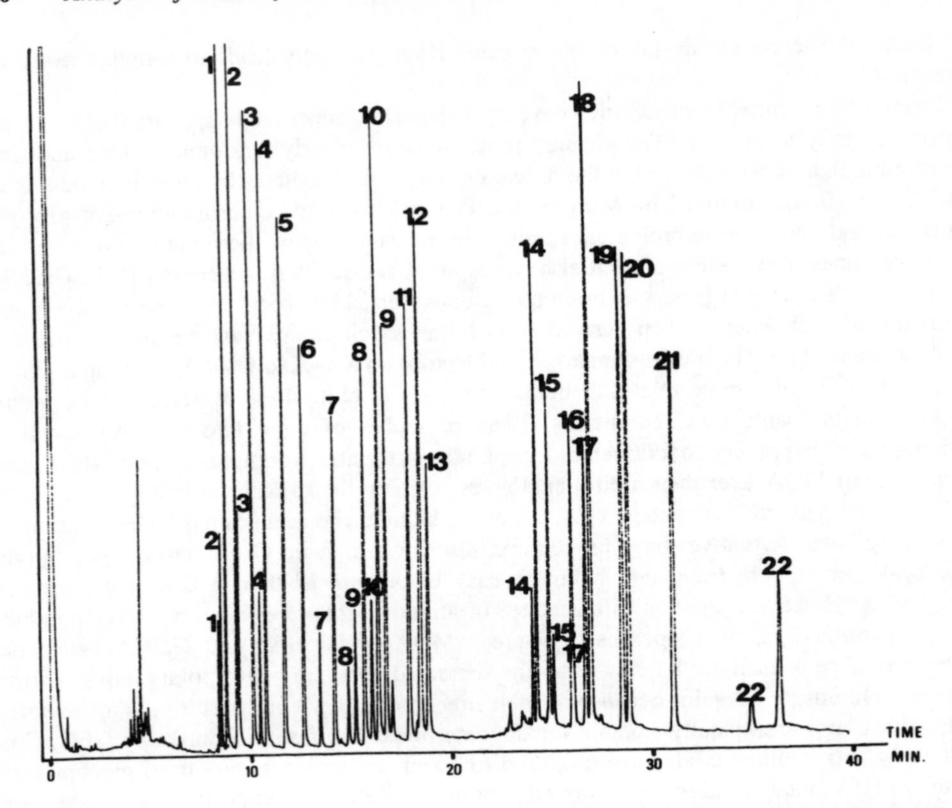

FIGURE 2. Gas chromatogram of *O*-methyloxime and alditol acetates from neutral and amino sugars separated on a fused-silica Carbowax® 20 *M* capillary column; coinjection of 2 × 0.7 μℓ containing an average of 0.05 μg of each sugar; initial temperature, 80°C; final temperature, 230°C. Peaks: 1, rhamnose; 2, fucose; 3, arabinose; 4, xylose; 5, arabinitol; 6, xylitol; 7, 3-*O*-methyl-glucose; 8, mannose; 9, galactose; 10, glucose; 11, mannitol; 12, *myo*-inositol; 13, galactitol; 14, 3-acetamido-3-deoxy-glucose; 15, GlcNAc; 16, GalNAc; 17, ManNAc; 18, GlcNAc-ol; 19, GalNAc-ol; 20, ManNAc-ol; 21, 6-acetamido-6-deoxyglucitol; 22, *N*-glycolylmannosamine. (From Neeser, J.-R., *Carbohydr. Res.*, 138, 189, 1985. With permission.)

the possibility illustrated above (to simultaneously determine both these kinds of derivatives in a single chromatographic run) has opened the door to new structural investigations regarding complex carbohydrates. In the past, Dutton et al. had already shown the possibility of determining the degree of polymerization of oligo- and polysaccharides, by separating glycitols and reducing sugars as TMS derivatives, after reduction of the terminal sugar and subsequent hydrolysis of the reduced oligo- or polymer.[38,39] However, this method was not widely used, probably because of the too-complex chromatograms obtained with TMS sugars. Later on, Morrison successfully subjected oligosaccharides to the sequence of reduction, hydrolysis, oximation, and acetylation and determined the degree of polymerization of the native oligosaccharides from the resulting ratio of aldononitrile and alditol acetates.[40] By an analogous sequence, Varma and Varma determined the degree of polymerization of glycosaminoglycans, which were released from their protein backbones by reductive cleavage.[41] It is also interesting to mention that identification of the reducing units of disaccharides has been performed via reduction of the corresponding sugar *O*-methyloximes with borane, followed by methanolysis, treatment of the mixture with ethylchloroformate, and finally TMS prior to GLC determination of both families of resulting products.[42]

The excellent GLC properties of both MOA and alditol acetates (which can be seen in Figure 2) have been further exploited in our laboratory, when studying carbohydrate chains released from glycoproteins by alkaline borohydride reduction.[37] Oligosaccharide cleavage in such conditions generates an alditol only from the reducing end; again, a subsequent acid hydrolysis followed by *O*-methyloximation/acetylation leads to a mixture of MOA and alditol acetates, the proportion of these last derivatives reflecting the efficiency of the alkaline cleavage. Terminal hexosamines involved in the linkage region could be distinguished from the other hexosamines, with the simultaneous separation and quantitation of GlcNAc, GalNAc, and of their alditols on a fused-silica capillary column.[37] More recently, the utility of the above methodology has been confirmed using packed columns.[43]

In conclusion, it clearly appears that the ideal GLC properties of MOA, along with their ease of preparation, offer possibilities for sophisticated investigations. For such purposes, MOA can favorably complement alditol acetates. Conditions for GLC analysis of MOA are listed on Table 1.

D. The *O*-Alkyloxime Trifluoroacetates (AO-TFA)

The third kind of volatile derivatives obtained from sugar *O*-alkyloximes for GC-MS determination, is the pertrifluoroacetates. After completion of the *O*-alkyloximation step in pyridine, *N*-methyl-bis-(trifluoroacetamid) can be added to the reaction mixture, which is subsequently heated at 70°C for 5 hr.[20] Alternatively, the use of trifluoroacetic anhydride mixed with ethyl acetate allows very mild reaction conditions, such as 12 hr at 4°C or 2 hr at room temperature.[21-29]

MO-TFA of di- and trisaccharides are sufficiently volatile to be separated on a packed column coated with OV-101 at 130°C and 160°C, respectively.[20] The above conditions were, however, not ideal, due to their long retention times (20 to 48 min).

Various AO-TFA have been prepared and analyzed by GC-MS by Schweer and associates.[21-26] Using a 50-m long capillary column coated with OV-225, they compared three different alkyl residues (methyl, 2-methyl-2-propyl, n-butyl) and found the AO-TFA to be versatile derivatives for separating *syn* and *anti* isomers derived from tetroses,[23] pentoses,[21] hexoses,[22] hexuloses,[24] and pentuloses.[25] The ease of trifluoroacetylation and the excellent selectivity of OV-225 towards these derivatives were seen by the authors as the main advantages. The *n*-butoxime trifluoroacetates have been further used to study complex carbohydrate mixtures by GC-MS, namely the formose sugars, arising from autocatalytic condensation of formaldehyde.[25,26]

The same authors have described (−)-menthyloxime per-trifluoroacetates[27,28] and (−)-bornyloxime per-trifluoroacetates[29] as new diastereomeric derivatives of enantiomeric sugars. Using such oximes, several racemic mixtures of D- and L-sugars could be separated. However, not all racemates could be fully resolved, each enantiomer giving rise to two chromatographic peaks.[27] GLC conditions for AO-TFA are mentioned in Table 1.

III. ANALYSIS OF *O*-ALKYLOXIME DERIVATIVES BY MASS SPECTROMETRY (MS)

The most widely used derivatives in MS of carbohydrates are the per-trimethylsilylethers, the per-acetates, and the alditol acetates. However, such species have shown a number of disadvantages in the electron-impact (EI) mode,[44] giving, for instance, molecular ions (M^+) of low abundance and being insensitive to stereochemical differences. Spectra of per-TMS sugars and of per-acetates are often dominated by ions stemming from the derivatization groups, the structurally significant ions being small and hidden in a complex spectrum. Alternative derivatives such as mercaptals, cyclic acetals, ketals, or cyclic boronates have not become very popular, probably because they are more difficult to prepare.[44]

Along with aldononitrile acetates and per-trifluoroacetates (Chapters 6 and 7, respectively), the *O*-alkyloxime derivatives appear to be very suitable alternative compounds for GC-MS

Table 2
OVERVIEW OF THE PRINCIPAL CARBOHYDRATE TYPES FOR WHICH MASS SPECTRAL DATA OF *O*-ALKYLOXIME DERIVATIVES HAVE BEEN REPORTED

	Derivatives and ref.	
Carbohydrates	MO-TMS	AO-TFA
Trioses and tetroses	7	26[a]
Aldopentoses	7, 16, 17	21[a-c]
Aldohexoses	6, 7, 15—17,45	22[a-c]
Aldoheptoses	7	26[a]
2- and 3- Pentuloses	17	25[a]
2- and 3-Hexuloses	7, 10, 16, 17	25[a]
2,5-Hexodiuloses	10	25[a]
Heptuloses	17	
Pentosuloses and hexosuloses	10, 11	
Hexosamines	16, 17, 30	
Neuraminic acids	34[d]	
Deoxypentoses	7	
Deoxyhexoses	10, 16, 17	
Aldonic acids	11, 17	
Alduronic acids	17	

[a] *O-n*-Butyloximes.
[b] *O*-2-Methyl-2-propyloximes.
[c] *O*-Methyloximes.
[d] And corresponding BO-TMS derivatives.

purposes, owing to their distinctive GC properties, their high volatility, and stability. Table 2 gives an overview of carbohydrate categories for which mass spectral data of *O*-alkyloxime derivatives have been reported. Only MO-TMS and AO-TFA have been used for a wide range of sugars. Surprisingly, no MS data are available for MOA, except one report mentioning the high relative abundance (28%) of the molecular ion from the MOA of GlcNAc,[31] and an indication from Laine and Sweeley that MOA exhibit fragmentation patterns similar to those of corresponding MO-TMS.[6] Similarly, BO-TMS have only been used for GC-MS in one single study[34] reporting the fragmentation pathways of derivatized *N*-acetylneuraminic and *N*-glycolyneuraminic acids. As might be expected, these patterns closely resembled those exhibited by the corresponding MO-TMS.

A. The per-Trimethylsilylated *O*-Methyloximes (MO-TMS)

The suitability of these derivatives for sugar analysis by GC-MS was first investigated by Laine and Sweeley.[6,7] The MO-TMS of glucose showed quite simple EI fragmentations, with abundant ions stemming from cleavage of the alditol chain. No cleavage occurs between C-1 and C-2, this bond being stabilized by the methyloxime group (Figure 3). The primary fragments can lose TMSOH to give a series of secondary ions, 90 mass units lower. Thus, the four-carbon fragment m/z 409 was not detected, but rather m/z 319 which constituted the base peak. No molecular ion could be observed at 40 eV ionizing energy. Instead, M^+-15, M^+-15-90, and M^+-15-2 × 90 indicated loss of CH_3 and of TMSOH. This interpretation was supported by the mass spectra of equally derivatized 1-($^2H_1$)-D-glucose and 6,6-($^2H_2$)-D-glucose.[6] In contrast to cyclic sugar derivatives, no rearrangements complicated the spectra.

This basic fragmentation pathway was also found in a series of aldoses with a three to seven carbon chain, and permitted certain identification of chain length.[7] Furthermore, it was possible to definitely locate the position of methyl groups in 2-*O*-methyl-D-ribose,

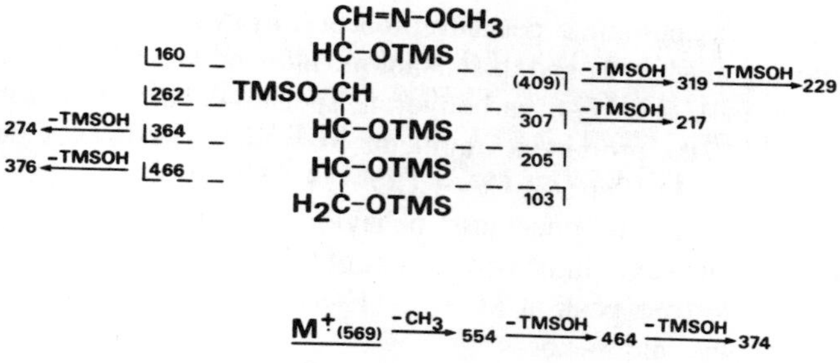

CH=N-OCH$_3$
└160 – – HC-OTMS – – (409)┐ —TMSOH→ 319 —TMSOH→ 229
└262 – TMSO-CH – – 307┐ —TMSOH→ 217
274 ←TMSOH– └364 – HC-OTMS – – 205┐
376 ←TMSOH– └466 – HC-OTMS – – 103┐
H$_2$C-OTMS

M$^+$(569) —CH$_3$→ 554 —TMSOH→ 464 —TMSOH→ 374

FIGURE 3. EI-fragmentation pattern of MO-TMS glucose showing masses of major ions. Masses in brackets were not observed (Adapted from Laine, R. A. and Sweeley, C. C., *Anal. Biochem.*, 43, 553, 1971.)

3-*O*-methyl-D-ribose, 2,3-di-*O*-methyl-D-ribose and 3-*O*-D-glucose.[7] The potential of MO-TMS derivatives for locating methyl groups was also demonstrated by the spectra of partially methylated glucose and galactose derived from permethylated lactose. However, isomeric trimethylderivatives could not be resolved sufficiently by GC, thus hindering linkage analysis of higher oligosaccharides.[7]

The fragmentation pathway outlined in Figure 3 is essentially the same as found for per-TMS non-*O*-alkylated sugar oximes,[8] ions containing the oxime group being 58 mass units larger. However, MO-TMS exhibit better volatility and great stability and, hence, appear preferable. In summary, since the *O*-methyloxime function effectively labeled the carbonyl carbon, the resulting mass spectra gave carbon chain lengths as well as location of substituents or linkages, and from the earliest studies of MO-TMS sugars it became evident that such derivatives were ideal for carbohydrate determination by combined GC-MS.[6,7]

Subsequently, MO-TMS derivatives have been applied for determining serum glucose by spiking serum with heptadeuterated glucose and determining the ratio of m/z 319 and 323 by GC-MS, after derivatization.[15] A convincing demonstration of the usefulness of MO-TMS sugars has been provided by Ehmann,[45] who established the structures of three isomeric *O*-(indole-3-acetyl)-D-glucoses isolated from Zea mays kernels. Here, direct TMS did not provide mass spectral data suitable for unambiguous structural assignments.

The fragmentation scheme outlined in Figure 3 can also be applied to other sugar types. All major mass peaks from MO-TMS of the three naturally occurring 2-acetamido-2-deoxy-aldoses have been assigned.[30] The spectra of GlcNAc, GalNAc, and ManNAc derivatives were almost the same, except the ratio between the C-3 to C-6 fragment and the C-1 to C-4 fragment was 2.9, 6.7, and 22.0, respectively.[30] Mawhinney et al.[34] have shown that MO-TMS of methyl and ethyl esters of *N*-acetylneuraminic and *N*-glycolylneuraminic acids also fragment in a pattern which is typical of their acyclic structure. Kito and co-workers[10,11] could identify several deoxy and dicarbonyl sugars arising from radiolytic decomposition of fructose. The pronounced stabilizing effect of the methyloxime group already mentioned, was particularly evident in 2,5-hexodiuloses, which gave only two major fragments stemming from cleavage between C-3 and C-4.

B. The *O*-Alkyloxime Trifluoroacetates (AO-TFA)

To our knowledge, these derivatives have been used exclusively in conjunction with chemical ionization (CI). This ionization technique avoids some limitations of EI-MS, such as the high thermal and electronic energy necessary for ionization which leads to low abundances of high mass ions. Schweer and associates[21,22,25,26] using methylpropane as the ionizing agent could identify *syn-* and *anti-*alkyloxime trifluoroacetates of a range of different sugar types (Table 2). In most cases, the ion m/z = M + 1 had the highest intensity: it results from addition of the methylpropyl ion $C_4H_9^+$ followed by loss of C_4H_8. Accordingly, clean chromatograms of *n*-butoxime per-trifluoroacetyl derivatives were obtained by using selected ion monitoring at M + 1, and this technique allowed identification and quantitation of complex mixtures of C-3 to C-6 carbohydrate species occurring in formose sugars.[26] Besides M + 1 and M^+, the prominent fragments from these derivatives stem from loss of trifluoroacetic acid (M = 114) and of CF_3–COO˙ (M = 113).

Glucose exhibited a curious fragmentation behavior; the *syn* isomers of its *n*-alkyloximes underwent a C–C bond cleavage, indicated by loss of CF_3–COO–CH_2.(M = 127). This gave rise to a glucose-specific base peak at M + 1-113-127 which permit detection of glucose in the presence of all other aldohexoses.[22]

REFERENCES

1. **Sweeley, C. C., Bentley, R., Makita, M., and Wells, W. W.,** Gas-liquid chromatography of trimethylsilyl derivatives of sugars and related substances, *J. Am. Chem. Soc.*, 85, 2497, 1963.
2. **Bourne, E. J., Tatlow, C. E. M., and Tatlow, J. C.,** Studies of trifluoroacetic acid. II. Preparation and properties of some trifluoroacetyl esters, *J. Chem. Soc.*, 1367, 1950.
3. **Gunner, S. W., Jones, J. K. W., and Perry, M. B.,** The gas-liquid partition chromatography of carbohydrate derivatives, *Can. J. Chem.*, 39, 1892, 1961.
4. **Sawardeker, J. S., Sloneker, J. H., and Jeanes, A.,** Quantitative determination of monosaccharides as their alditol acetates by gas liquid chromatography, *Anal. Chem.*, 37, 1602, 1965.
5. **Mason, B. S. and Slover, H. T.,** A gas chromatographic method for the determination of sugars in foods, *J. Agric. Food Chem.*, 19, 551, 1971.
6. **Laine, R. A. and Sweeley, C. C.,** Analysis of trimethylsilyl *O*-methyloximes of carbohydrates by combined gas-liquid chromatography - mass spectrometry, *Anal. Biochem.*, 43, 553, 1971.
7. **Laine, R. A. and Sweeley, C. C.,** *O*-Methyl oximes of sugars. Analysis as *O*-trimethylsilyl derivatives by gas-liquid chromatography and mass spectrometry, *Carbohydr. Res.*, 27, 199, 1973.
8. **Petersson, G.,** Gas-chromatographic analysis of sugars and related hydroxy acids as acyclic oxime and ester trimethylsilyl derivatives, *Carbohydr. Res.*, 33, 47, 1974.
9. **Funcke, W. and Von Sonntag, C.,** *Syn* and *anti* forms of some monosaccharide *O*-methyl oximes: a ^{13}C-n.m.r. and g.l.c. study, *Carbohydr. Res.*, 69, 247, 1979.
10. **Kito, Y., Kawakishi, S., and Namiki, M.,** Radiation — induced degradation of D-fructose in anaerobic condition, *Agric. Biol. Chem.*, 43, 713, 1979.
11. **Kito, Y., Kawakishi, S., and Namiki, M.,** Radiation-induced degradation of D-fructose in aerated condition, *Agric. Biol. Chem.*, 45, 1999, 1981.
12. **Dizdaroglu, M., Henneberg, D., Schomburg, G., and Von Sonntag, C.,** Radiation chemistry of carbohydrates, VI; γ-radiolysis of glucose in deoxygenated N_2O saturated aqueous solution, *Z. Naturforsch.*, 30b, 416, 1975.
13. **Den Drijver, L. and Holzapfel, C. W.,** Separation and quantitative determination of the radiolysis products of D-fructose as their *O*-benzyloximes, *J. Chromatogr.*, 363, 345, 1986.
14. **Adam, S. and Jennings, W. G.,** Gas chromatographic separation of silylated derivatives of disaccharide mixtures on open tubular glass capillary columns, *J. Chromatogr.*, 115, 218, 1975.
15. **Björkhem, I., Blomstrand, R., Falk, O., and Oehman, G.,** The use of mass fragmentography in the evaluation of routine methods for glucose determination, *Clin. Chim. Acta,* 72, 353, 1976.
16. **Storset, P., Stokke, O., and Jellum, E.,** Monosaccharides and monosaccharide derivatives in human seminal plasma, *J. Chromatogr.*, 145, 351, 1978.
17. **Pelletier, O. and Cadieux, S.,** Glass capillary or fused-silica gas chromatography - mass spectrometry of several monosaccharides and related sugars: improved resolution, *J. Chromatogr.*, 231, 225, 1982.
18. **Murphy, D. and Pennock, C. A.,** Gas chromatographic measurement of blood and urine glucose and other monosaccharides, *Clin. Chim. Acta,* 42, 67, 1972.
19. **Zegota, H.,** Separation and quantitative determination of fructose as the *O*-methyloxime by gas-liquid chromatography using glass capillary columns, *J. Chromatogr.*, 192, 446, 1980.
20. **Schwind, von H., Scharbert, F., Schmidt, R., and Kattermann, R.,** Gaschromatographische Bestimmung von Di- und Trisacchariden, *J. Clin. Chem. Clin. Biochem.*, 16, 145, 1978.
21. **Schweer, H.,** Gas chromatography - mass spectrometry of aldoses as *O*-methoxime, *O*-2-methyl-2-propoxime and *O*-*n*-butoxime pertrifluoroacetyl derivatives on OV-225 with methylpropane as ionization agent. I. Pentoses, *J. Chromatogr.*, 236, 355, 1982.

22. **Schweer, H.,** Gas chromatography - mass spectrometry of aldoses as *O*-methoxime, *O*-2-methyl-2-propoxime and *O*-*n*-butoxime pertrifluoroacetyl derivatives on OV-225 with methylpropane as ionization agent. II. Hexoses, *J. Chromatogr.*, 236, 361, 1982.
23. **Decker, P. and Schweer, H.,** Gas-liquid chromatography on OV-225 of tetroses and aldopentoses as their *O*-methoxime and *O*-*n*-butoxime pertrifluoroacetyl derivatives and of C_3-C_6 alditol pertrifluoroacetates, *J. Chromatogr.*, 236, 369, 1982.
24. **Decker, P. and Schweer, H.,** Gas-liquid chromatography of hexoses as *O*-methoxime and *O*-*n*-butoxime pertrifluoroacetyl derivatives on silicone OV-225, *J. Chromatogr.*, 243, 372, 1982.
25. **Schweer, H.,** G.L.C.-M.S. of the oxidation products of pentitols and hexitols as *O*-butyloxime trifluoroacetates, *Carbohydr. Res.*, 111, 1, 1982.
26. **Decker, P., Schweer, H., and Pohlmann, R.,** Identification of formose sugars, presumable prebiotic metabolites, using capillary gas chromatography/gas chromatography - mass spectrometry of *n*-butoxime trifluoroacetates on OV-225, *J. Chromatogr.*, 244, 281, 1982.
27. **Schweer, H.,** Gas chromatographic separation of carbohydrate enantiomers as (−)-menthyloxime pertrifluoroacetates on silicone OV-225, *J. Chromatogr.*, 243, 149, 1982.
28. **Schweer, H.,** G.l.c.-m.s. of mixtures of enantiomeric tetruloses and pentuloses as trifluoroacetylated *O*-(−)-menthyloxime derivatives, *Carbohydr. Res.*, 116, 139, 1983.
29. **Schweer, H.,** Gas chromatographic separation of enantiomeric sugars as diastereomeric trifluoroacetylated (−)-bornyloximes, *J. Chromatogr.*, 259, 164, 1983.
30. **Orme, T. W., Boone, C. W., and Roller, P. P.,** The analysis of 2-acetamido-2-deoxyaldose derivatives by gas-liquid chromatography and mass spectrometry, *Carbohydr. Res.*, 37, 261, 1974.
31. **Mawhinney, T. P., Feather, M. S., Ricardo Martinez, J., and Barbero, G. J.,** A rapid, convenient method for the determination of hexosamines as *O*-acetylated-*O*-methyloximes by gas-liquid chromatography, *Carbohydr. Res.*, 75, C 21, 1979.
32. **Mawhinney, T. P., Feather, M. S., Barbero, G. J., and Ricardo Martinez, J.,** The rapid, quantitative determination of neutral sugars (as aldononitrile acetates) and amino sugars (as *O*-methyloxime acetates) in glycoproteins by gas-liquid chromatography, *Anal. Biochem.*, 101, 112, 1980.
33. **Hicks, R. E. and Newell, S. Y.,** An improved gas chromatographic method for measuring glucosamine and muramic acid concentrations, *Anal. Biochem.*, 128, 438, 1983.
34. **Mawhinney, T. P., Madson, M. A., Rice, R. H., Feather, M. S., and Barbero, G. J.,** Gas-liquid chromatography and mass-spectral analysis of per-*O*-trimethylsilyl acyclic ketoxime derivatives of neuraminic acid, *Carbohydr. Res.*, 104, 169, 1982.
35. **Neeser, J.-R., and Schweizer, T. F.,** A quantitative determination by capillary gas-liquid chromatography of neutral and amino sugars (as *O*-methyloxime acetates), and a study on hydrolytic conditions for glycoproteins and polysaccharides in order to increase sugar recoveries, *Anal. Biochem.*, 142, 58, 1984.
36. **Guerrant, G. O. and Moss, C. W.,** Determination of monosaccharides as aldononitrile, *O*-methyloxime, alditol, and cyclitol acetate derivatives by gas chromatography, *Anal. Chem.*, 56, 633, 1984.
37. **Neeser, J.-R.,** G.L.C. of *O*-methyloxime and alditol acetate derivatives of neutral sugars, hexosamines, and sialic acids: "one-pot" quantitative determination of the carbohydrate constituents of glycoproteins and a study of the selectivity of alkaline borohydride reductions, *Carbohydr. Res.*, 138, 189, 1985.
38. **Dutton, G. G. S., Gibney, K. B., Jensen, G. D., and Reid, P. E.,** The simultaneous estimation of polyhydric alcohols and sugars by gas-liquid chromatography. Applications to periodate oxidized polysaccharides, *J. Chromatogr.*, 36, 152, 1968.
39. **Dutton, G. G. S., Reid, P. E., Rowe, J. J. M., and Rowe, K. L.,** Determination of the degree of polymerisation of oligo- and polysaccharides by gas-liquid chromatography, *J. Chromatogr.*, 47, 195, 1970.
40. **Morrison, I. M.,** Determination of the degree of polymerisation of oligo- and polysaccharides by gas-liquid chromatography, *J. Chromatogr.*, 108, 361, 1975.
41. **Varma, R. and Varma, R. S.,** A simple procedure for combined gas chromatographic analysis of neutral sugars, hexosamines and alditols. Determination of degree of polymerization of oligo- and polysaccharides and chain weights of glycosaminoglycans, *J. Chromatogr.*, 139, 303, 1977.
42. **Chaves das Neves, H. J., Bayer, E., Blos, G., and Frank, H.,** A gas-chromatographic method for identification of the reducing units of disaccharides via reduction of the methoximes with borane, *Carbohydr. Res.*, 99, 70, 1982.
43. **Mawhinney, T. P.,** Simultaneous determination of *N*-acetylglucosamine, *N*-acetylgalactosamine, *N*-acetylglucosaminitol and *N*-acetylgalactosaminitol by gas-liquid chromatography, *J. Chromatogr.*, 351, 91, 1986.
44. **Radford, T. and De Jongh, D. C.,** Carbohydrates, in *Biochemical Applications of Mass Spectrometry (First Supplementary Volume)*, Waller, G. R. and Dermer, O. C., Eds., John Wiley & Sons, New York, 1980, 255.
45. **Ehmann, A.,** Identification of 2-*O*-(indole-3-acetyl)-D-glucopyranose, 4-*O*-(indole-3-acetyl)-D-Glucopyranose and 6-*O*-(indole-3-acetyl)-D-glucopyranose from kernels of Zea mays by gas-liquid chromatography-mass spectrometry, *Carbohydr. Res.*, 34, 99, 1974.

Chapter 9

LINKAGE STRUCTURE OF CARBOHYDRATES BY GAS CHROMATOGRAPHY-MASS SPECTROMETRY (GC-MS) OF PARTIALLY METHYLATED ALDITOL ACETATES

Nicholas C. Carpita and Elaine M. Shea

TABLE OF CONTENTS

I. INTRODUCTION

Not unexpectedly, this decade has witnessed an evergrowing interest in the chemical structure of carbohydrates. Their function as the structural determinants of the cell walls of bacteria, fungi, and plants and their role as a broad range of storage polymers are well established. Enormous potential for chemical and structural diversity is provided by the variety of ways each sugar can be linked, so it is not surprising that oligosaccharide moieties of glycoproteins and glycolipids participate in more subtle roles in cell recognition, surface sensing, and, perhaps, an even wider range of functions of unrealized dimension. A fundamental step in understanding the three-dimensional structure of the carbohydrate is to determine how the sugar components are linked; the most widely used method for determining linkage structure of polysaccharides is methylation analysis by gas chromatography (GC) of partially methylated alditol acetate derivatives of neutral and amino sugars. Despite considerable advances in technologies which allow derivatization and detection of minute amounts of material, the basic strategy has remained the same. The free hydroxyl groups of polymerized sugars are completely methylated, and upon subsequent hydrolysis of the polymer, the unmasked hydroxyls indirectly reveal the position of the former linkage. Chemical reduction of the anomeric carbons to their respective alcohols permits acetylation of all remaining positions, establishing which carbons participate in the ring and linkage. With the advent of combined GC and mass spectrometry (MS), techniques for analysis have been so refined that as little as 50 μg of material is more than enough to provide quantitative linkage information on complex polysaccharides. Indeed, microscale preparations designed to maximize recovery, coupled with selective ion monitoring MS, has pushed limits of successful analysis to 1 μg.[1-3] Such analyses are performed routinely in many laboratories, and several excellent reviews have documented improved methods in preparation and separation of derivatives.[4-10] The principles of electron impact mass spectrometry (EI-MS) of permethylated sugars pioneered by Chizhov[11] and Lindberg[6-8,12] and their associates have also been updated frequently,[9,10,13] but a comprehensive practical guide to current methodologies is needed in sample preparation and derivatization and use of MS for analysis of a wide range of oligo- and polysaccharides. The intention of this review is to provide a straightforward operational guide to the formation of partially methylated alditol acetate derivatives, to provide the well known theoretical and practical bases for MS analysis of derivatives, and a more complete summation of MS data on a wide class of polymers. There is no one method that can be described, because the chemistry of carbohydrates differs markedly, so attention is given to modifications of common procedures to optimize recovery of special polymers. Figure 1 summarizes the basic techniques and some of the specific problems that can be encountered. The goal of this chapter is to provide a research technician, lacking experience in sugar chemistry, with a systematic approach to successful preparation of derivatives and a body of examples of fragmentation analyses that allow deduction of sugar linkage composition for a wide range of glycoproteins and polysaccharides.

II. PREPARATION OF MATERIALS

A. General

The information provided by linkage analysis obviously depends on homogeneity of the material, and thus, considerable effort should be devoted to purification of the polysaccharides and oligomers. The major limitation of linkage analysis is that little sequence information can be deduced directly. With oligomers, additional chemical or enzymic degradation can be used in tandem with linkage analysis to deduce sequence information, but with polysaccharides, one must accept these limitations. The linkage analysis will directly provide only backbone composition and the linkage position and frequency of substitution of other constituents. If the polymer has a repeating unit structure, then its digestion with specific enzymes may be of help. Subsequent purification of the digestion products and additional linkage information could then provide considerable sequence information. This approach, when used with the newer technologies of ^{1}H- and ^{13}C-NMR spectroscopy or FAB-MS (see Chapter 10), can provide an enormous amount of information. Purification of material and extensive chemical and enzymic modification of carbohydrates is beyond the scope of this guide, but a brief review of the potential of these techniques in deducing fine structure is provided.

B. Protection Against Reducing End "Peeling"

Successful permethylation of oligo- and polysaccharides depends equally on their purity and ability to disperse in the solvents used for methylation. While many carbohydrates are freely soluble in dimethylsulfoxide (DMSO), others become so only upon addition of alkali. Some more crystalline materials, such as cellulose, resist such dissolution, and stronger reagents compatible with the methylating reagents are recommended. Oligomers made of neutral and amino sugars present little problem. Polysaccharides are isolated typically by solvent extraction, often with alkali. Because the methylation itself requires introduction of alkali, all materials should be protected from end "peeling" by addition of $NaBH_4$ or $NaBD_4$ to the alkali extraction solution.[14,15] With oligomers, additional information is provided upon reduction of the reducing end to the corresponding alditol. For other material purified in the absence of alkali, 1 *M* NH_4OH containing 1 mg/mℓ $NaBH_4$ or $NaBD_4$ is added to the carbohydrate, and the solution is incubated at 40°C for 90 min to complete the reduction. Material is then neutralized with acetic acid and either dialyzed against deionized water or precipitated twice in 80% ethanol at −70°C for 2 hr. Oligomers can be desalted by gel chromatography on small columns of polyacrylamide gel. Carbohydrate-based gels should be avoided, particularly if only small amounts of material are available, as considerable column bleed can contaminate the preparation. Desalted material is suspended in water and lyophilized.

C. Reduction of Glycosyluronic Acids

Polymers containing glycosyluronic acid, particularly the rhamnogalacturonans of plant pectins and the uronic acid-rich bacterial polysaccharides, pose additional problems. Alkaline solutions used either in extraction or in the methylation reaction can cause β-elimination,

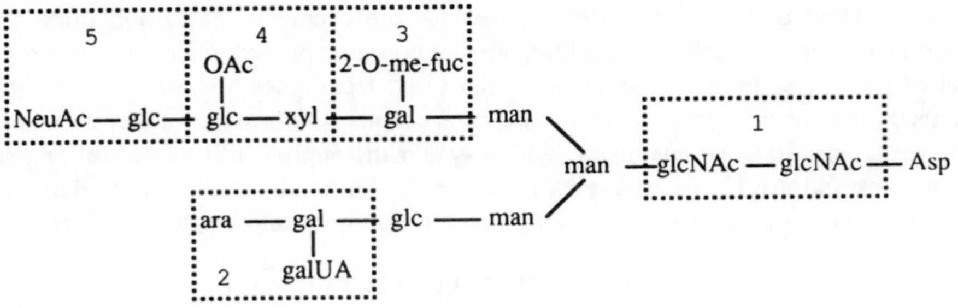

5 4 3
OAc 2-O-me-fuc
NeuAc — glc — glc — xyl — gal — man
1
man — glcNAc — glcNAc — Asp
ara — gal — glc — man
2 galUA

a

FIGURE 1. a. Model oligosaccharide illustrating potential problems to be overcome for successful methylation analysis. 1. Permethylation of amino sugars and subsequent hydrolysis can result in *N*-deacetylation, rendering the sugar resistant to hydrolysis; alternatives of acetolysis and methanolysis are discussed. 2. While furanose sugars, such as arabinose, are easily hydrolyzed by acid, glycosyluronic acids are quite resistant to hydrolysis, and the aldobiouronic acids would be lost; their identification is based on chemical reduction of the carboxyl group with $NaBD_4$ generating the corresponding aldose. The 6,6-dideuterium label permits quantitative determinations by MS. The positions of the uronic acids can also be determined by β-elimination reactions after permethylation, yielding sequence information when done in parallel with permethylation of carboxyl-reduced polymers. 3. Some sugars are endogenously methylated, and the positions of these groups are revealed by use of CD_3I or CH_3CH_2I instead of CH_3I, producing unique derivatives that are also quantified by MS. 4. Positions of endogenous acetylations can also be determined by chemical derivatization in reactions compatible with the methylation analysis. 5. Positions of neuraminic acid (best determined as methyl ester instead of alditol acetate) and other sugars can be determined by enzymic or mild-acid hydrolysis to provide additional linkage information. b. Basic preparation of partially methylated alditol acetate derivatives. Carbohydrate is dissolved in a strong nonaqueous alkali solution to form alkoxyl ions of all free hydroxyl units. Methyl ethers are formed upon addition of CH_3I, and after purification of the now lipophilic permethylated polymer, monosaccharides are generated by acid hydrolysis. The sugars are reduced with $NaBD_4$ to form their respective alditols, tagging the C-1 carbon with deuterium, and the remaining hydroxyl groups are acetylated. The family of derivatives is separated by GLC, and the positions of the methyl and acetyl substituents are deduced unequivocally by EI-MS.

thus fragmenting the chains, particularly with 4-*O*-substituted uronic acids.[12] β-Elimination is not a likely problem during the methylation itself because the solution is neutralized rapidly upon formation of the methyl ethers.[16] Further, the uronic acids are generally quite resistant to acid hydrolysis and, instead, form aldobiouronic acids. Hence the uronic acids, and neutral sugars to which they are attached, are lost. This potential problem is best solved by chemical reduction of an activated ester of the carboxyl with $NaBD_4$. The deuterium label ensures that the carboxyl ester is reduced to the 6,6-dideuterio primary alcohol that, with MS, is clearly differentiated from native neutral sugar. Several strong reagents, such as $LiAlH_4$ (and $LiAlD_4$) in hot ethyl ether,[17-20] have been used for reduction, but more recently the carbodiimide-activated reduction has proven to be both gentle and efficient and has gained widespread use.[21,22] The procedure listed below is tailored to larger amounts of material, but the procedure can be scaled down to much smaller volumes. About 50 mg of material are dissolved in 25 mℓ of H_2O, and the solution is adjusted to pH 4.75 with 0.01 *M* HCl. While continuously stirring and monitoring the pH, 0.5 g of 1-cyclohexyl-3-(2-morpholino-ethyl)-carbodiimide-metho-p-toluenesulfonate (CMC) solid powder is added. The pH will slowly rise as the mixed anhydride is formed; pH is maintained by drop-wise addition of 0.05 *M* HCl. After about 2 hr, the pH will stabilize as adjunct formation is complete. A drop or two of *n*-octanol is added to diminish foaming, and reduction is accomplished by drop-wise addition of 10 mℓ of 2 *M* $NaBD_4$. The pH will rise quickly to above neutrality. Taylor and Conrad[21] recommended that the pH be maintained at 7 with an automated titrimeter, or by drop-wise additions of 0.1 *M* HCl, because the adjunct is sensitive to alkali and possibly to avoid β-elimination before reduction is complete. Addition of HCl results

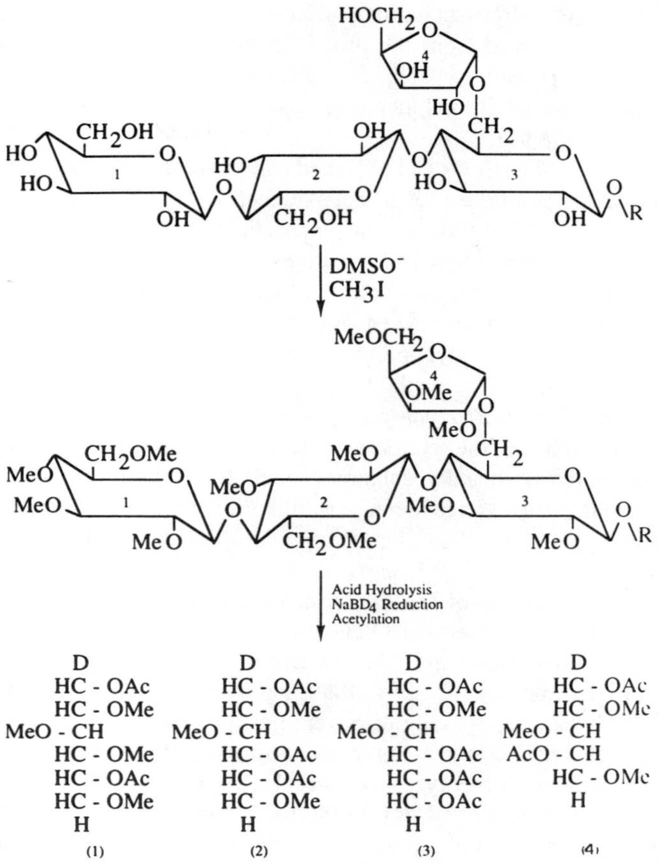

DMSO⁻
CH_3I

Acid Hydrolysis
$NaBD_4$ Reduction
Acetylation

FIGURE 1b

in extensive destruction of much of the $NaBD_4$. The procedure can be modified to include a neutral buffer,[22] but more simply 1 mℓ of the 2 m*M* $NaBD_4$ is added drop-wise by pipet and pH carefully maintained at neutrality with a second pipet containing 0.1 *M* HCl. Sufficient borate has now formed to buffer the solution; the remainder of the $NaBD_4$ solution is added, and the pH allowed to rise to about 8.5 to 9. The mixture is stirred for an additional hour, then chilled on ice and titrated to pH 5.5 with glacial acetic acid and dialyzed against deionized water. Borate, which binds strongly to some sugars, can be removed from preparations of polymeric material by dialysis against 10% ethylene glycol and then deionized water, or from oligomers and sugars by successive evaporations with 10% acetic acid in methanol followed by methanol. To check yield, samples of the material should be assayed for uronic acid before and after reduction by either the method of Blumenkrantz and Asboe-Hansen[23] or with carbazole[24] as modified by addition of sulfamate to inhibit neutral sugar interference.[25] If uronic acids are not reduced to their respective sugars, the material should at least be passed through a H^+-cation exchange column to protonate the carboxyl groups, rendering the polymers more soluble in the DMSO mixtures.[26]

D. Glycoproteins and Glycolipids

Glycolipids require little additional preparation, but with glycoproteins, the cleavage of the carbohydrate moieties from the peptides may be necessary for several reasons. First, some glycoproteins display only slight solubility in solutions of DMSO used for methylation;

second, because different oligosaccharides often occupy multiple sites on a single peptide, the sugars should be removed from the protein and purified by conventional liquid chromatography. Sulfated, phosphorylated, and other substituted derivatives can add to the complexity, but also offer additional means of separation by electrophoretic or ion-exchange methods.

Formerly, hydrolysis of both *O*- and *N*-linked carbohydrates was achieved by elimination in alkaline-borohydride solutions.[27-29] Extensive destruction of the protein was inevitable, but, because the borohydride reduces the anomeric reducing end to form an alditol or aminohexitol, alkaline degradation of the oligosaccharide was prevented and additional linkage information was provided as a bonus. More recently, however, enzymic hydrolysis is favored over alkaline digestions when possible, both to preserve the protein (valuable when only minute amounts of glycoprotein can be obtained) and to ensure collection of native sugar. Two commercially available enzymes now used extensively are endo-β-glycosaminidases (endo H from *Streptomyces grieus* and endo F from *Flavobacterium meningosepticum*) which cleave the chitobiosyl residues of the core-glycosylated *N*-linked proteins.[30,31] The protein is usually denatured in SDS or Na thiocyanate to relieve steric hindrance of the sugar groups and to increase availability of the chitobiosyl groups to enzymic attack.[32] The chitobiosyl groups are often further substituted with α-fucosyl residues which suppress endo H activity.[33] Endo F is capable of cleaving the chitobiosyl residue when the fucosyl is attached to the *O*-6 of the GlcNAc residue, but its activity is reduced toward hybrid structures containing bisecting GlcNAc residues, or tri- and tetra-antennary structures.[31] Two new commercially available enzymes, "*N*-glycanase" and "*O*-glycanase" (Genzyme/Koch-Light), are significant in that they hydrolyze the entire carbohydrate from the protein.[34,35] *N*-Glycanase, actually discovered as a contaminant of endo F preparations, is an amidase that releases the amino sugar reducing end and converts asparagine to aspartic acid,[34] whereas *O*-glycanase hydrolyzes the disaccharide gal-galNAc from serine or threonine residues of mucin-type proteins.[35] Considering the complexity based on elaboration of the disaccharide core structure now recognized among this class of glycoproteins,[36] the *O*-glycanase may be of more limited use. The intact *N*-linked oligosaccharides are hydrolyzed by *N*-glycanase from simple, hybrid, and complex glycoprotein structures. Regardless of the method of cleavage of the carbohydrate moieties from protein, the reducing (amino) sugar end should be reduced to its corresponding alditol (aminohexitol) to prevent alkaline degradation during methylation. By enzymic degradation techniques, the reduction can be done conveniently with $NaBD_4$ to produce asymmetry necessary to deduce linkage structure.

All deionized materials prepared for methylation should be lyophilized for ease of dispersion in DMSO. If only small amounts of the material are available, they should be prepared directly in the reaction vial. Lyophilized materials in reaction vials should be stored complete with magnetic stir-bar and stopples over P_2O_5 in a vacuum desiccator at least overnight to ensure minimal moisture, with the vials covered with paper towelling to avoid contamination. A drying tube filled with $CaSO_4$ desiccant (with indicator) should be used when reintroducing air to the vacuum desiccator.

III. METHYLATION OF CARBOHYDRATES

A. A Brief History of Methylation and Linkage Analysis

Use of methylated sugars for analysis of linkage structure has a long history. Analysis was done by tedious chemical degradation and relied extensively on recrystallization of products, determination of melting points, and colorimetric and titrimetric assays. Needless to say, separation of a mixture of derivatives from complex polysaccharides was not explored until some time later, and hence, rather simple polymers were deduced. With the advent of paper chromatography, the diversity of products was realized. Analysis still heavily relied upon methylation, but until the advent of gas-liquid chromatography (GLC), reactions required rather large amounts of sample, and rather incomplete methylation was often overcome by continued remethylations, as many as four or five times. With the advancement in GLC combined with MS analysis, the complexity of many seemingly simple polymers became apparent, as did the observation of the limits of the analysis and the potential for destruction of alkali-sensitive, partially methylated sugars. While many of those limitations have not been overcome, complete methylation can be accomplished in one step which avoids potential degradation by repeated contact with the strongly alkaline reagents. While many methods have become available, they all share the same strategy by which the polymer is solubilized in a strong nonaqueous alkali to completely ionize the hydroxyl, amino, and carboxyl groups, and a methyl halogen is introduced to form the stable methyl ether or methyl amino linkages.

Nevertheless, the method of Purdie and Irvine,[37] using silver oxide and methyl iodide, and later, the method of Haworth,[38] using silver oxide and a mixture of methyl sulfate and NaOH, were used extensively for nearly 50 years without modification. One of the first modifications recognized the importance of nonaqueous solvents in facilitating methylation in which Ag_2O and CH_3I were used in *N,N*-dimethylformamide,[39] and a comparison of methodologies demonstrated its superiority over the Haworth and Purdie methylations.[40] Both reports tested only monosaccharides and oligomers, however. The most significant development was the demonstration by Corey and Chaykovsky[41] that methylsulfinylmethanide, or "dimsyl" ion, was a useful anion in organic synthesis. The anion could be made from DMSO and NaH or KH, but not from LiH, and also suggested that *tert*-butoxide or *n*-butyllithium could also generate the dimsyl anion. Many possible uses for the anion were suggested, but methylation of sugars was not one of them. Price and Whiting[42] described reactions of the methanide ion with several alcohols, but it was the classic description by Hakomori[43] that documented the dimsyl ion as a superior methylation reagent for sugars and polysaccharides. Anderson and Cree[44] also demonstrated that the dimsyl ion was superior to Haworth and Purdie methylations for complete methylation of acidic polysaccharides. Further technical refinements by Sandford and Conrad[45] made preparation of the reagent routine. Steiner and Gilbert,[46] and later Hiller,[47] demonstrated that the Na^+ anion induced stoichiometrically complete ionization of many alcohols as judged by titrimetric detection of free methanide with triphenylmethane.[41,48] The K^+ anion was actually less effective by their estimation, although more recently many have found the K^+ anion easier to prepare and more reactive with sugars and polysaccharides. By virtue of the low solubility of K^+ salts in chloroform or other organic solvents, it produces derivative preparations with significantly less contamination.[2,49,50] Although no systematic studies have been done, it is possible that contamination of commercial preparations of either DMSO, KH, or NaH could have resulted in these apparent differences.[50,51]

B. Laboratory Equipment

CAUTION: *n-Butyllithium, NaH, and KH are extremely flammable and potentially explosive reagents. Exercise extreme care to avoid introduction of water or air to dry hydrides, not to heat the DMSO solution excessively during evolution of hydrogen, and not to exceed recommended volumes of preparations. Wear protective gloves throughout preparation of reagents and all subsequent steps in the preparation of permethylated material.*

Because of the dangers involved in forming the methylsulfinylmethanide anion from KH or NaH and the necessity in maintaining anhydrous conditions for preparation of the reagent, some special attention should be given to preparing appropriate glassware and laboratory conditions. A stainless steel fume hood devoted to preparation of the reagents should be free of organic solvents to minimize fire danger, free of water and steam baths that could inadvertently introduce water vapor to the dry hydrides during initial preparation of the anion, and free of paper or plastic flammable materials. To keep the reaction mixtures under an inert atmosphere, and for evaporation of acids and solvents, a regulated supply of dry N_2 or Ar is necessary. Several devices are commercially available with flow rates of about 1 to 100 mℓ/min of cylinder gas through as many as 24 individual jets, each with a needle valve control and a Luer-Lok® fitting so that individual flow rates can be controlled. To give greater flexibility so that reaction mixtures can be stirred continuously, needles from disposable syringe tips are removed with pliers, and 1/2-in. latex tubing stretched over the fitting. The Luer-Lok® end is connected to the nitrogen delivery; the opposite end of 12 to 18 in. of tubing is stretched over a second intact disposable 18-gauge syringe needle, hence nitrogen, controlled by needle valve, can be introduced into reaction vials sealed with rubber stopples or crimp-on septa. Nitrogen should be of exceptional purity, since lower-grade commercial cylinders can have significant water vapor. A trap of $CaSO_4$ (with indicator) and molecular seive can also be used if so desired. Harris et al.[2] have also suggested that argon, being denser than air, is preferable to N_2. All glassware should be rinsed with acetone or methanol and oven dried just before use. For added safety, a Plexiglas® shield such as those used in radioisotope work should be used during preparation of the methanide ion.

C. Preparation of Reagents

1. Preparation of the Sodium Methylsulfinylmethanide Anion

The sodium anion is an excellent reagent, but is slow to form, requiring heating at 50°C for about an hour. It can be prepared and stored frozen under N_2 in sealed ampules and used when needed. The presence of the active anion can be assayed by formation of a deep red adjunct with triphenylmethane,[41,48] and the normality of the reagent determined by titration of an aliquot to phenolphthalein endpoint with 0.1 *M* HCl.

The NaH is commercially available as a 60 to 80% suspension in mineral oil. Because batches of reagent vary substantially in purity, some attention to its appearance should be noted upon obtaining new or different batches of reagent. The suspension should be homogeneous and beige to light gray in color. Due to incomplete formation of the hydride, Na metal is sometimes a significant contaminant. Any reagent containing substantial amounts of visible impurities should be returned, since it is unsuitable for use in these reactions. Wear gloves when handling the NaH and during all subsequent steps of the preparation.

Suggested amounts of anion vary among investigators. In a recent study, it was suggested that three to six equivalents per exchangeable H^+ (an average of three per sugar) was optimal and higher amounts slightly inhibited subsequent methylation.[53] Therefore, 1 mg of hexose polysaccharide needs only 100 μmol of the anion. Others have experienced little problem when mg amounts of polysaccharide were suspended in 1 to 1.5 *M* anion, and this is most conveniently added as an equal volume to a suspension of the sugar in DMSO. Hence, 2 or 3 *M* anion is prepared, but titration of the reagent should be made because yields of methanide ion can sometimes be considerably less than anticipated as a result of impurities in the reagents or moisture, particularly if the prepared reagent is deep yellow or greenish yellow. To guard against excessive alkali, many now add recommended equivalents of anion, and withdraw a sample of the mixture to test for presence of the methanide ion by triphenylamine. Rauvala[48] has suggested that a slightly positive methanide reaction rarely results in undermethylation, but there are reports to the contrary.[2]

The dense NaH usually settles out to form a firm cake in the mineral oil, and some time should be spent to disperse this cake into a homogeneous suspension by gently swirling the solution. Keep the bottle sealed under N_2 atmosphere to prevent unwanted contamination with oxygen, carbon dioxide, and water vapor. Take a 150-mℓ round-bottom, triple-neck flask containing a magnetic stir bar upon a top-loading balance. Using a large bore disposable pipet, quickly dispense 8 g of the homogeneous suspension of 60% NaH in mineral oil into it. Quickly place the suspension in an N_2 atmosphere by inserting the syringe needle carrying

N_2 flow from the delivery source into a rubber stopple secured on one side neck of the flask. Delivery of about 60 mℓ/min should be satisfactory, just enough to detect flow when the needle is held about 2 in. from one's lips. With a ring clamp, secure the flask in a heating mantle resting above a magnetic stirrer. The stock reagent should be flushed with N_2 or Ar and the bottle resealed tightly. Add about 50 mℓ of hexane to the suspension, and stir the NaH up into the hexane for about 30 sec. The dense NaH will settle out within a few min. Carefully pour off the hexane containing the dissolved oil out the open side neck so as not to disturb the sedimented NaH. Repeat three more times to ensure complete removal of the oil, pouring off as much of the hexane as possible in the last wash to hasten drying. The remainder of the hexane can be dried by gently stirring the cake under the stream of N_2. Alternatively, a vacuum pump can be used to accelerate the drying, but be careful not to introduce water vapor into the sample. Further, quickly inducing vacuum can spatter the NaH-hexane suspension. Small portable vacuum pumps that contain needle valves to control vacuum are best. Attach the pump via rubber tubing to a glass tubing adapter with a check valve at the center neck of the flask, and seal the side necks securely with rubber stopples. Simultaneously begin pumping and withdraw the syringe needle with N_2 or Ar flow. Gradually increase vacuum by the needle valve until the hexane just begins to bubble. One may want to practice this step with a mℓ of hexane without NaH. When the hexane has evaporated, then boost the vacuum, continuously stirring the NaH to break up the cake. When dry, the NaH should have a fine sandy appearance. If chunks of material persist, gently tap the sides of the flask so that the stir bar disperses them. Evaporation of the hexane takes about 15 min. When the vacuum pump is turned off, the flask will self-seal by means of the check valve in the tubing adapter. Reintroduce N_2 slowly to prevent dispersion of the powder. Remove the rubber tubing and vacuum pump and vent the flask with N_2 or Ar flow. With a syringe, add 100 mℓ of anhydrous, silylation-grade DMSO (obtained commercially) through the stopple in the side valve. Molecular sieve can be added to the commercial DMSO to ensure dryness, but the flask should then be sealed with a septum. The DMSO is withdrawn by syringe with N_2 introduced to replace the volume taken and to keep the solution under pressure of inert gas. Gently stir and continuously purge the suspension of NaH and DMSO with N_2 or Ar flow and then heat to about 50°C until gentle evolution of H_2 is perceived. After about 1 hr, a pale yellow to greenish yellow clear solution is formed. The color of the reagent is variable at times and subject to some myths. It varies from nearly colorless[51,52] to yellow, yellow green, deep green, or even gray green.[6,12,41] Exner and Steiner[51] and Rose et al.[52] prepared essentially colorless methanide from purified metal hydrides and dry, distilled DMSO. Rose and colleagues[52] also noted that yellowing, or further darkening of the solutions, resulted from overheating or entry of traces of air or water. Indeed, the heated Na^+ methanide preparations are always darker than other preparations made at ambient temperatures. Allow the methanide solution to cool to ambient temperature and dispense by pipet into individual ampules in desired volumes. Top each ampule with N_2 or Ar, and seal either with rubber stopples or with septa that can be crimped on tightly. The reagent can be stored for months at −20°C without apparent loss of reactivity.

2. Preparation of the Potassium Methylsulfinylmethanide Anion

More recently, many have preferred the use of the K^+ methylsulfinylmethanide anion over the Na^+ species. Because KH is a more reactive species than NaH,[51] additional caution should be exercised in its preparation and use. Preparation is much easier, so that fresh reagent can and should be made just prior to use. As was the case with Na^+ ion, 2 to 3 *M* reagent is prepared and added as an equal volume to the suspension of carbohydrate in DMSO or as a 3 to 6× stoichiometric amount of exchangeable protons.[53] Samples of the mixture should be tested for the presence of excess methanide ion by triphenylmethane.[41,48] Because KH is more reactive than the sodium species, only up to 10 mℓ of reagent is made at one time. The reaction proceeds at ambient temperature, so additional heating is not required or recommended. *The triple-neck flask must be about five volumes the amount of reagent prepared because considerable foaming occurs beginning about 30 sec after addition of the DMSO*. Pipet 2.3 g of a homogeneous suspension of 35% KH in mineral oil into a 50-mℓ triple-neck, round-bottom flask and purge with either N_2 or Ar. Wash with hexane and dry as described for the NaH. Chilling the stock anhydrous DMSO on ice a few minutes before addition to the KH is also recommended to slow the reaction initially and to prevent unnecessary heating of the reagent. Contamination of KH with potassium metal contributes an additional danger because it can spontaneously spark upon addition of DMSO, igniting the evolving hydrogen. This problem can be avoided easily by ensuring that the chilled DMSO, when injected, is so directed as to completely coat the gently stirred KH powder as quickly as possible. After complete addition of the DMSO and submersion of the KH, the potential for fire is quenched. If for some reason the N_2 flow is not sufficient to disperse the foam, it can usually be broken up with gentle swirling. Some of the incomplete reaction product occasionally may spew from the flask; do not attempt to wipe up this material as it will ignite paper toweling. Instead, simply cover spills with mineral oil, and neutralize them later with ethanol. The reaction is complete in about 30 min, and a clear pale yellow to slightly greenish yellow solution free of sediment is obtained.

3. Other Anions

Corey and Chaykovsky[41] demonstrated that potassium *tert*-butoxide could also form the methanide ion from DMSO, and considering the relative ease and safety of preparation, it also has been proposed as an effective reagent for the methylation of polysaccharides.[9,54] Because this reagent is not as reactive as the metal salts,[51] its ease of formation is compromised by incomplete methylation of certain polysaccharides.[2,53] Further, Brauman and colleagues[55] pointed out that the reaction actually produces equal amounts of *tert*-butanol and methanide ion from DMSO and *tert*-butoxide; this mixture could interfere with alkoxide formation of carbohydrates.[51] Although the reagent may be adequate in some applications, it cannot be recommended for general use.

Excessive heating required to form the Li^+ methanide ion from LiH resulted in extensive decomposition of the ion,[41] but *n*-butyllithium offers a reasonable alternative to the LiH. The reagent evolves butane exothermically, so a potential fire danger must be avoided by adequate purging with inert gas and slow introduction of the reagent into the dry DMSO. The *n*-butyllithium is commercially available as a 2.5 *M* solution in hexane, and can be used directly with dry DMSO. Blakeney and Stone[56] have found that the methanide ion can be easily made at near ambient temperatures, producing a pale yellow solution free of sediment, acceptable for methylation of polysaccharides. Parente et al.[57] also demonstrated its usefulness in *N*-methylation of amino sugars of glycoproteins, and from our own recent experience, complete methylation of plant pectin and hemicellulosic polysaccharides was consistently obtained with this reagent.

Preparation of the methanide ion is simple. Dry DMSO (8 mℓ) is injected into a 50-mℓ, round-bottom flask purged with N_2 or Ar as usual, and 6.4 mℓ of the 2.5 *M* *n*-butyllithium in hexane is added slowly by syringe into the DMSO, which is slowly stirred magnetically. A cloudy solution is incubated at 40°C for about 30 min until a pale yellow clear solution is obtained. The hexane slowly evaporates in an additional 30 min, but the last traces must be removed by gentle vacuum evaporation using the precautions described for drying the metal hydrides. A further simplification of this method suggests addition of 0.4 mℓ of the hexane solution of 2.5 *M* *n*-butyllithium directly to a 1 mℓ suspension of polymers sonicated in DMSO, permitting simultaneous methanide and polyalkoxide formation as the hexane evaporates.[57,58]

The most recent addition to the list of anions is perhaps the simplest — the addition of powdered KOH and NaOH to DMSO. Ciucanu and Kerek[53] recently questioned whether or not the methanide is the active species in alkoxide formation, even in preparations of the Na^+ and K^+ methanide anion. Although Na^+ methylsulfinylmethanide preparations were effective, anion formation with *tert*-butoxide was poor, and efficient methylations were possible only if NaOH was added to the mixture. They further showed that 2.5 to 24% NaOH added for only a few minutes to a DMSO solution of sugar resulted in 98.8% ± 2% yield in permethylated sugar upon addition of CH_3I. They suggest that OH^- or H^- are the active species. This work only considered simple sugars, and only a few additional works have explored use of this simple technique and so its acceptance for general use will need testing.[59-61] Although one would intuitively expect that the preparation is too wet, the effectiveness is based on the fact that KOH and NaOH are only slightly soluble in DMSO, and the remaining insoluble alkali absorbs the water hygroscopically. Its preparation will be discussed later in relation to the methylation of polysaccharides.

D. Dissolution of Polymers

Lyophilized material is more easily dispersed in DMSO, but solubility varies considerably. Many materials become soluble upon addition of the alkaline methanide ion. Still others, like cellulose for example, have strong crystalline structures that might require dissociation with stronger solvents. Processing of 1 mg or more of material can be accomplished in 15-mℓ Corex® tubes with snug-fitting rubber stopples, whereas lesser amounts are best processed in 1-mℓ conical reaction vials sealed with silicon/Teflon® septa with either screw caps or with crimp-on adapters to ensure vapor-tight seals. Anhydrous DMSO can be distilled from above $CaCl_2$ and stored over molecular sieve or purchased as "silylation grade", which is commercially prepared as essentially anhydrous and packed under N_2. The DMSO is withdrawn with a syringe, and the volume is replaced with dry N_2. One mℓ is added to the sealed tubes containing sugar and magnetic stir bar. The tubes are incubated in a sonic bath in a small amount of water, and a vacuum pump with rubber tubing fitted with a syringe needle is used to evacuate each tube for about 30 sec to facilitate "wetting" of the carbohydrate with DMSO. The tube is gently tapped against the bottom of the sonic bath to dislodge bubbles formed during the degassing. Gentle sonication for up to 4 hr will result in heating to about 50°C, and this will aid dissolution.

With polysaccharides such as cellulose or plant cell walls, solubility is a major problem. Some authors recommend that simply dry milling the material to a fine powder will aid swelling and dispersion in the DMSO,[62] while others have sought better solvents compatible with the methylation reagents.[63-69] Narui and colleagues[69] found that mixtures of 1,1,3,3-tetramethyl urea hastened formation of anions, perhaps by dissociation of hydrogen bonds, and suggested that cellulose could be successfully methylated. Another recent evaluation of cellulose solvents has shown that while several are compatible with methylation reagents, a SO_2-diethylamine-methylsulfoxide solvent dispersed cellulose without degradation and resulted in high yields of methylation when used with powdered NaOH.[66,67] N_2O_2-methylsulfoxide and LiCl-*N,N*-dimethylacetamide are perhaps good solvents,[64] but resulted in poor yields and caused noticeable depolymerization, respectively.[65] 4-Methylmorpholine *N*-oxide was a reasonable solvent,[63] although some depolymerization was also reported.[66,68]

Concentrated SO_2-methylsulfoxide is made by bubbling 15 g of SO_2 (dried over $CaCl_2$) into 50 mℓ of anhydrous DMSO, gently stirred and resting on a top-loading balance. The SO_2 should be bubbled slowly so that the heat from this exothermic reaction can be dissipated. The solution should be stable for several months. Injected into 1 mℓ of the DMSO suspension of the material is 40 μℓ of this mixture, followed by 20 μℓ of the diethylamine. The solution should clear in a few minutes and is then ready for addition of the anion.

Regardless of the technique of solubilization, to prepare for addition of the methanide ion or powdered alkali, the tubes containing the material are flushed continuously with N_2 or

Ar (about 5 mℓ/min) by insertion of syringe needles into the stopple- or septum-sealed vials connected to the inert gas supply as described. After pressure has developed, introduce a second syringe needle for escape flow. A thawed preparation of the Na^+ methanide ion is centrifuged at about 1500 *g* to remove a sediment that forms during preparation, but the fresh anion preparations should be clear and used directly. An equal volume of 2 *M* reagent is injected directly into the DMSO suspension, taking care not to run it down the sides of the tube. Initially yellow, the mixture will turn greenish yellow after continued incubation for up to 4 hr. For addition of dry KOH or NaOH, technical grade alkali is recommended because it is shipped as flakes, rather than pellets, and can be pulverized quickly and vigorously in a mortar and pestle so that absorption of moisture is minimized. A scoop of about 0.25 g of the powder is added directly to the DMSO sonication mixture. The tube is purged of air by vacuum, and then flushed with inert gas. Reaction times for alkoxide formation vary, depending on the methanide preparation and the solubility of the carbohydrate. Oligosaccharides should take no longer than 1 hr with either the Li^+ or K^+ preparation, but polysaccharides may require several hours to completely dissolve in the alkaline reagent. Unless the ion is titrated, there is no certain way to tell the exact time, but reactions can be varied from a few minutes[53] to several hours,[6,45] and tested empirically with standards.

Problems with ''undermethylation'' often result from the manner by which CH_3I is added, rather than by improper formation of the anion. The commercial silver-stabilized liquid CH_3I should be stored chilled in doubly sealed containers. In some special preparations (discussed later), CD_3I or CH_3CH_2I may be used in place of CH_3I, and the following procedures are equally appropriate for the use of these reagents. The CH_3I is a potent mutagen, but forms innocuous amounts of methanol and iodine upon contact with water. Nevertheless, the reagent should be treated with extreme caution and only opened in the fume hood. The solution should be clear and colorless; an orange tint indicates the significant conversion of CH_3I to iodine and methanol from introduction of water. With many materials, CH_3I is injected directly into the mixture in molar amounts equal to, or slightly higher than, the anion. For reaction mixtures containing 2 mℓ of 1 *M* methanide ion, about 0.5 mℓ is sufficient. With some oligosaccharides, and many polysaccharides, however, rapid formation of methyl ethers at some positions inhibits subsequent methylation of adjacent residues. Furthermore, the partially permethylated polymers become insoluble and the remainder of the polymer may precipitate before complete methyl ether formation. Some authors recommend addition of $^1/_{10}$ of the CH_3I intended, incubation for a few minutes, and then slow addition of the remainder. To slow the neutralization, tubes containing the DMSO suspension are put into crushed ice for about 15 min until well frozen. The entire volume of CH_3I is added, and the methylation mixture left on ice for a few minutes before stirring is resumed. When the K^+ anion is used, the solution clears quickly to a yellow orange, but with the Na^+ ion, a milky orange solution is produced upon melting that requires about 30 min to clear. Occasionally, a few samples fail to clear unless additional CH_3I is added. The CH_3I is added directly to the warm Li^+ alkoxide for it is particularly slow to react, and extended incubation time, perhaps overnight, is necessary for complete methylation.[56,57] Because addition of powdered KOH or NaOH is less quantitative, care must be taken to add sufficient CH_3I to neutralize the alkaline mixture. Upon clearing, the N_2 flushing can be halted, and the sample processed directly or left overnight with gentle stirring. The KOH or NaOH samples, or the anion preparations, can be processed similarly. The effectiveness of the methylation can sometimes be estimated by looking closely at the sample. Fully permethylated polymers are quite soluble in DMSO, whereas visible, finely suspended material may indicate substantial undermethylation. Detection of R–OH groups by IR spectroscopy has also been exploited,[43] but it is doubtful whether the sensitivity of most instruments is sufficient when considering that only mg quantities of the material, or less, are now used. The actual extent of undermethylation can be judged best from direct quantitation of derivatives by GLC which will be described later.

E. Purification of Methylated Polymers and Sugars

Material can be purified from the methylation reaction mixtures in several ways. Formerly, the mixture was simply dialyzed against running deionized water overnight, and the remaining emulsion or suspension partitioned twice into about 2 mℓ of chloroform, followed by 2 mℓ of chloroform/methanol (2:1, v/v). The combined chloroform and chloroform/methanol phases were collected in 1-dram (about 4 mℓ) screw-cap vials, and the solvents were evaporated in a stream of N_2. Because this technique was unsuitable for oligo- and monosaccharides, and because of considerable losses expected when using small samples, the material can also be partitioned directly in organic phases without dialysis.[2] A mixture of chloroform and methanol is added directly to the methylation reaction, and the DMSO and additional salts removed by at least five washes with water. The organic phase is then evaporated. Because a small amount of water carried over can slow the evaporation, a small amount of 2,2-dimethoxypropane (6.8 mℓ/mℓ of water) is added, which reacts with the water to form methanol and acetone.[2]

Column chromatography can also be used with some limitations. Organic phase columns of LH-20 have been prepared,[70-72] but these have been replaced primarily with mini columns of C_{18} reverse-phase material (Sep-Paks®; Waters, available through Millipore) which fit standard Luer-Lok® 5-mℓ glass syringe barrels.[3,73] The columns are washed twice with 5 mℓ of methanol and then equilibrated with deionized water, pushing gently with the syringe plunger to produce a flow of about 10 mℓ/min. Care should be taken to disconnect the Sep-Pak® from the syringe to avoid air contact with the material when filling the syringe with fresh solvent during further preparation or, alternatively, use a syringe with a three-way valve. An equal volume, or greater, of water is added to the methylation mixture to increase the polarity of the solvent so that the permethylated polymers bind tightly to the reverse-phase material; a milky suspension quite often forms, but with small amounts of material or with small molecules, the solution may remain clear. The mixture is loaded onto the column through the syringe, and washed with about 15 mℓ of water. The iodine often remains on the column and will require low concentrations of methanol or acetonitrile in water to remove it. The permethylated material varies in its tenacity of binding. Mono- and oligosaccharides, including fructans of about DP 25, will begin to elute in about 15% methanol, whereas most polysaccharides require at least 40% methanol to displace them from the column. Higher concentrations of acetonitrile than methanol are generally needed to displace the sugars, and it may be the solvent of choice for oligosaccharides. To test the elution profile when experimenting with new material, elute, in 2-mℓ volumes, in 10 to 15% increments of methanol or acetonitrile to 100% and finally with chloroform/methanol (1:1, v/v). Sugar in each fraction can be determined by a microversion of the phenol-sulfuric acid method,[74] where 200 μℓ of the sample is mixed with 200 μℓ of 5% phenol, 2 mℓ of concentrated H_2SO_4 is injected, and the reaction mixture vortex agitated thoroughly. The yellow-orange reaction product is measured spectrophotometrically at 490 nm. This will give the safe washing concentration of eluant; the material, in subsequent samples, should be eluted in 100% methanol or acetonitrile to hasten subsequent evaporation of solvent. However, water can be removed safely by addition of 2,2-dimethylpropane.

IV. PREPARATION OF ALDITOL ACETATES

A. Acid Hydrolysis

1. Neutral Sugars

Generally, ease of hydrolysis and the propensity of degradation of the permethylated monosaccharides released upon hydrolysis is similar to that of the native sugars. One important difference is that the permethylated polymers are less soluble in aqueous solutions, so additional hydrolysis times are warranted. Many early studies used quite strong acids to cleave the sugar residues, and much degradation was possible. One popular method used hot H_2SO_4, followed by addition of $BaCO_3$ or $AgCO_3$, to neutralize the acid; insoluble sulfates were sedimented and the remaining solvent evaporated to dryness. Conrad et al.[75] advised against concentrated aqueous HCl or H_2SO_4, since either acid caused too much degradation. The use of H_2SO_4 can complicate matters further by sulfation of some of the carbohydrate hydroxyl groups.[4] Formolysis in 98% formic acid is particularly destructive and can even induce demethylation.[76] Slightly diluted formic acid (90%) for up to 6 hr at 100°C achieved quite satisfactory hydrolysis with only minimal degradation,[76] and is well recommended for permethylated material.[6,7,12,71] A more common method for carbohydrates containing primarily neutral sugar is hydrolysis in 2 *M* TFA, for 60 to 90 min at 120°C.[77] Internal standards are usually introduced at this step; *myo*-inositol is the most common standard and, for mg amounts of sugar, 1 μmol is typically used. For samples of 1 mℓ or less, hydrolysis can be done safely in 1-dram glass vials with screw caps lined with Teflon®. Nitrogen can also be bubbled through the solution briefly before sealing the vials to minimize oxidation. A heating block is a reasonable alternative to autoclaving as originally described.[77] Either formic acid or TFA is volatile and can be removed by evaporation, reducing losses of material. Permethylated monosaccharides, particularly the trimethyl pentoses and deoxyhexoses, are also reasonably volatile, so losses can occur during evaporation of the TFA. It is recommended that the reaction mixture be cooled to 30°C and maintained there while the acid is evaporated until just dry.

Unfortunately, considerable variation exists in the rates of hydrolysis and, after hydrolysis, the reactive anomeric carbons are more susceptible to degradation. The furanosyl linkages are easily hydrolyzed, and some loss of arabinosyl is expected under conditions required for complete hydrolysis of hexopyranosyl units.[75] With fructans, the fructofuranosyl linkages are completely hydrolyzed in 2 *M* TFA at ambient temperatures in 30 min or less; heating to 120°C completely destroys fructose. Fucose, galactofuranose, and 3,6-dideoxysugars of many bacterial cell surface polysaccharides are equally acid sensitive.[75,78-80]

2. *Glycoproteins and Glycolipids*

The complexity of carbohydrates of glycoproteins and glycolipids offer several unique problems with respect to quantitative recovery of all sugars in a single step. The acid-labile 3,6-dideoxyhexose and fucose, often found as units of oligosaccharide chain, are easily lost under conditions required to hydrolyze the more resistant amino sugars. In fact, upon methylation of the amino sugar, subsequent acid hydrolysis can result in loss of the amino acetyl group, generating the aminomethyl cation and producing a derivative extremely resistant to acid hydrolysis.[8,9] In a systematic study of this problem, Stellner et al.[81] found that among 0.5 to 2 *M* H_2SO_4, 90% formic acid + 0.15 *M* H_2SO_4, methanolysis in 0.5 *M* HCl, TFA in acetic acid, TCA in acetic acid, and boron trifluoride in DMSO, none afforded satisfactory hydrolysis of such linkages, but acetolysis followed by hydrolysis was quite superior. The permethylated material is suspended in 0.3 mℓ of 0.25 *M* H_2SO_4 in 95% acetic acid (100% acetic acid resulted in decomposition) and incubated overnight (16 to 18 hr) at 80°C to achieve acetolysis. Hydrolysis is induced by addition of 0.3 mℓ of H_2O, and incubation is continued at 80°C for 5 hr. To remove sulfate, the mixture is applied to a small (0.5 mℓ) column of Amberlite® IR-45 in acetate form prepared by incubation in 1 *M* sodium acetate, followed by extensive washing with H_2O. The sugars are then eluted with a few milliliters of methanol, and the acetic acid and methanol are evaporated in a stream of N_2. This procedure is particularly useful for hydrolysis of glycolipids or glycoproteins in which the oligosaccharides remain attached to the noncarbohydrate moiety. However, in cases where the aminohexitol was formed either by alkaline borohydride degradation of glycoproteins or by reduction of enzymic digestion products, this methylaminoalditol may be lost. The relatively harsh conditions of acetolysis-hydrolysis can result in some demethylation, although there is some confusion as to whether loss is from the C-1 methoxyl[82] or the C-2 aminomethyl group.[9,83,84]

Chambers and Clamp[85] suggest that methanolysis in 1 or 2 *M* methanolic HCl is more acceptable for several reasons. Methanolysis results in the formation of stable methyl glycosides and, under these conditions, the usually acid-sensitive fucosyl, xylosyl, and mannosyl were all stable for at least 24 hr at 100°C. A 4 to 6 hr incubation is also generally sufficient to release the amino sugars. Methanolic HCl is made in a well-ventilated fume hood. Dry methanol is prepared either by reflux for 1 hr over Mg turnings and I_2 followed by distillation, or by incubation of dry, distilled commercial analytical-grade methanol over molecular sieve. HCl gas is then slowly bubbled into the methanol in a small bottle tared on a top-loading balance until the desired concentration is reached. Upon completion of methanolysis, the methanolic HCl should not be evaporated directly, since concentration of the less volatile HCl could then cause decomposition of the sugars. Chambers and Clamp[85] suggest addition of $AgCO_3$ to neutralize the extract and, following sedimentation of AgCl and excess $AgCO_3$, evaporation of the methanol in a stream of N_2. Alternatively, *tert*-butanol can be added to the hydrolysate to provide a solvent with a higher vapor pressure than HCl, and safe coevaporation in a stream of N_2 can be achieved.[86] The permethylated methyl glycosides can be chromatographed directly, although the α- and β-isomers will be resolved and can confound quantitative analysis. Alditol acetates can also be formed from these glycosides, but the methyl glycosides must first be hydrolyzed with TFA before reduction of the anomeric carbon.

B. Reduction and Acetylation Reactions

The principal advantages of the formation of permethylated alditol acetate derivatives are the simplification of the derivative formation and stability of those derivatives. Reduction of the anomeric carbon, to form the corresponding alditol, eliminates the asymmetry that would otherwise result in equivocal determination of some derivatives. In so doing, reduction with $NaBD_4$ introduces a single deuterium atom onto the C-1, thus tagging the former anomeric carbon. This deuterium is essential in differentiating between symmetrical derivatives that often are not resolved chromatographically. In a widely used procedure, the sugar residues are dissolved in 1 mℓ of 1 *M* NH_4OH containing 3 mg/mℓ of $NaBD_4$ and incubated at ambient temperature overnight[77] or at 40°C for 90 min.[87] Excess borodeuteride is destroyed by concentrated acetic acid (added dropwise until the "fizzing" produced by the decomposition of the borohydride has ceased). Because borate interferes substantially with subsequent acetylation, it is removed by repeated evaporation as volatile trimethyl borate in 10% acetic acid in methanol followed by 100% methanol. This tedious operation requires several hours, and substantial loss of trimethyl pentitol and deoxyhexitol derivatives often occurs. The derivatives are then dried over P_2O_5, and acetylation is achieved by incubation in 200 μℓ of acetic anhydride for 90 min at 120°C. Because the permethylated sugars are less soluble in aqueous solutions, a mixture of pyridine and acetic anhydride are used for acetylation.[7,8]

A simplification of the reduction and acetylation procedure predicated on the ability of 1-methylimidazole to catalyze the acetylation, has been introduced and is highly recommended for general use.[88] Sugar residues are dissolved in 100 μℓ of 1 *M* NH_4OH. After adding $^1/_2$ mℓ of DMSO containing 10 mg of $NaBD_4$, the mixture is incubated at 40°C for 90 min. The excess borodeuteride is destroyed with 100 μℓ of concentrated glacial acetic acid; 100 μℓ of anhydrous 1-methylimidazole is added, followed by 0.5 mℓ of acetic anhydride. Acetylation is complete in 10 min at ambient temperature. Water (1.5 mℓ) is added to destroy unreacted acetic anhydride, and the reaction mixture is cooled to ambient temperature. Dichloromethane (1 mℓ) is added, and the permethylated alditol acetates are partitioned into this organic phase by gentle shaking for a few seconds. Other slightly polar solvents, such as ethyl acetate, are also suitable if chlorinated solvents much be avoided for other reasons. The mixtures can be centrifuged gently to achieve complete separation of the phases. The dichloromethane phase is carefully pipetted into a new vial, and the aqueous phase reextracted with another 1 mℓ of dichloromethane. The combined dichloromethane phases are washed with additional water to remove traces of DMSO and evaporated in a stream of N_2 at ambient temperature. A small amount of 2,2-dimethoxypropane can be added to hasten evaporation by eliminating traces of water that may accompany the organic phase. The derivatives are dissolved in up to 200 μℓ of dichloromethane or other preferred solvent and are ready to inject into a GC column.

V. GAS-LIQUID CHROMATOGRAPHY (GLC)

A. Separation Technology

One of the first comprehensive practical guides to methylation analysis contains a summary of relative retention times of many derivatives using the four popular stationary phases of that period.[6] Since then, there have been considerable advances in capillary chromatography to yield better resolution, in bonded-phase construction to increase column life, and in some newer high-polarity phases necessary to improve separation of the alditol acetate derivatives. Considering the complexity of polysaccharide structure now evident, investigation of new polymers essentially demands the use of capillary columns for resolution and MS for unequivocal analysis of individual derivatives. Data tabulated represent only a small fraction of the data now available in the literature, but some of the more complete summaries are represented.

For neutral sugars, Bjorndal et al.[89] described separation of partially methylated alditol acetates on packed columns using a copolymer of ethylene glycol succinate polyester and nitrile silicone (ECNSS-M) (Table 1). This high-polarity phase gave good resolution of several closely related derivatives, but persistent column bleed was a problem. Subsequently, ECNSS-M was replaced with a methyl-, phenyl-, and cyanopropyl-substituted silicone (OV-225) which was more stable at temperatures required to elute the more fully acetylated sugars.[90] With the advance of capillary chromatography, several other phases were described that vary considerably in polarity, all available commercially as fused (vitreous)-silica (WCOT) capillary columns (Table 1).[91-95] The vitreous silica columns are extremely versatile and permit connection directly into the MS source. Some high-polarity phases do not coat the capillary walls as well as they do the glass columns and can exhibit substantial bleed. For those who have invested in deactivated glass-capillary column material and prepare their own phases, deactivated vitreous silica "ends" can be attached to the glass column with blunt-end connectors, to give the flexibility necessary for low dead-volume connection to the MS source. Bonded-phase columns,[96,97] which offer thermally stable phases that can actually be cleaned by solvent flushing, are also available now with high-polarity phases. Bacic et al.[98] described separations on BP-75, a bonded-phase equivalent of OV-275, and showed that separation order, with only one unexplained exception, was identical to the unbonded phase[92] (Table 1).

Partially methylated amino-hexitol acetates are more reactive than alditols and decompose on high-polarity columns. It is recommended that use of metal inserts should be avoided at injection ports. Geyer et al.[1,99] compared several columns with a wide range of polarities with respect to their ability to resolve partially methylated derivatives typically found in glycoproteins. For such applications, low-polarity columns are superior for resolving aminohexitol derivatives, but at the expense of loss of resolution of the alditols (Table 2); a multiple column approach, low polarity for aminohexitols and high polarity for alditols is recommended. Separation of end-reduced di- and trisaccharides, including those containing aminohexitol groups, has also been described.[100-102] Permethylated oligomers larger than these have been separated by GLC, but these higher oligosaccharides are best run by FAB-MS techniques (see Chapter 10). One must also bear in mind that not only will sugar derivatives be separated chromatographically, but many contaminants as well. Plasticizers contaminating the solvents used are the most significant problem; phthalate esters can also be leached from Eppendorf® pipet tips by organic solvents and one is often detected as a small peak eluting near 2,3,4,6-tetra-methyl-1,5-di-*O*-acetylglucitol on SP-2330.[131] EI-MS reveals it by a diagnostic m/z 149. In fact, because these phthalate esters and other contaminants can be such a problem, particularly with micropreparations, several of the commonly encountered ones have been documented.[103]

B. Standards

Because MS does not differentiate the various epimers of the pentosyl, hexosyl, and aminohexosyl units, much of the analysis still relies on chromatographic separation. Hence, an appropriate bank of standards is necessary. While many purified and well-characterized polysaccharides and oligosaccharides, and even high-mannose and complex glycoproteins, are available commercially, a full range of standards is actually prepared simply by intentionally "undermethylating" monosaccharide standards. Use of monosaccharides directly permits formation of both pyranose and furanose sugars in solution, important for some sugars like L-arabinose which occur naturally in both ring forms. Xylose, deoxysugars, and most hexoses exist in the pyranose form, so *p*-nitrophenyl α- or β-glycosides are used to help retain the native configuration, simplifying the standard analysis. Milligram quantities of the free sugar or glycoside are dissolved in DMSO and methanide ion of choice. Formation of polyalkoxide ions should take no more than a few minutes. One-third to two-thirds equimolar amounts (relative to exchangeable proton) of CH_3I are added directly to the warm, vigorously stirred methanide solution to ensure random methoxylation. The partially methylated sugars are then purified by partitioning,[2] and alditol acetates prepared as desired.[77,88]

VI. MASS SPECTROMETRY (MS)

A. Electron Impact Mass Spectrometry (EI-MS) and Rules of Fragmentation

In EI-MS, the derivatives exit the chromatography column into an electron beam that produces fragments in a predictable manner from each derivative. From these fragments, one can deduce the original distribution of the methoxylated and acetoxylated carbons on the carbon skeleton. While standard published spectra could be used as "fingerprints" to identify specific derivatives,[6] any spectrum may be deceptive considering the near coelution of many derivatives. Furthermore, many symmetrical equivalents, such as 3,4- and 2,3-methyl xylitol, are unresolved on many columns, and identification must rely on the C-1 deuterium introduced upon reduction of the permethylated sugar. Hence, one derivative may be "hidden" in the spectrum if the other predominates. In this regard, chemical ionization MS, although less sensitive than EI-MS, can be used to generate molecular ions. Little fragmentation information is provided, but the technique is quite useful to initially screen the family of derivatives and identify the various linkage groups.[104,105] In EI-MS the general theoretical rules of fragmentation are quite straightforward, and it must be stressed that each spectrum should be scrutinized rigorously using these rules. Based on the early work of Golovkina et al.[106] and Bjorndal et al.[107,108] describing the fragmentation of permethylated alditol acetates, more comprehensive reviews of the theoretical bases of fragmentation for a wide variety of derivatives are available.[6-9,12,13] One should consult these reports for a more complete description of the physico-chemical bases for primary and secondary fragmentations. Nevertheless, one should be familiar enough with these rules of fragmentation to be able to recognize the major primary and secondary fragments of a derivative and the relative abundance expected for each. One should also be able to recognize the fragments that reflect contamination with additional material.

Table 1
COMPARISON OF ELUTION ORDER OF NEUTRAL SUGAR PARTIALLY METHYLATED ALDITOL ACETATES ON STATIONARY PHASES OF VARIOUS POLARITY

Derivative	Deduced linkage	(1)[a]	(2)[b]	(3)[c]	(4)[d]	(5)[e]	(6)[f]	(7)[g]	(8)[h]	(9)[i]	(10)[j]
2,3-4-Me_3 Ara	t-Ara*p*	0.73	0.54	0.778	0.671	0.807	0.812	0.363	—	0.679	0.797
2,3,5-Me_3 Ara	t-Ara*f*	0.48	0.41	0.651	0.564	0.700	0.726	0.232	0.59	0.597	0.666
2,3-Me_2 Ara	5-Ara*f*	—	1.07	1.258	1.125	1.090	1.195	0.738	1.32	0.886	1.131
2,4-Me_2 Ara	3-Ara*p*	1.40	1.10	1.270	1.205	1.105	1.213	0.771	—	0.901	1.161
2,5-Me_2 Ara	3-Ara*f*	1.10	0.84	1.105	0.977	1.004	1.067	0.572	1.09	0.806	1.028
3,4-Me_2 Ara	2-Ara*p*	1.38	—	1.298	1.171	1.111	1.435	—	—	0.901	1.174
3,5-Me_2 Ara	2-Ara*f*	0.91	0.80	0.980	0.935	0.943	0.981	0.570	0.94	0.798	0.946
2-Me Ara	3,5-Ara*f*	—	1.93	1.747	1.815	1.329	1.688	1.161	2.23	1.089	1.450
3-Me Ara	2,5-Ara*f*	—	1.48	1.821	1.936	1.366	1.787	1.259	—	1.124	1.490
4-Me Ara	2,3-Ara*p*	—	—	1.804	—	1.355	—	—	—	1.099	1.496
5-Me Ara	2,3-Ara*f*	—	—	1.403	—	1.168	1.335	0.910	—	0.971	1.581
Ara-$(OAc)_5$	—	—	—	2.092	2.623	1.493	2.164	1.615	—	1.259	1.650
2,3-4-Me_3 Xyl	t-Xyl*p*	0.68	0.54	0.801	0.686	0.856	0.841	0.365	0.74	0.662	0.823
2,3,5-Me_3 Xyl	t-Xyl*f*	—	—	0.707	—	—	—	—	—	—	—
2,3-Me_2 Xyl	4-Xyl*p*	1.54	1.19	1.390	1.208	1.183	1.319	0.766	1.48	0.890	1.241
2,4-Me_2 Xyl	3-Xyl*p*	1.34	1.06	1.255	1.154	1.123	1.198	0.723	—	0.869	1.154
2,5-Me_2 Xyl	3-Xyl*f*	—	—	1.255	—	—	—	—	—	—	—
3,4-Me_2 Xyl	2-Xyl*p*	1.54	1.19	1.390	1.208	1.183	1.319	0.766	1.48	0.890	1.241
3,5-Me_2 Xyl	2-Xyl*f*	1.08	—	1.097	—	—	—	—	—	—	—
2-Me Xyl	3,4-Xyl*p*	2.92	2.15	1.982	1.988	1.439	1.972	1.222	2.70	1.108	1.592
3-Me Xyl	2,4-Xyl*p*	2.92	2.15	1.959	1.988	1.431	1.952	1.241	2.65	1.116	1.582
4-Me Xyl	2,3-Xyl*p*	2.92	2.15	1.982	1.988	1.439	1.972	1.222	—	1.108	1.592
5-Me Xyl	2,3-Xyl*f*	—	—	1.699	—	—	—	—	—	—	—
Xyl-$(OAc)_5$	—	—	—	2.551	3.023	1.651	2.868	1.739	—	1.306	1.811

1,3,4,6-Me_4 Man	t-Man*f*	0.72	—	0.804	—	—	—	—	—	—	—
2,3,4,6-Me_4 Man	t-Man*p*	1.00	0.99	0.987	1.000	0.955	0.977	1.000	0.98	1.010	0.985
1,3,4-Me_3 Man	6-Man*f*	—	—	1.505	—	—	—	—	—	—	—
2,3,4-Me_3 Man	6-Man*p*	2.48	2.19	1.676	1.807	1.312	1.630	1.880	2.16	1.331	1.504
2,4,6-Me_3 Man	3-Man*p*	2.09	1.90	1.472	1.803	1.234	1.449	1.683	—	1.276	—
3,4,6-Me_3 Man	2-Man*p*	1.95	1.82	1.505	1.621	1.222	1.439	1.593	1.81	1.245	1.346
2,3,6-Me_3 Man	4-Man*p*	2.20	2.03	1.589	1.765	1.267	1.540	1.621	2.01	1.257	1.406
2,3-Me_2 Man	4,6-Man*p*	4.83	3.69	2.253	3.218	1.563	2.471	2.671	—	1.561	—
2,4-Me_2 Man	3,6-Man*p*	5.44	4.51	2.352	3.691	1.618	2.703	2.967	4.41	1.627	—
2,6-Me_2 Man	3,4-Man*p*	3.35	—	1.934	2.603	1.435	1.986	2.163	—	1.433	—
3,4-Me_2 Man	2,6-Man*p*	5.37	4.36	2.426	3.276	1.618	2.766	—	4.56	1.601	—
3,6-Me_2 Man	2,4-Man*p*	4.15	3.67	2.186	2.907	1.522	2.351	2.428	—	1.502	—
4,6-Me_2 Man	2,3-Man*p*	3.29	2.92	1.903	2.603	1.425	1.956	2.262	—	1.460	1.593
2-Me Man	3,4,6-Man*p*	7.00	5.65	2.640	4.793	1.731	3.282	3.461	—	1.734	—
3-Me Man	2,4,6-Man*p*	8.80	6.80	2.974	5.488	1.836	4.125	3.942	—	1.825	—
4-Me Man	2,3,6-Man*p*	8.80	6.80	2.974	5.488	1.836	4.125	3.942	—	1.825	—
6-Me Man	2,3,4-Man*p*	4.48	4.10	2.207	3.529	1.556	2.412	2.832	—	1.601	—
Man-$(OAc)_6$	—	—	—	2.985	6.630	1.873	4.201	4.481	—	1.914	2.047
2,3,4,6-Me_4 Gal	t-Gal*p*	1.25	1.19	1.140	1.143	1.055	1.102	1.115	1.19	1.061	1.097
2,3,5,6-Me_4 Gal	t-Gal*f*	1.15	1.10	1.102	—	1.016	—	—	—	1.021	—
2,3,4-Me_3 Gal	6-Gal*p*	3.41	2.89	1.980	2.379	1.444	2.003	2.134	—	1.422	1.612
2,3,6-Me_3 Gal	4-Gal*p*	2.42	2.22	1.730	1.842	1.329	1.688	1.618	—	1.257	1.484
2,4,6-Me_3 Gal	3-Gal*p*	2.28	2.03	1.561	1.912	1.281	1.541	1.744	1.97	1.302	1.407
2,5,6-Me_3 Gal	3-Gal*f*	2.25	1.95	1.605	—	—	—	—	—	1.302	—
3,4,6-Me_3 Gal	2-Gal*p*	2.50	2.15	1.686	—	1.320	—	—	—	—	1.468
3,5,6-Me_3 Gal	2-Gal*f*	—	—	1.536	—	—	—	—	—	—	—
2,3-Me_2 Gal	4,6-Gal*p*	5.68	4.70	2.523	3.620	1.661	2.972	2.845	—	1.606	1.831
2,4-Me_2 Gal	3,6-Gal*p*	6.35	5.10	2.536	—	1.684	3.132	—	5.28	1.679	1.848
2,6-Me_2 Gal	3,4-Gal*p*	3.65	3.14	2.038	2.736	1.478	2.143	2.203	3.19	1.447	1.655
3,4-Me_2 Gal	2,6-Gal*p*	5.37	5.50	2.676	—	1.719	—	—	—	1.673	1.887
3,6-Me_2 Gal	2,4-Gal*p*	4.15	—	2.232	3.037	1.522	2.428	2.411	—	1.506	1.729
4,6-Me_2 Gal	2,3-Gal*p*	3.64	—	2.036	—	1.487	2.143	—	—	1.496	1.655
2-Me Gal	3,4,6-Gal*p*	8.10	—	2.801	5.176	1.782	3.686	3.655	—	1.779	1.956
3-Me Gal	2,4,6-Gal*p*	11.10	—	3.181	6.185	1.897	4.643	4.167	—	1.875	2.101
4-Me Gal	2,3,6-Gal*p*	11.10	—	3.181	6.185	1.897	4.643	4.167	—	1.875	2.101
6-Me Gal	2,3,4-Gal*p*	5.10	—	2.342	4.069	1.616	—	2.918	—	1.629	1.781
Gal-$(OAc)_6$	—	—	—	3.186	7.419	1.928	4.719	4.780	—	1.970	2.134

Table 1 (continued)
COMPARISON OF ELUTION ORDER OF NEUTRAL SUGAR PARTIALLY METHYLATED ALDITOL ACETATES ON STATIONARY PHASES OF VARIOUS POLARITY

Derivative	Deduced linkage	(1)[a]	(2)[b]	(3)[c]	(4)[d]	(5)[e]	(6)[f]	(7)[g]	(8)[h]	(9)[i]	(10)[j]
1,3,4,6-Me_4 Glc	t-Glc*f*	0.72	—	0.809	—	—	—	—	—	—	—
2,3,4,6-Me_4 Glc	t-Glc*p*	1.00[k]	1.00[k]	1.000[k]	1.000[k]	1.000[k]	1.000[k]	1.000[k]	1.00[k]	1.000[k]	1.000[k]
1,3,4-Me_3 Glc	6-Glc*f*	—	—	1.492	—	—	—	—	—	—	—
2,3,4-Me_2 Glc	6-Glc*p*	2.49	2.22	1.668	1.931	1.343	1.665	1.863	2.19	1.327	1.553
2,3,6-Me_3 Glc	4-Glc*p*	2.50	2.32	1.770	1.889	1.363	1.759	1.674	2.38	1.279	1.493
2,4,6-Me_3 Glc	3-Glc*p*	1.95	1.82	1.449	1.680	1.246	1.428	1.595	1.73	1.245	1.317
3,4,6-Me_3 Glc	2-Glc*p*	1.98	1.83	1.513	—	1.261	1.480	—	—	1.230	1.347
2,3-Me_2 Glc	4,6-Glc*p*	5.39	4.50	2.452	3.417	1.637	2.850	2.724	4.75	1.582	1.786
2,4-Me_2 Glc	3,6-Glc*p*	5.10	4.21	2.332	3.417	1.604	2.648	2.803	4.25	1.597	1.727
2,6-Me_2 Glc	3,4-Glc*p*	3.83	3.38	2.115	2.707	1.509	2.240	2.219	3.39	1.454	1.613
3,4-Me_2 Glc	2,6-Glc*p*	5.27	4.26	2.457	—	1.637	—	2.814	—	1.597	1.844
3,6-Me_2 Glc	2,4-Glc*p*	4.40	3.73	2.222	2.893	1.546	2.420	2.368	3.82	1.498	1.707
4,6-Me_2 Glc	2,3-Glc*p*	4.02	3.49	2.153	2.893	1.526	2.298	2.355	—	1.487	1.613
2-Me Glc	3,4,6-Glc*p*	7.90	6.60	2.865	5.023	1.784	3.808	3.538	—	1.762	—
3-Me Glc	2,4,6-Glc*p*	9.60	7.60	3.031	5.399	1.832	4.259	3.800	—	1.816	1.970
4-Me Glc	2,3,6-Glc*p*	11.50	8.40	3.258	—	1.899	—	3.996	—	1.846	1.970
6-MeGlc	2,3,4-Glc*p*	5.62	5.00	2.538	—	1.663	—	2.897	4.99	1.625	—
Glc-$(OAc)_6$	—	—	—	3.485	6.430	1.973	5.434	4.618	—	1.945	2.191
2,3,4-Me_3 Rhm	t-Rhm	0.46	0.35	0.597	0.540	0.669	0.686	0.390	0.54	0.694	0.630
2,3-Me_2 Rhm	4-Rhm	0.98	0.92	1.015	0.945	0.965	1.000	0.776	—	0.911	0.983
2,4-Me_2 Rhm	3-Rhm	0.99	0.94	0.995	1.000	0.965	0.968	0.831	—	0.930	—
3,4-Me_2 Rhm	2-Rhm	0.92	0.87	0.977	0.888	0.938	0.926	—	—	0.911	0.956
2-Me Rhm	3,4-Rhm	1.52	1.37	1.304	1.403	1.138	1.260	1.137	—	1.074	—
3-Me Rhm	2,4-Rhm	1.94	1.67	1.503	1.547	1.222	1.448	1.300	—	1.139	1.328
4-Me Rhm	2,3-Rhm	1.72	1.57	1.413	1.470	1.185	1.355	1.237	—	1.118	1.267
Rhm-$(OAc)_5$	—	—	—	1.520	1.892	1.255	1.480	1.554	—	1.228	1.343

2,3,4-Me_3 Fuc	t-Fuc	0.65	0.58	0.758	0.649	0.778	0.792	0.497	0.69	0.749	0.784
2,3-Me_2 Fuc	4-Fuc	1.18	—	1.171	1.027	1.039	1.114	0.832	—	0.930	1.095
2,4-Me_2 Fuc	3-Fuc	1.12	1.02	1.084	1.076	1.008	1.066	0.898	—	0.958	1.054
3,4-Me_2 Fuc	2-Fuc	—	—	1.212	—	1.064	1.147	0.898	—	0.966	1.130
2-Me Fuc	3,4-Fuc	1.67	1.43	1.380	1.462	1.168	1.324	1.153	—	1.076	1.255
3-Me Fuc	2,4-Fuc	2.05	—	1.589	—	1.261	1.524	1.319	—	1.148	—
4-Me Fuc	2,3-Fuc	2.08	1.71	1.559	—	1.253	—	—	—	1.154	1.371
Fuc-$(OAc)_5$	—	—	—	1.594	2.081	1.296	1.577	1.530	—	1.247	1.412

[a] Packed column 3% ECNSS-M, 2 m × 3 mm, 170°C isothermal. N_2 carrier flow at 40 mℓ/min.[6]

[b] Packed column 3% OV-225, 1.8 m × 1.5 mm, 155°C or 170°C isothermal (± 3%). N_2 carrier flow at 15 to 20 mℓ/min.[6]

[c] Glass-capillary column OV-275 (Chrompack, Middlebury, The Netherlands), 25 m × 0.25 mm, 165°C to 215°C at 2°C/min.[92]

[d] Glass-capillary column SP-1000 (Erba Science, Swindon, UK), 20 m × 0.3 mm, 60°C for on-column injection, then raised to 206°C for isothermal elution. Helium carrier at 60 kPa.[93]

[e] Vitreous-silica column BP-75 (bonded phase OV-275) (SGE, Melbourne, Australia), 25 m × 0.22 mm, 150°C to 250°C at 4°C/min. Helium carrier flow at 0.78 mℓ/min.[98]

[f] Fused-silica column CP-Si188 (Chrompack,® Middlebury, UK), 50 m × 0.33 mm, 210°C isothermal. Helium carrier at 130 kPa.[94]

[g] Fused-silica column BP-1 (bonded phase OV-1) (SGE), 50 m × 0.33 mm, 0.5 μm film, 195°C isothermal. Helium carrier at 130 kPa.[94]

[h] Glass-capillary column Silar®-10C, 30 m × 0.28 mm, 150°C to 190°C at 1°C/min.[95]

[i] Fused-silica column SP-2100 (Supelco, Bellefonte, Pa.), 15 m × 0.24 mm, 0.25 μm film thickness, 120°C to 200°C at 2°C/min. Helium carrier flow of 0.78 mℓ/min.[91]

[j] Fused-silica column SP-2330 (Supelco, Bellefonte, Pa.), 52 m × 0.25 mm, 0.20 μm film thickness, 160°C to 210°C at 2°C/min, then 210°C to 240°C at 5°C/min. Helium carrier flow 180 kPa.[131]

[k] Retention times are relative to 2,3,4,6-Me_4 Glc.

Table 2
COMPARISON OF ELUTION ORDER OF NEUTRAL AND AMINO SUGAR PARTIALLY METHYLATED ALDITOL ACETATE DERIVATIVES USING STATIONARY PHASES OF VARIOUS POLARITIES

Derivative	Deduced linkage	(1)[a]	(2)[b]	(3)[c]	(4)[d]	(5)[e]
2,3,4-Me_3 Fuc	t-Fuc	0.815	0.743	0.759	0.867	0.853
Fuc-$(OAc)_5$	—	—	1.227	1.241	1.110	—
2,3,4,6-Me_4 Gal	t-Gal	1.096	1.094	1.057	1.025	1.025
2,3,4-Me_3 Gal	6-Gal	1.444	1.587	1.364	1.180	1.195
2,4,6-Me_3 Gal	3-Gal	1.282	1.367	1.267	1.129	1.140
3,4,6-Me_3 Gal	2-Gal	1.339	1.396	1.267	1.129	1.140
2,4-Me_2 Gal	3,6-Gal	1.634	1.883	1.575	1.277	1.305
3,6-Me_2 Gal	2,4-Gal	1.526	1.638	1.435	1.212	1.232
4,6-Me_2 Gal	2,3-Gal	1.452	1.587	1.428	1.205	1.223
Gal-$(OAc)_6$	—	—	2.129	1.840	1.389	—
2,3,4,6-Me_4 Man	t-Man	1.009	1.000	1.011	1.003	1.000[f]
2,3,4-Me_3 Man	6-Man	1.321	1.431	1.290	1.146	1.157
2,3,6-Me_3 Man	4-Man	1.302	1.345	1.231	1.112	1.120
2,4,6-Me_3 Man	3-Man	1.241	1.328	1.244	1.121	1.130
3,4,6-Me_3 Man	2-Man	1.258	1.308	1.221	1.107	1.115
2,4-Me_2 Man	3,6-Man	1.573	1.818	1.536	1.264	1.288
2,6-Me_2 Man	3,4-Man	1.428	1.532	1.377	1.184	1.199
3,4-Me_2 Man	2,6-Man	1.592	1.802	1.513	1.253	1.276
3,6-Me_2 Man	2,4-Man	1.515	1.662	1.435	1.212	1.232
4,6-Me_2 Man	2,3-Man	1.412	1.532	1.399	1.195	1.211
2-Me Man	3,4,6-Man	1.662	1.909	1.623	1.303	1.332
Man-$(OAc)_6$	—	—	2.066	1.798	1.371	—
2,3,4,6-Me_4 Glc	t-Glc	1.000[g]	1.000[g]	1.000[g]	1.000[g]	—
2,4,6-Me_3 Glc	3-Glc	1.231	1.283	1.221	1.107	—
3,4,6-Me_3 Glc	2-Glc	1.247	1.292	1.202	1.102	—
Glc-$(OAc)_6$	—	—	2.183	1.822	1.380	—
3,4,6-Me_3 GalNAc	t-GalNAc	1.897	2.362	1.758	1.383	1.420
Gal-NAc-$(OAc)_5$	—	—	3.093	2.218	1.558	—
1,3,5,6-Me_4 GlcNAc	4-GlcNAcOH	1.452	1.662	1.364	1.195	1.208
3,4,6-Me_3 GlcNAc	t-GlcNAc	1.782	2.184	1.652	1.343	1.373
3,4-Me_2 GlcNAc	6-GlcNAc	—	2.761	1.969	1.470	1.513
3,6-Me_2 GlcNAc	4-GlcNAc	2.002	2.466	1.824	1.421	1.460
4,6-Me_2 GlcNAc	3-GlcNAc	—	2.650	1.969	1.466	1.510
3-Me GlcNAc	4,6-GlcNAc	—	2.864	2.111	1.525	1.574
6-Me GlcNAc	3,4-GlcNAc	—	2.746	2.042	1.502	1.549
GlcNAc-$(OAc)_5$	—	—	2.986	2.156	1.534	—

[a] Glass-capillary column Silar® 9CP (WGA, Dusseldorf, FRG), 22 m × 0.25 mm, 10 min hold 100°C, then to 230°C at 1°C/min. H_2 carrier at 0.8 bar.[99]

[b] Glass-capillary column Dexsil® 410 (Supelco, Pa.), 20 m × 0.22 mm, 130°C to 240°C at 2°C/min. H_2 carrier at 0.4 bar.[99]

[c] Glass-capillary column SE-30 (LKB, Bromma, Sweden), 25 m × 0.25 mm, 100°C to 240°C at 2°C/min. H_2 carrier at 0.5 bar.[99]

[d] Glass-capillary column OV-101 (WGA, Dusseldorf, FRG), 85 m × 0.25 mm, 100°C to 240°C at 3°C/min. H_2 carrier at 1.5 bar.[99]

[e] Fused-silica bonded Dura-Bond® DB-1 (ICT, Frankfurt, FRG), 60 m × 0.25 mm, 100°C to 250°C at 2°C/min. Helium carrier at 2.8 bar.[1]

[f] Retention times relative to 2,3,4,6-Me_4 Man.

[g] Retention times relative to 2,3,4,6-Me_4 Glc.

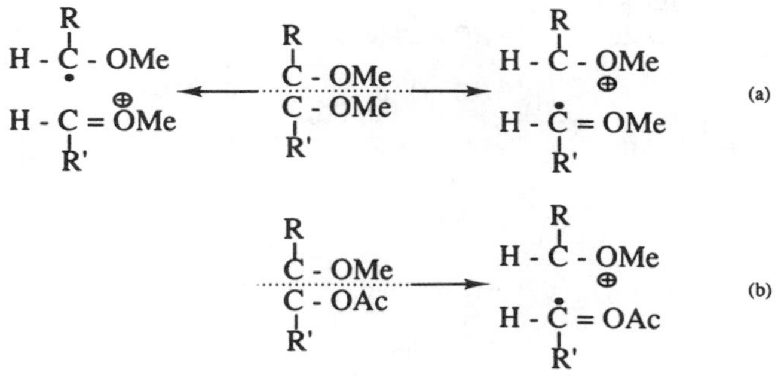

FIGURE 2. Primary fragmentation. Cleavage of the chain between contiguous methoxylated carbons is favored over cleavage between acetoxylated and methoxylated carbons. (a) If R and R′ are similar, then fragmentation between the contiguous methoxylated carbons generates cations of either fragment with about equal frequency. (b) However, cleavage between acetoxylated and methoxylated carbons produces cations only from the fragment bearing the methoxylated carbon.

Energy produced by the electron beam is sufficient to produce fragments primarily from the carbon-carbon bonds of the alditol, and subsequent secondary fragments from consecutive loss of their methoxyl and acetoxyl groups. All fragments reported from either magnetic or quadrupole MS instruments are positive ions. Cleavage between the contiguous alditol carbons bearing methoxyl groups is greatly favored over cleavage between methoxylated and acetoxylated carbons. The methyl ether groups exhibit little electronic influence on the alditol carbon; hence, if the R-groups are similar, then cleavage between contiguous methoxylated carbons will produce positive carbonium ions of either fragment with about equal frequency (Figure 2a). The keto group of the acetyl substitution, however, exhibits a strong electron withdrawing function that results in cation formation exclusively by the methoxylated carbon (Figure 2b). Fission between contiguous acetoxylated carbons is rare, except for fully acetylated derivatives or those bearing a single methoxyl group.

The molecular ions are rarely detected with EI-MS. Loss of the acetylium ion (CH_3CO^+; m/z 43) predominates and is the base peak of all derivatives of neutral sugars and most methyl amino sugars. Cleavage of the alditol chain usually results in detectable two-, three-, four-, and occasionally, five-carbon fragments bearing methoxyl and acetoxyl groups. Secondary fragmentations result from fission of these groups resulting in loss of acetic acid or methanol. An extremely important diagnostic tool is the observation that fission of the methyl or acetyl substituents of the primary fragment is primarily from the β-carbon to the formal charge of the carbonium ion (Figure 3a, b). Loss of acetic acid is always favored over loss of methanol when fission of the substituent of the α- or γ-carbon occurs. Additional significant secondary fragments form by elimination of formaldehyde (− m/z 30) or ketene (− m/z 42). Although confirmatory, these observations alone are insignificant as diagnostic tools (Figure 3c, d).

B. Derivatives of Nonreducing Terminal Sugars

Derivatives of nonreducing terminal pyranosyl and furanosyl units are quite distinctive. The terminal pyranosyl pentitol derivative is a symmetrical 2,3,4-tri-*O*-methyl alditol acetate comprising two facile cleavages between contiguous methoxylated carbons (Figure 4a). Reduction of the permethylated sugar with $NaBD_4$ introduces asymmetry that results in "doublets" of each major ion. The furanosyl pentitol is acetylated at the C-4 rather than the C-5 position, and so only a single cleavage between contiguous methoxylated carbons is possible yielding m/z 118 and 161 as primary fragments (Figure 4b). Secondary fission

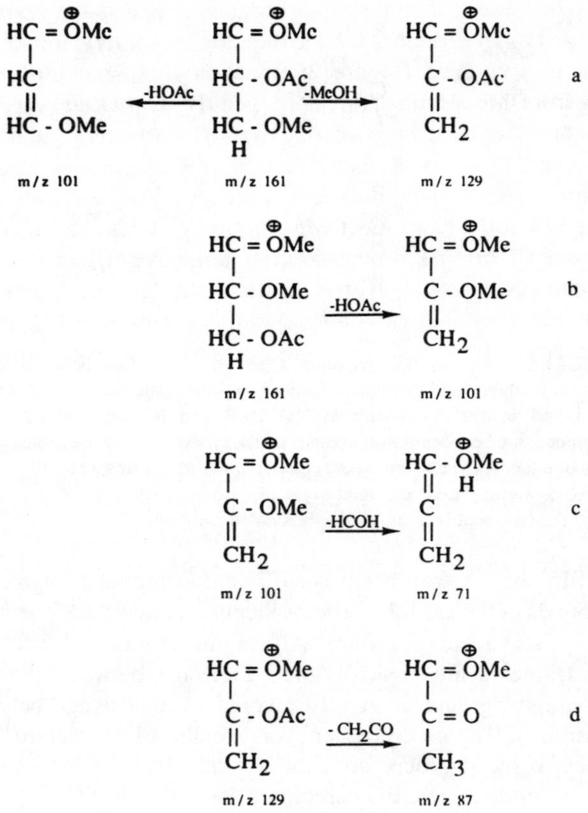

(a)

$$HC{=}\overset{\oplus}{O}Me-CH{=}CH-OMe \;(m/z\ 101) \xleftarrow{-HOAc} HC{=}\overset{\oplus}{O}Me-CH(OAc)-CH_2-OMe \;(m/z\ 161) \xrightarrow{-MeOH} HC{=}\overset{\oplus}{O}Me-C(OAc){=}CH_2 \;(m/z\ 129)$$

(b)

$$HC{=}\overset{\oplus}{O}Me-CH(OMe)-CH_2-OAc \;(m/z\ 161) \xrightarrow{-HOAc} HC{=}\overset{\oplus}{O}Me-C(OMe){=}CH_2 \;(m/z\ 101)$$

(c)

$$HC{=}\overset{\oplus}{O}Me-C(OMe){=}CH_2 \;(m/z\ 101) \xrightarrow{-HCOH} HC{-}\overset{\oplus}{O}Me{=}CH{=}CH_2 \;(m/z\ 71)$$

(d)

$$HC{=}\overset{\oplus}{O}Me-C(OAc){=}CH_2 \;(m/z\ 129) \xrightarrow{-CH_2CO} HC{=}\overset{\oplus}{O}Me-C(=O)-CH_3 \;(m/z\ 87)$$

FIGURE 3. Secondary fragmentation. Further fission of the primary fragments is also predictable. Loss of the (a) methoxyl group (as methanol) or (b) acetoxyl group (as acetic acid) occurs principally from the β-position to the formal carbonium charge, however, some loss of acetic acid from the α-position is also detected when the β-position is methoxylated. These secondary fragments can undergo further fission by loss of (c) formaldehyde (− m/z 30) or (d) ketene (− m/z 42).

of the m/z 161 fragment well illustrates how β-elimination yields m/z 129 by loss of methanol in excess to m/z 101 produced by loss of acetic acid from the α-position to the formal carbonium charge (Figure 4b). When the α-position to the carbonium ion is methoxylated, as in the example of the pyranosyl pentitol, secondary ions are almost exclusively from β-elimination (Figure 4a). The less frequent m/z 162 from cleavage between C-3 and C-4 of the furanosyl derivative, yields m/z 102 by loss of the β-acetyl group, the proton needed for acetic acid formation arising from C-2 rather than C-1 leaving the deuterium to contribute to the mass (Figure 4b). The methoxylated terminal C-5 is also conspicuous by the relative abundance of m/z 45 from cleavage of this terminal carbon.

The derivative of the nonreducing terminal hexopyranosyl unit introduces additional complexity. Two facile cleavages exist between contiguous methoxylated carbons. One cleavage produces about equal amounts of m/z 161 and 162, and secondary fragmentation by β-elimination yields primarily m/z 129 from m/z 161 fragment and m/z 102 from m/z 162. Hence, the relative positions of the single acetyl substituents of each 3-carbon fragment are deduced (Figure 4c). Note that significant amounts of m/z 101 are produced by loss of acetic acid from the α-position to the carbonium ion of m/z 161, but analogous loss of methanol from the α-position of m/z 162 is essentially undetected (Figure 4c). Loss of acetic acid is favored here mostly because fission results in formation of a stable derivative in which the formal charge is shared upon formation of a symmetric mesomeric ion, a situation untenable when the β-carbon is acetylated (Figure 2a, b).[13] A significant difference between this derivative and that from the terminal furanosyl pentose is that cleavage between the C-2, C-3 contiguous methoxylated carbons of the hexosyl derivative yields m/z 118 and the 4-carbon fragment m/z 205. Loss of acetic acid from β-elimination results in an abundant m/z 145 (Figure 4c). Other minor cleavages yield m/z 206 and 249 and their principal secondary ions m/z 174 and 217, respectively, but may or may not be detected depending on how the MS is tuned. Terminal 6-deoxyhexitol derivatives (from rhamnosyl and fucosyl constituents) are quite distinctive by virtue of the methyl group. The two facile cleavages between contiguous methoxylated carbons yield unique ions m/z 175 and 131, the former producing the secondary ion m/z 115 from loss of acetic acid by β-elimination (Figure 4d).

C. Derivatives of Linked and Branched Pentoses

Linkage of a sugar in polymers results in an additional acetylation that is the primary determinant in retention time upon chromatography. Establishing its position is key to identification, and reduction of the anomeric carbon with $NaBD_4$ becomes of great importance. Some examples below will illustrate this point further. The *O*-2, *O*-3, and *O*-5 positions are the possible linkage carbons of a pentofuranose, while the *O*-2, *O*-3, and *O*-4 constitute the possible linkage positions for pentopyranose units. Note that the derivatives from 5-linked pentofuranose and 4-linked pentopyranose will produce identical fragmentations, and identification of these derivatives must rely primarily on chromatographic separation. Linkage at the *O*-2 of a pentofuranose is obvious by its subsequent acetylation, yielding m/z 190 (Figure 5a). A 3-carbon primary fragment bearing two acetyl substitutions and one methyl ether is m/z 189, and the additional mass unit contributed by the C-1 deuterium establishes the symmetry. The *O*-1 acetyl is understood, but the position of the second acetyl must be at *O*-2 for two reasons — acetyl substitution at *O*-3 should yield m/z 118, absent from this spectrum, and cleavage at the C-3–C-4 bond could only result in formation of an undetectable negative ion. The m/z 161 establishes that the other 3-carbon fragment contains a single acetyl substitution, whose position is deduced by virtue of its loss of methanol by β-elimination to form the predominant m/z 129 (Figure 5a).

The 3-linked pentofuranose is identified more by the absence of fragments than by their presence. The m/z 118 indicates that the *O*-2 was methylated, but the absence of m/z 162 suggests that an acetyl was at the *O*-3 position (Figure 5b). The ring carbon remains to be established. The primary fragment m/z 233 indicates that only one primary fragment exists excluding the C-1, and it comprises a 4-carbon fragment with two acetyl and two methoxyl groups. The absence of m/z 117 of about equal intensity as m/z 118, indicates that the remaining acetyl unit must be at the *O*-4 and not the *O*-5 position (Figure 5b). The symmetrical derivative of 3-linked pentopyranose would also yield nearly equal intensities of m/z 233 and 234, as well as m/z 117 and 118 (Figure 5c).

The 4-linked pentopyranose and 5-linked pentofuranose produce nearly identical fragmentations, hence, identification must rely on chromatographic separation. With the major cleavage at the only pair of adjacent methoxylated carbons, the spectrum is easily deduced. The m/z 118 indicates that the C-2 must be methoxylated, and the m/z 189 indicates that, analogous to the derivative of 2-linked pentofuranose, the acetyl cannot be at *O*-3 (Figure 5d). The m/z 129 is produced by loss of acetic acid by β-elimination of m/z 189, rather than loss of methanol from m/z 161, but the secondary fragment is nevertheless the same. The derivatives of 2-linked pentopyranosyl units are symmetrical equivalents of the 4-linked units and are unresolved chromatographically. The amounts of each must be estimated from MS. The asymmetry provided by the C-1 deuterium yields primary fragments m/z 117 and 190, rather than m/z 118 and 189 (Figure 5e). The amounts of each are determined, with

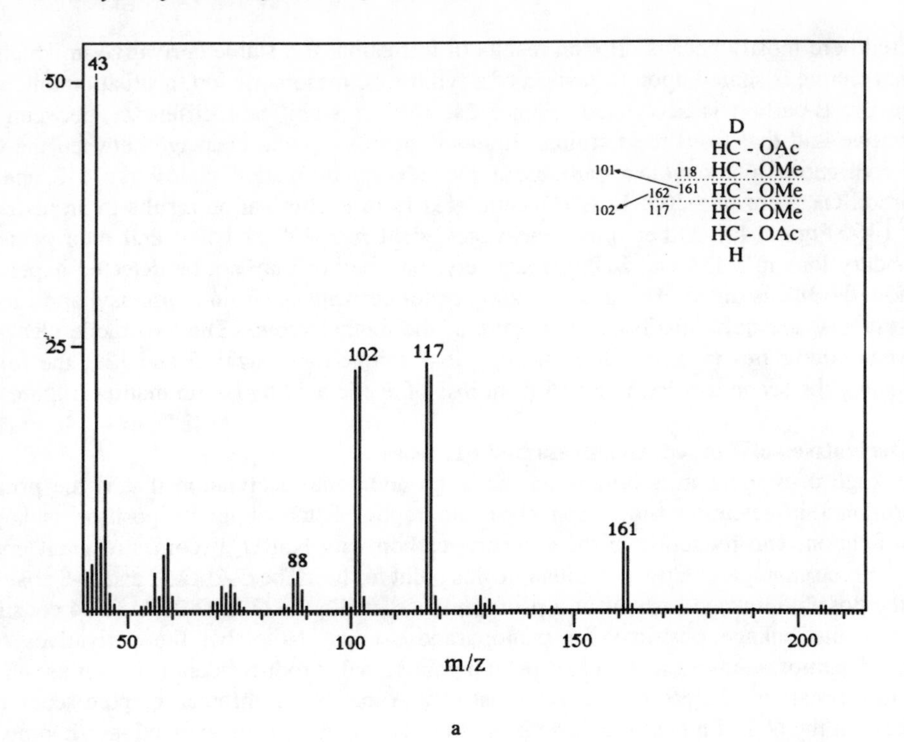

a

b

FIGURE 4. Spectra of derivatives of nonreducing terminal sugars: a, 1,5-di-*O*-acetyl-(1-deuterio)-2,3,4-tri-*O*-methyl pentitol; b, 1,4-di-*O*-acetyl-(1-deuterio)-2,3,5-tri-*O*-methyl pentitol; c, 1,5-di-*O*-acetyl-(1-deuterio)-2,3,4,6-tetra-*O*-methyl hexitol; d, 1,5-di-*O*-acetyl-(1-deuterio)-2,3,4-tri-*O*-methyl-6-deoxy-hexitol.

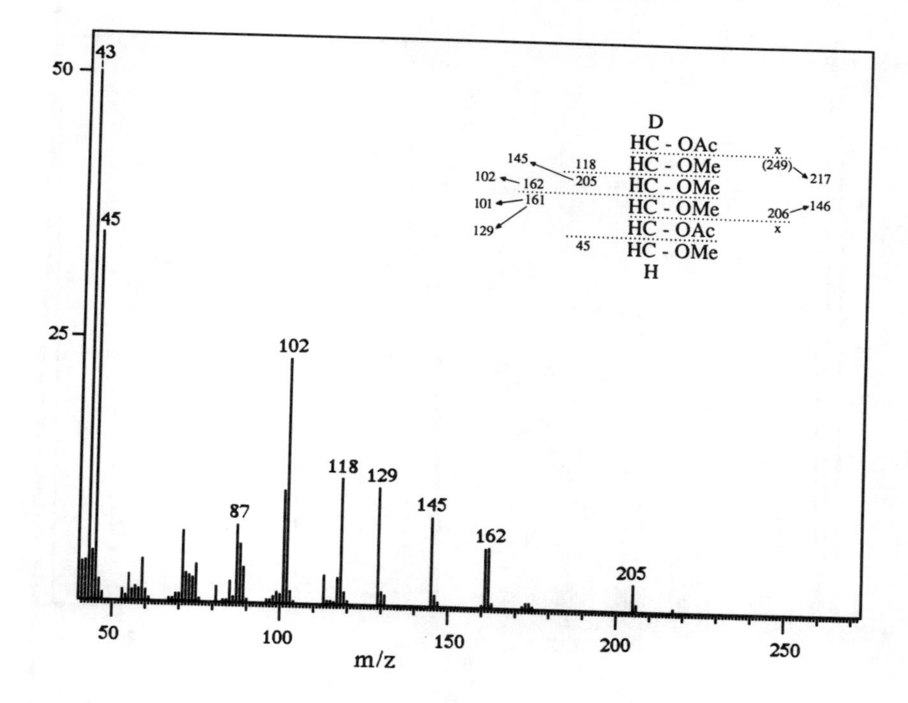

FIGURE 4c

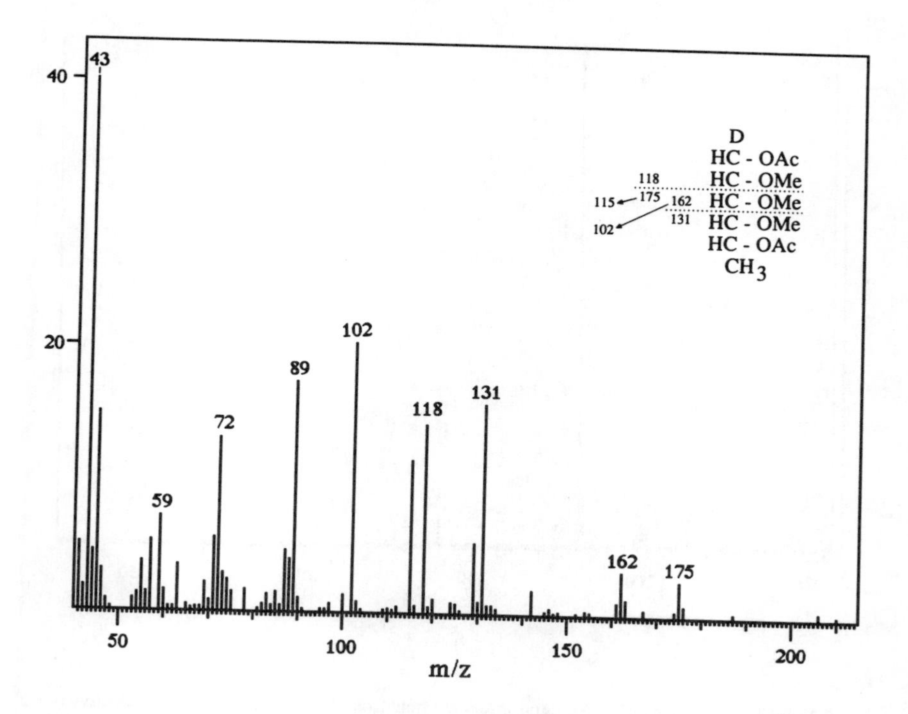

FIGURE 4d

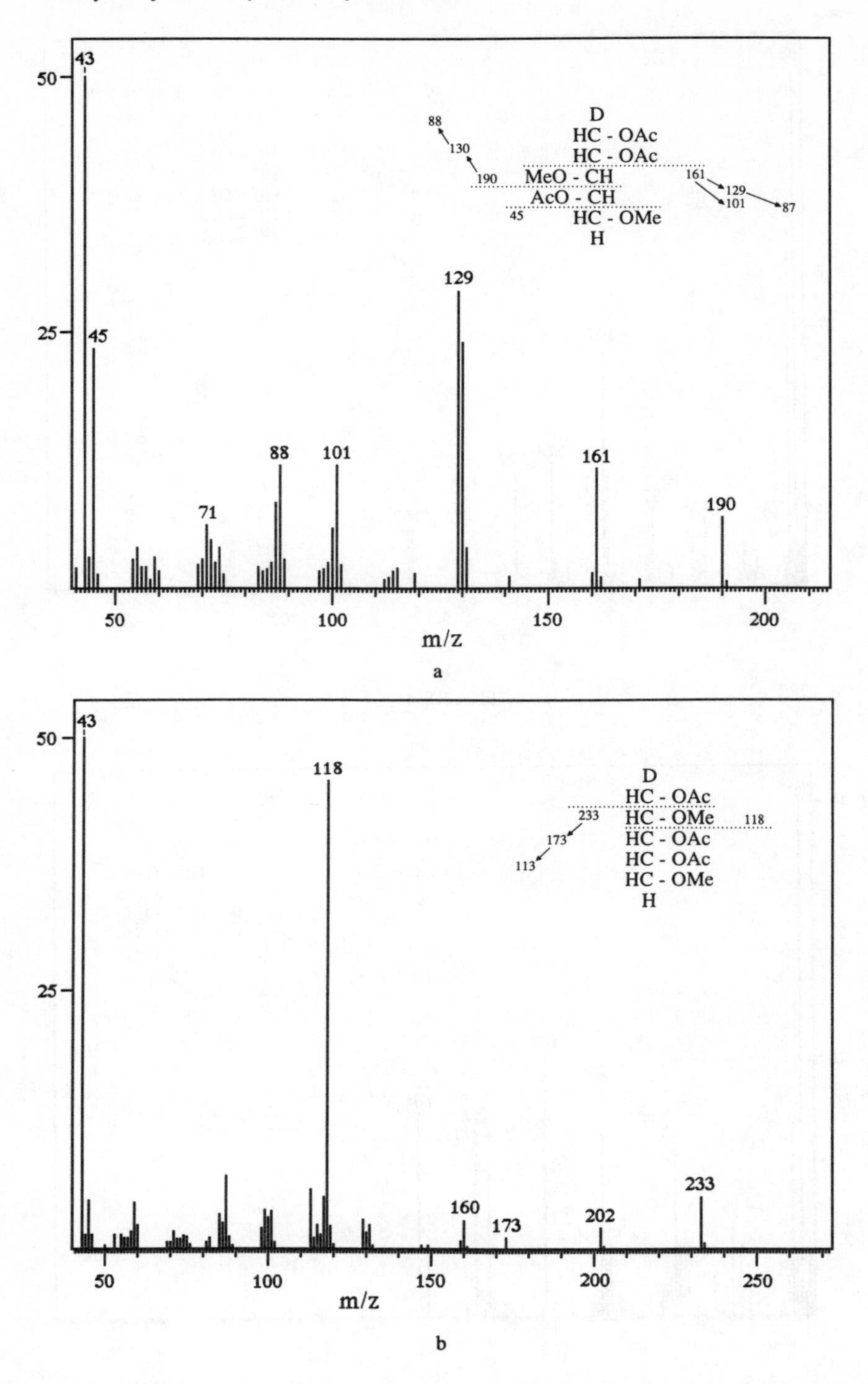

FIGURE 5. Spectra of derivatives of linked and branched pentoses: a, 1,2,4-tri-*O*-acetyl-(1-deuterio)-3,5-di-*O*-methyl pentitol; b, 1,3,4-tri-*O*-acetyl-(1-deuterio)-2,5-di-*O*-methyl pentitol; c, 1,3,5-tri-*O*-acetyl-(1-deuterio)-2,4-di-*O*-methyl pentitol; d, 1,4,5-tri-*O*-acetyl-(1-deuterio)-2,3-di-*O*-methyl pentitol; e, mixture of (d) and 1,2,5-tri-*O*-acetyl-(1-deuterio)-3,4-di-*O*-methyl pentitol; f, 1,2,4,5-tetra-*O*-acetyl-(1-deuterio)-3-*O*-methyl pentitol; g, 1,3,4,5-tetra-*O*-acetyl-(1-deuterio)-2-*O*-methyl pentitol; h, Mixture of (g) and 1,2,3,5-tetra-*O*-acetyl-(1-deuterio)-4-*O*-methyl pentitol.

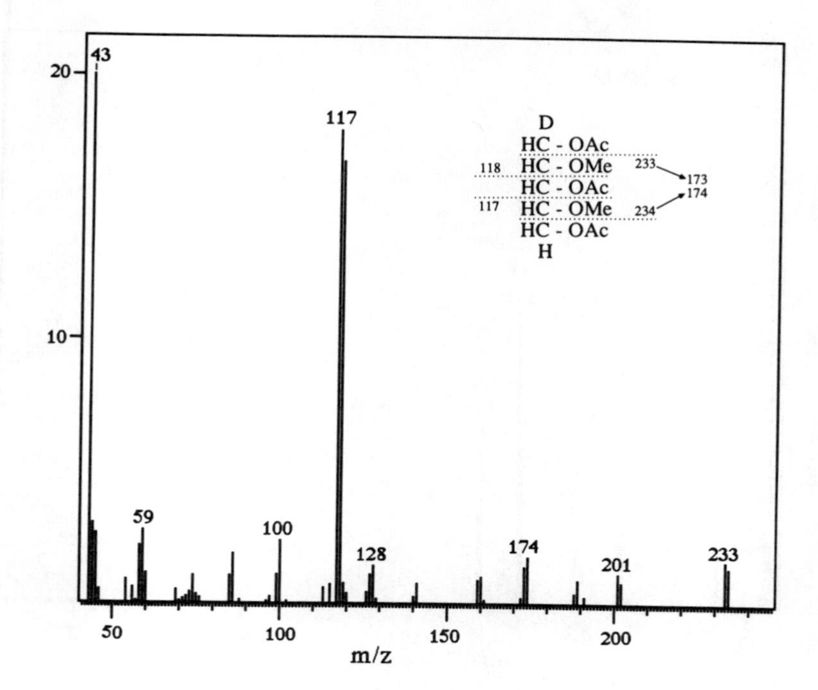

FIGURE 5c

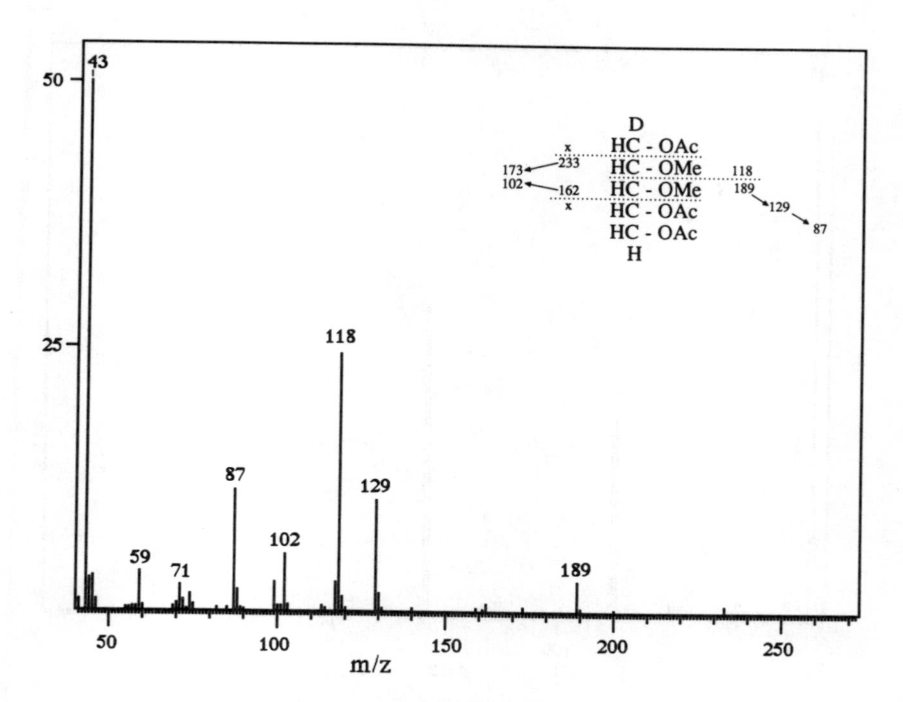

FIGURE 5d

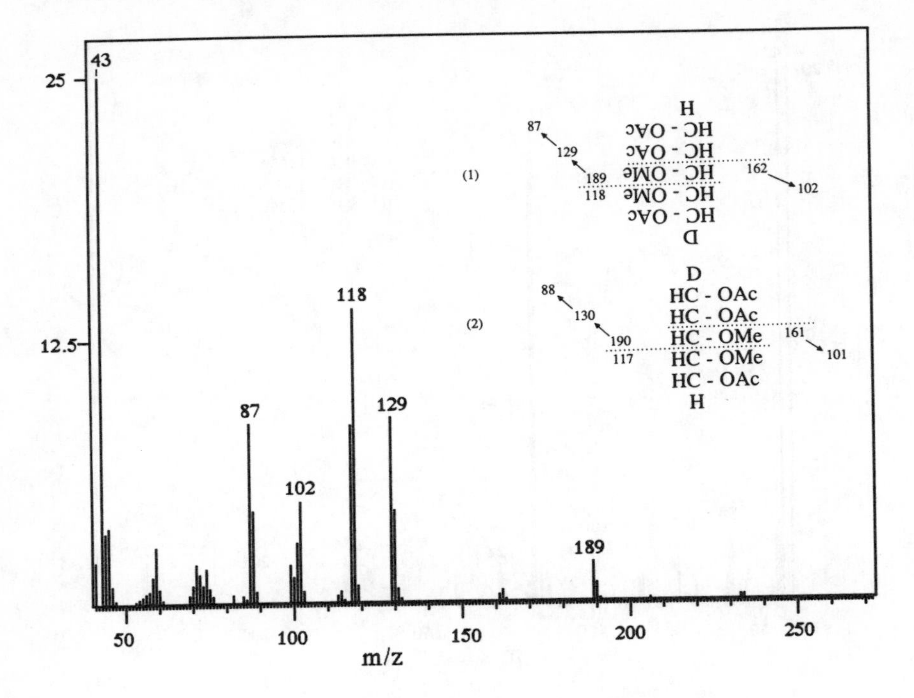

FIGURE 5e

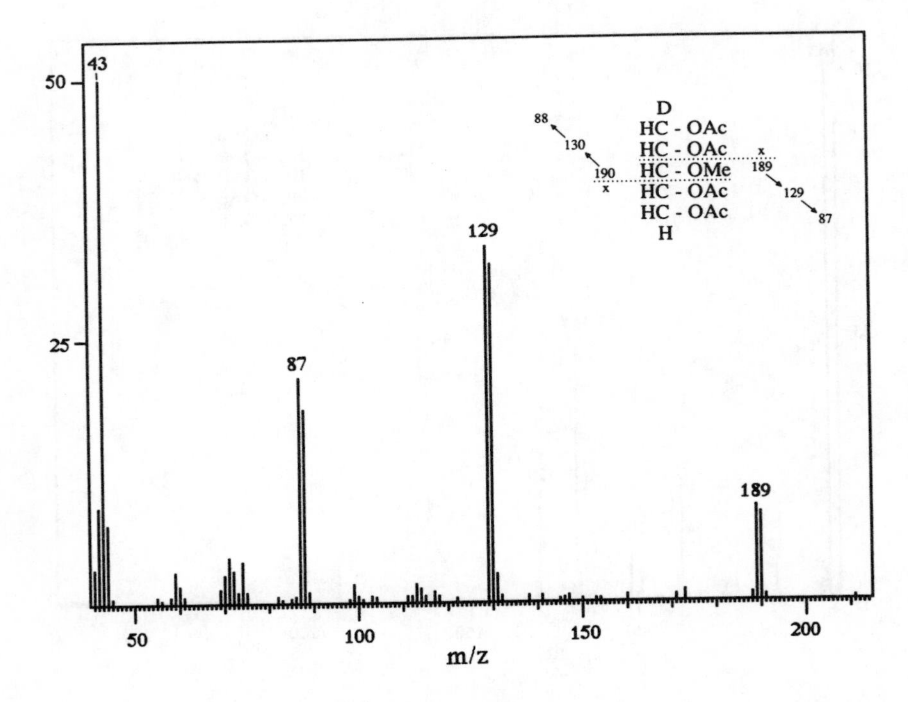

FIGURE 5f

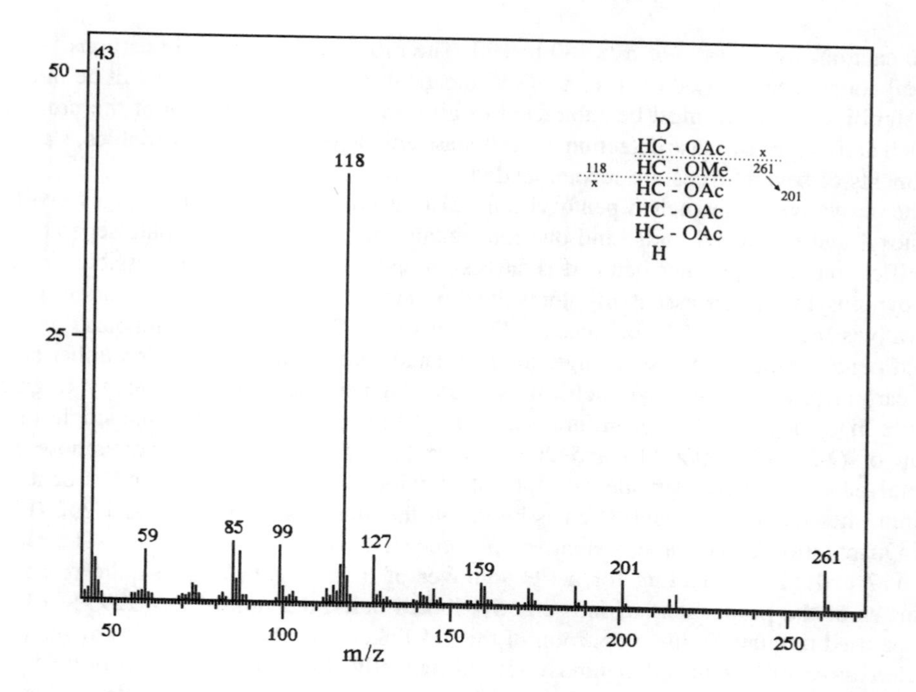

FIGURE 5g

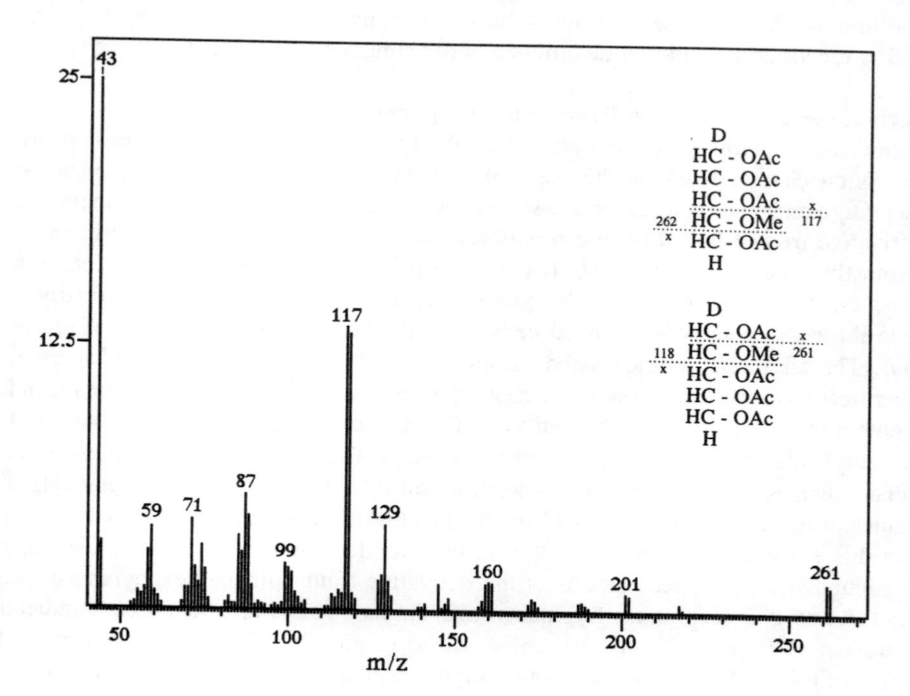

FIGURE 5h

some caution, by the ratio of m/z 189 to 190. The m/z 189 contains eight carbons, and ^{13}C natural abundance[109] based on 1.11% of ^{12}C means that 8.88% of m/z 189 will be m/z 190, this significant amount must be subtracted to give an accurate estimation of the proportion of each derivative. Because detection of each mass unit depends on many variables, statistical treatments of several trials are recommended.

The derivatives of branched pentosyl units also result in ambiguity in linkages involving furanosyl and pyranosyl units, and one must again rely on chromatographic separation for identification. On the other hand, derivatives representative of all three possible branched xylopyranosyl units are essentially unresolved on most columns, and identification of these derivatives must rely solely on spectra. Derivatives of 2,5-linked pentofuranose and 2,4-linked pentopyranose possess a single methoxylated carbon, and cleavage on either side of this carbon produces two symmetrical primary fragments and four secondary fragments (Figure 5f). For 3,5-linked pentofuranose and 3,4-linked pentopyranose, the single methyl group of *O*-2 yields m/z 118 and 261 (Figure 5g). The 2,3-linked xylopyranose is an unresolved symmetrical equivalent of 3,4-linked xylopyranose, but again, the C-1 deuterium distinguishes it, and its quantitation is based on the amounts of m/z 117 and 262 (Figure 5h). Quantitation of all three derivatives is somewhat complicated, however. The ratio of m/z 117 and 118, correcting for a ^{13}C spillover of 5.55%, should give quite reasonable estimates of the proportions of 2,3- and 3,4-linked units; the sum of m/z 129 + 130 can also be used to quantify the proportion of the 2,4-linked unit, but because fragmentation of the derivative of 2,4-linked pyranosyl is substantially different, the *proportion* of these diagnostic masses to the base peak m/z 43 must be determined for each derivative with standards that can be resolved chromatographically, such as those of 2,5- and 3,5-linked arabinofuranosyl units. These relative values can change substantially depending on how the MS is tuned, and should be determined under similar run conditions.

D. Derivatives of Linked and Branched Hexopyranoses

Fragmentation of many of the derivatives of hexosyl units resemble those of pentoses, but are often characterized on the basis of abundant 4-carbon, and sometimes 5-carbon primary fragments and their respective secondary fragments. For example, fragmentation of the derivative from 2-linked hexose resembles that of the derivative of 2-linked pentofuranose, mostly because m/z 190 and 161 are the principal primary fragments for each — cleavage on either side of the methoxylated carbon of the pentofuranose derivative and between the contiguous methoxylated carbons of the hexopyranose derivative (Figures 5a and 6a). The additional methoxylated carbon within the chain of the hexosyl derivative, however, results in the formation of 4-carbon fragments m/z 205 and 234 and their secondary fragments m/z 145 and 174, each from loss of acetic acid by β-elimination (Figure 6a).

The acetyl substitution at the *O*-3 results in a unique m/z 234, in addition to m/z 118 and 161, that indicates that the C-2 and C-4 positions must be methoxylated. The relatively high abundance of m/z 234 partially results from the fact that no contiguous methoxylated carbons exist in this derivative, the only case for all hexosyl derivatives (Figure 6b). Cleavage of the contiguous methoxylated carbons of the derivative from 4-linked hexopyranose results in m/z 118 and 233 as primary fragments, and the m/z 162 is also diagnostic, indicating that both the C-2 and C-3 must be methoxylated (Figure 6c). The fragmentation of the derivative of 6-linked hexopyranosyl units is quite similar to that of the 4-linked derivative, with one significant difference. The m/z 189 indicates that the *O*-6 must bear the acetyl substitution; substitution at the *O*-4, and subsequent cleavage between the C-3 and C-4, cannot result in formation of the carbonium cation (Figures 6c, d).

Identification of branched hexopyranosyl units requires detection of primary fragments of even higher masses because of the appearance now of additional contiguous acetoxylated carbons. A 4-carbon fragment bearing three acetoxylated and one methoxylated carbon is m/z 261. Identification of m/z 262 indicates that such a fragment bearing the C-1 deuterium must have a single methyl group at the *O*-4 position to result in a detectable carbonium cation (Figure 7a). Further, the detection of m/z 161 is the sole remaining information necessary to complete the analysis of this derivative of 2,3-linked hexopyranosyl unit. The 2,4- and 2,6-linked hexopyranosyl units each bear *O*-2 acetyl and *O*-3 methyl that yield a diagnostic m/z 190, but the contiguous acetoxylated carbons of the 2,4-hexose derivative can yield only the m/z 233 as a second primary fragment (Figure 7b). The 2,6-hexose derivative is symmetrical, yielding "doublets" of m/z 189 and 190, and their respective secondary fragments m/z 129 and 130, from loss of acetic acid by elimination (Figure 7c).

Other branch linkages involving the C-3 are 3,4- and 3,6-hexopyranosyl units, and these derivatives are quite common in many polysaccharides and glycoproteins. The three contiguous acetoxylated carbons of the derivative of the 3,4-linked unit result in detection of the rare 5-carbon primary fragment m/z 305, whereas *O*-4 methylation of the derivative of 3,6-hexosyl units makes possible the primary fragments m/z 189 and 234 that substantiate the linkage (Figures 7d, e). The 4,6-hexopyranosyl linkage, perhaps the most abundant branched hexose in nature, yields contiguous methoxylated carbons that yield the diagnostic pair m/z 118 and 261 (Figure 7f).

E. Derivatives of Hexofuranoses

Although most hexoses exist in the pyranose form, the most obvious exceptions are the keto sugars, such as fructose, which is the principal sugar of inulins, bacterial levans, and other plant fructans. The predicted fragmentation of C-2 deuterioreduced ketoses has been described, and is straightforward.[110,111] The nonreducing terminal sugar is symmetrical and yields nearly equal amounts of m/z 161 and 162 as major primary fragments and m/z 205 and 206 as quite diagnostic minor fragments. Loss of methanol by β-elimination results in m/z 129 from either m/z 161 or 162, a rare demonstration that the proton, and in this case deuterium, to neutralize the MeO^- comes from the penultimate carbon (Figure 8a). Unlike aldoses, reduction of keto sugars results in formation of both possible epimers. In the case of fructose, reduction of the C-2 anomeric carbon will yield either mannitol or glucitol, but ultimately the fragmentation will be identical. Without the C-2 deuterium, the 1- and 6-linked fructofuranosyl units usually found are symmetrical and would yield identical fragments. With the C-2 deuterium, the derivative from 1-linked hexofuranose yields m/z 190 and 161 as major fragments and m/z 234 and 205 as minor fragments, whereas the derivative from 6-linked hexofuranose would yield major fragments m/z 189 and 162 and minor fragments m/z 233 and 206 (Figure 8b, c). Although the mannitol and glucitol derivatives are resolved chromatographically, the respective derivatives of 1- and 6-linked fructofuranose are not. Many levans and fructans from bacterial and plant sources, including inulin, comprise mixtures of these two linkages (Figure 8d). The amounts of each derivative can be determined by ratio analysis of m/z 190 and 189 or m/z 161 and 162 after correction of spillover of ^{13}C (7.77% of m/z 161 and 8.88% of m/z 189). One unusual feature of the derivatives from terminal- and 6-linked fructosyl units is that methylation of the *O*-1 directs formation of predominantly the mannitol epimer upon reduction, so both peaks must be summed to obtain an accurate estimation of the proportion of 1- and 6-linked units.[112] Some fructans contain 1,6-linked branch points, and their derivatives are characterized by m/z 189 and 190 as the major primary fragments and m/z 233 and 234 as minor fragments (Figure 8e). Ostensibly, secondary fragmentation should be identical to that of the 2,6-hexopyranosyl derivative, but loss of deuterium from the C-2, upon β-elimination of acetic acid from m/z 190, yields only m/z 129.

F. Derivatives of Deoxyhexitols

Deoxysugars are quite abundant components of certain carbohydrates, most notably plant pectins (rhamnose) and slime secretions (fucose), some glycoproteins (fucose), and bacterial

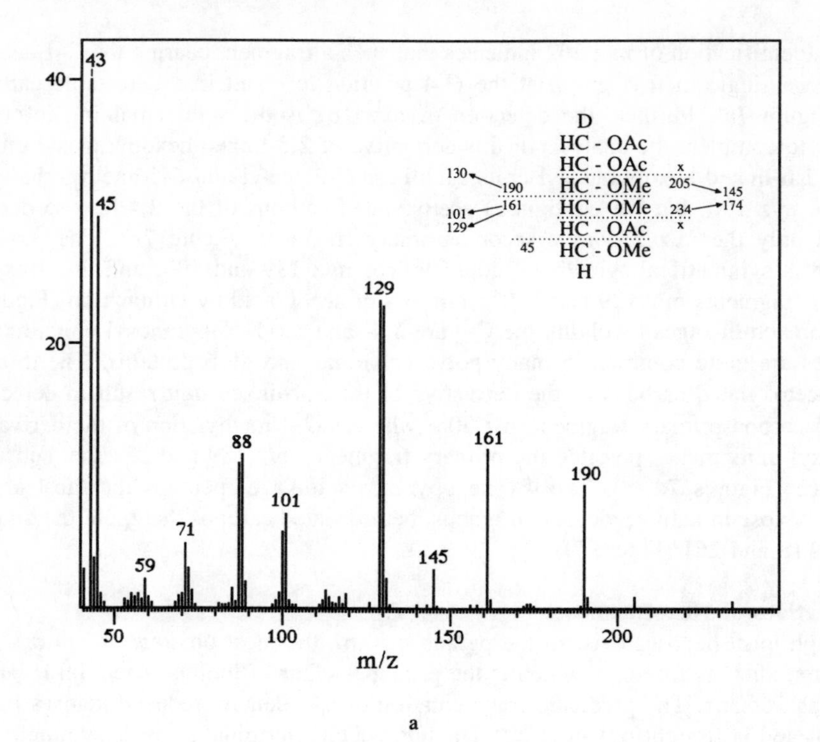

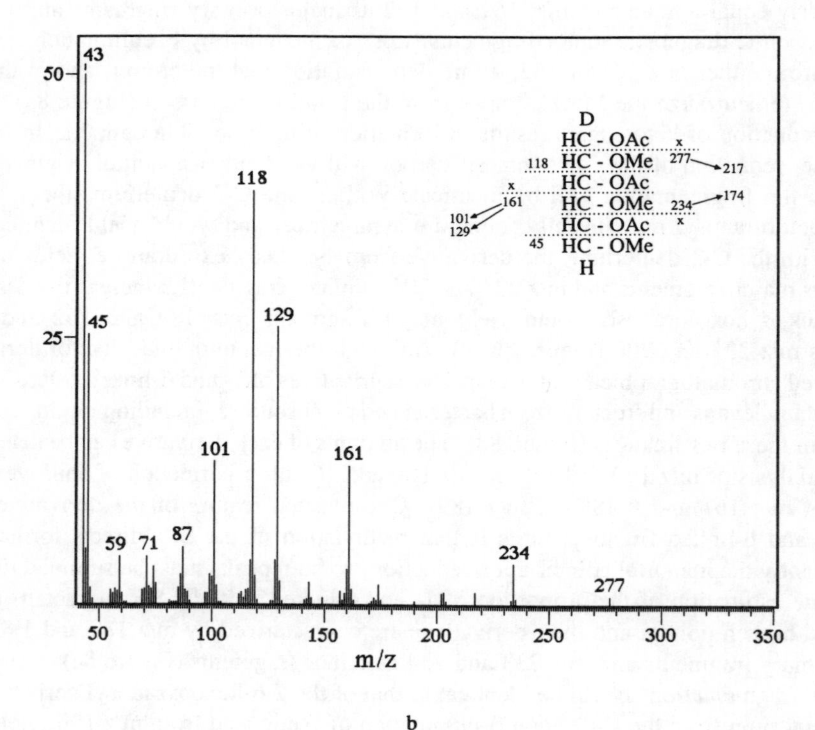

FIGURE 6. Spectra of derivatives of linked hexopyranoses: a, 1,2,5-tri-*O*-acetyl-(1-deuterio)-3,4,6-tri-*O*-methyl hexitol; b, 1,3,5-tri-*O*-acetyl-(1-deuterio)-2,4,6-tri-*O*-methyl hexitol; c, 1,4,5-tri-*O*-acetyl-(1-deuterio)-2,3,6-tri-*O*-methyl hexitol; d, 1,5,6-tri-*O*-acetyl-(1-deuterio)-2,3,4-tri-*O*-methyl hexitol.

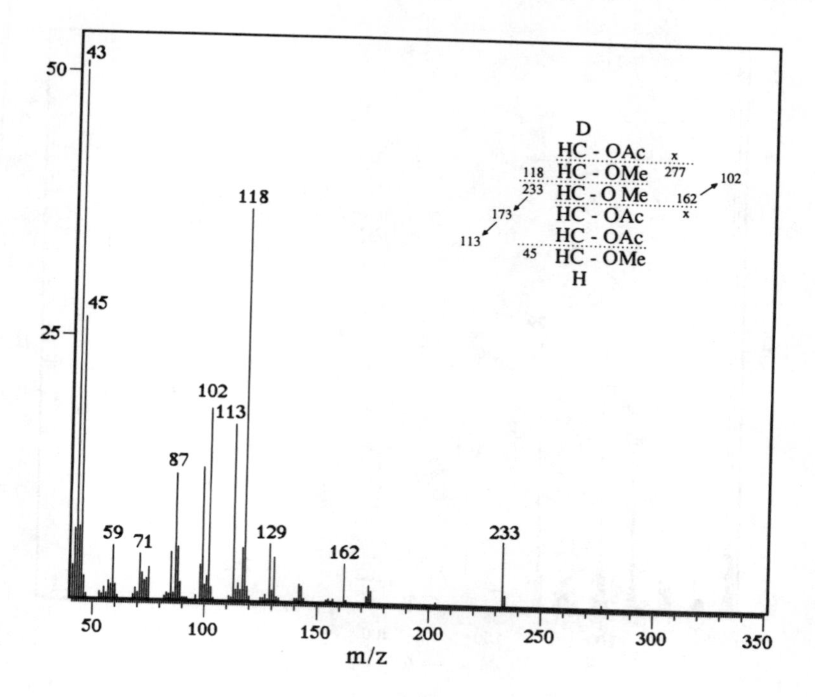

FIGURE 6c

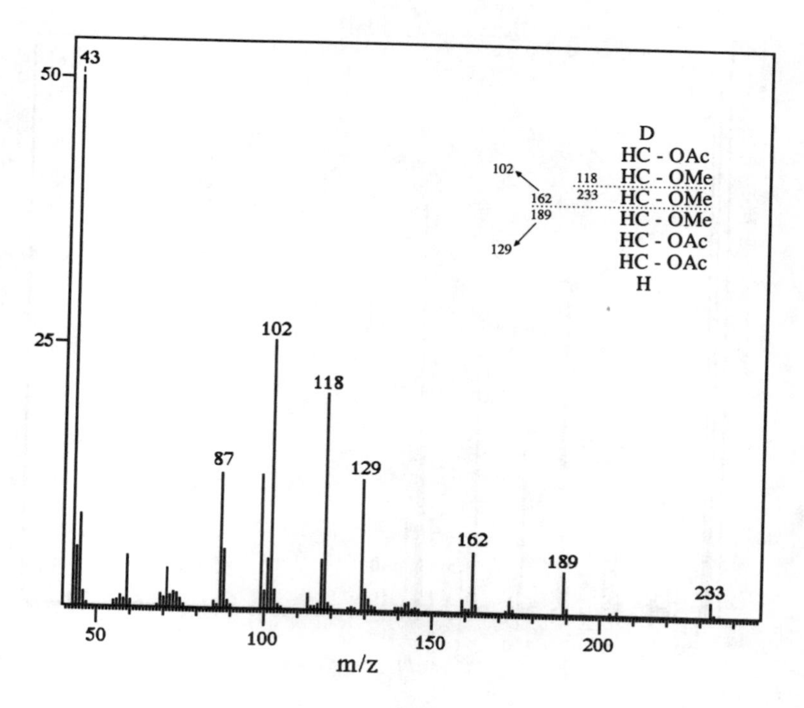

FIGURE 6d

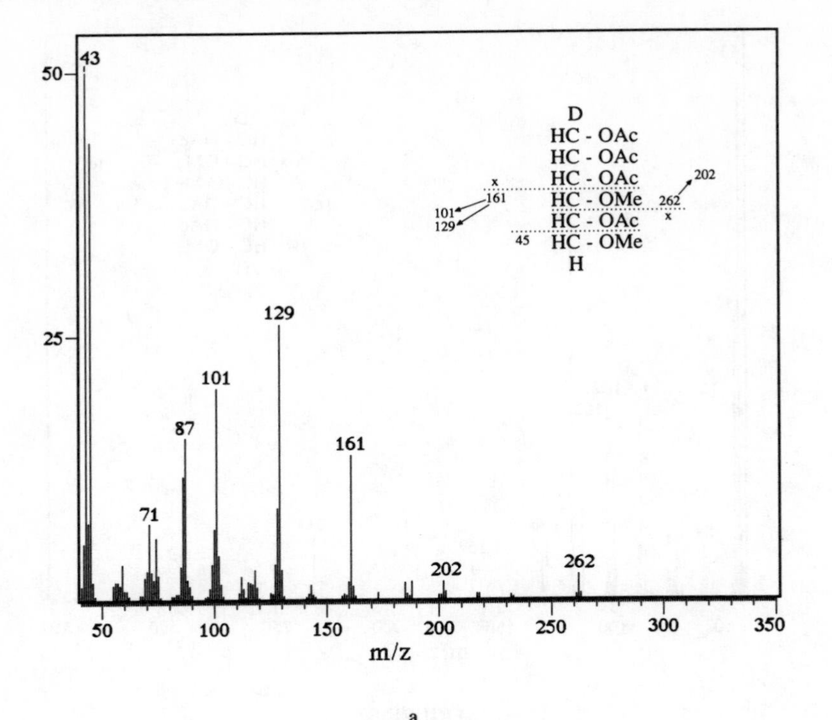

a

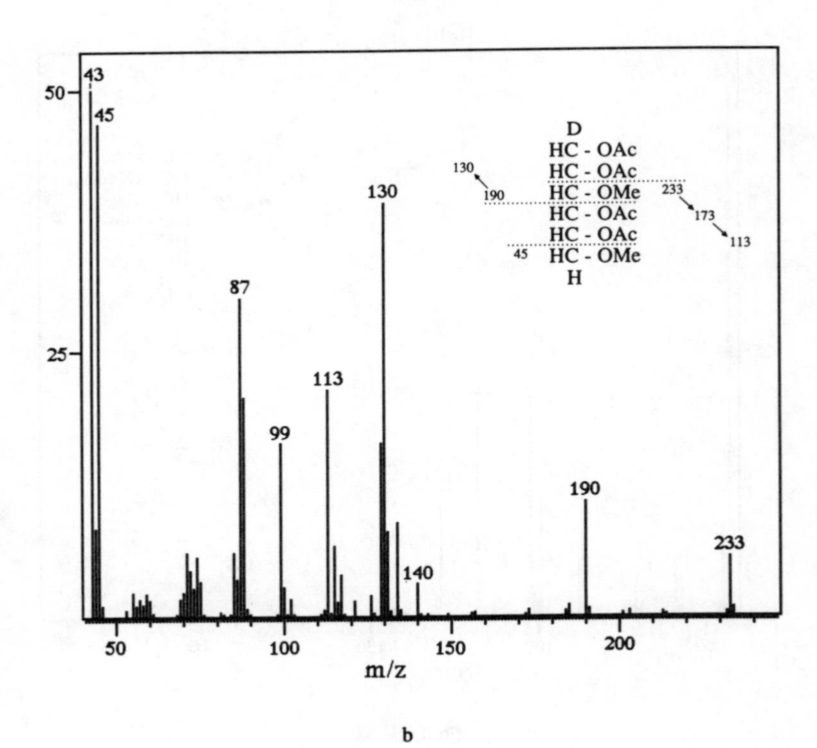

b

FIGURE 7. Spectra of derivatives of branched hexopyranoses: a, 1,2,3,5-tetra-*O*-acetyl-(1-deuterio)-4,6-di-*O*-methyl hexitol; b, 1,2,4,5-tetra-*O*-acetyl-(1-deuterio)-3,6-di-*O*-methyl hexitol; c, 1,2,5,6-tetra-*O*-acetyl-(1-deuterio)-3,4-di-*O*-methyl hexitol; d, 1,3,4,5-tetra-*O*-acetyl-(1-deuterio)-2,6-di-*O*-methyl hexitol; e, 1,3,5,6-tetra-*O*-acetyl-(1-deuterio)-2,4-di-*O*-methyl hexitol; f, 1,4,5,6-tetra-*O*-acetyl-(1-deuterio)-2,3-di-*O*-methyl hexitol.

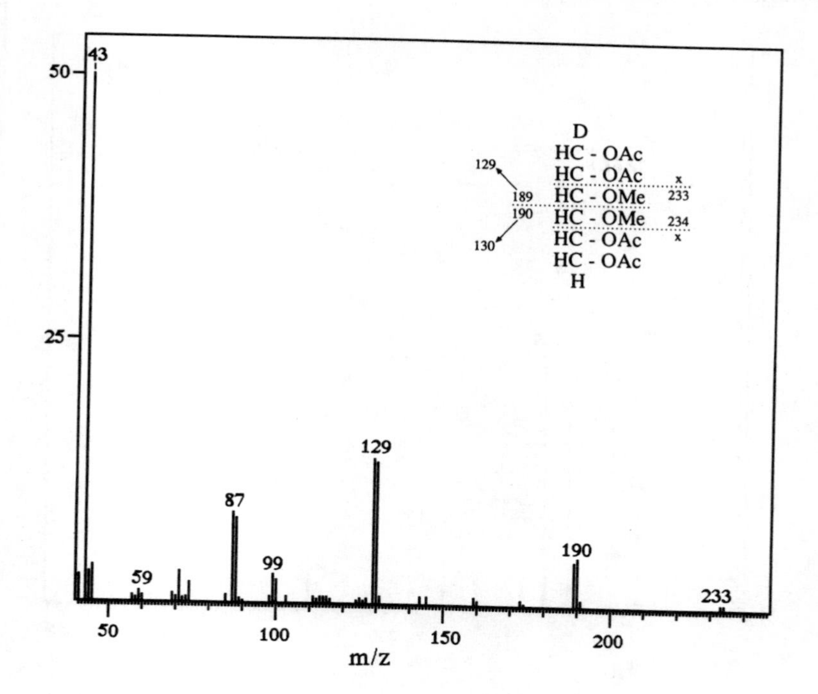

FIGURE 7c

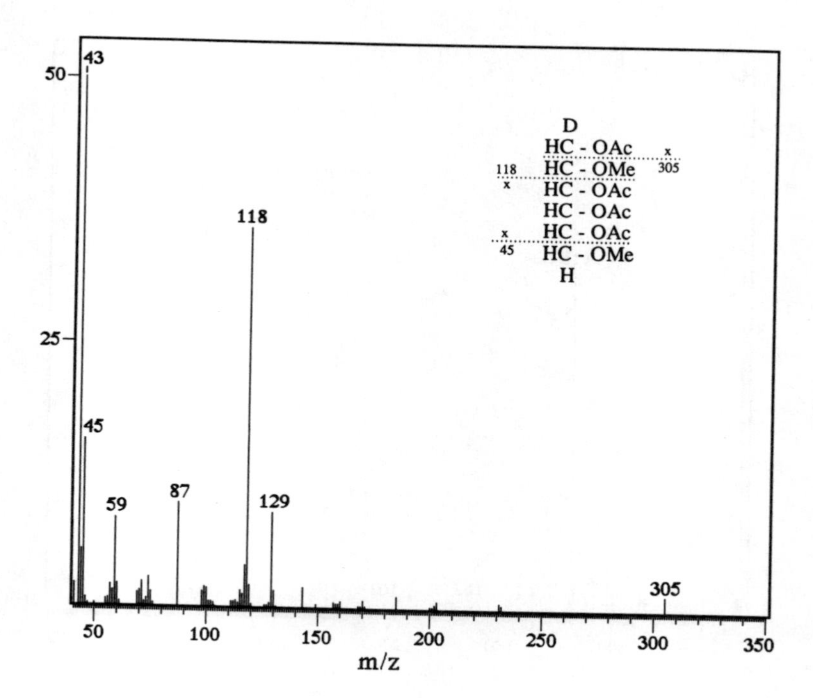

FIGURE 7d

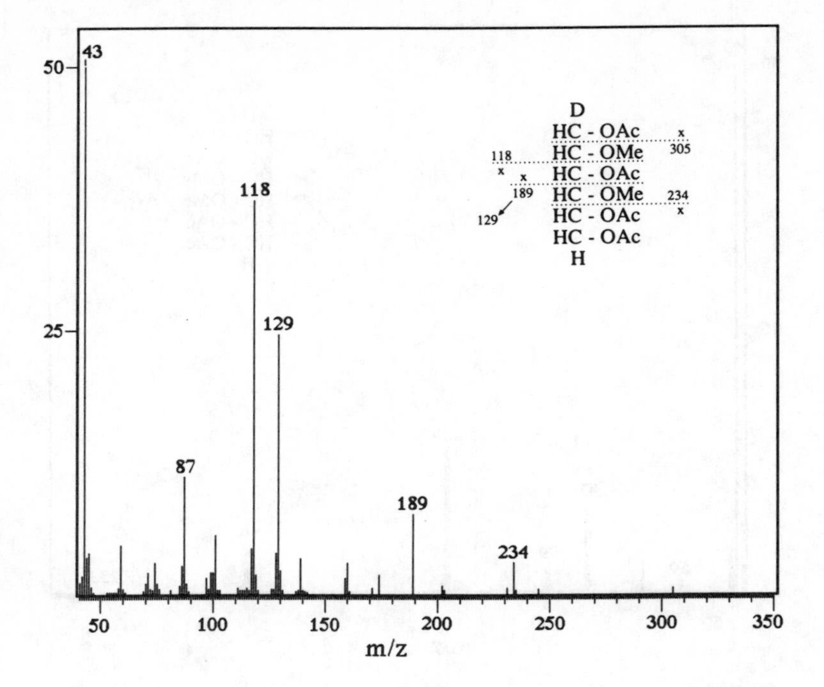

FIGURE 7e

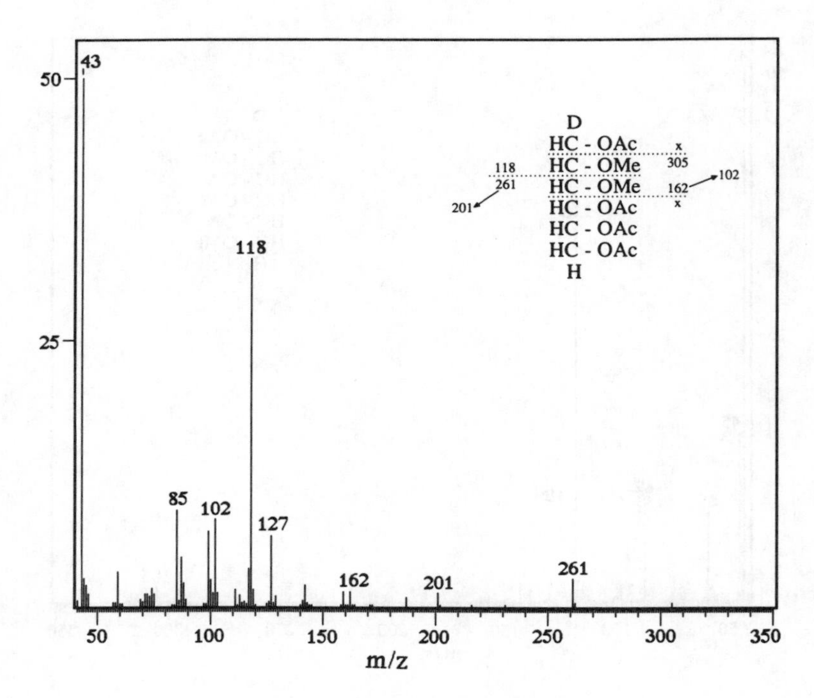

FIGURE 7f

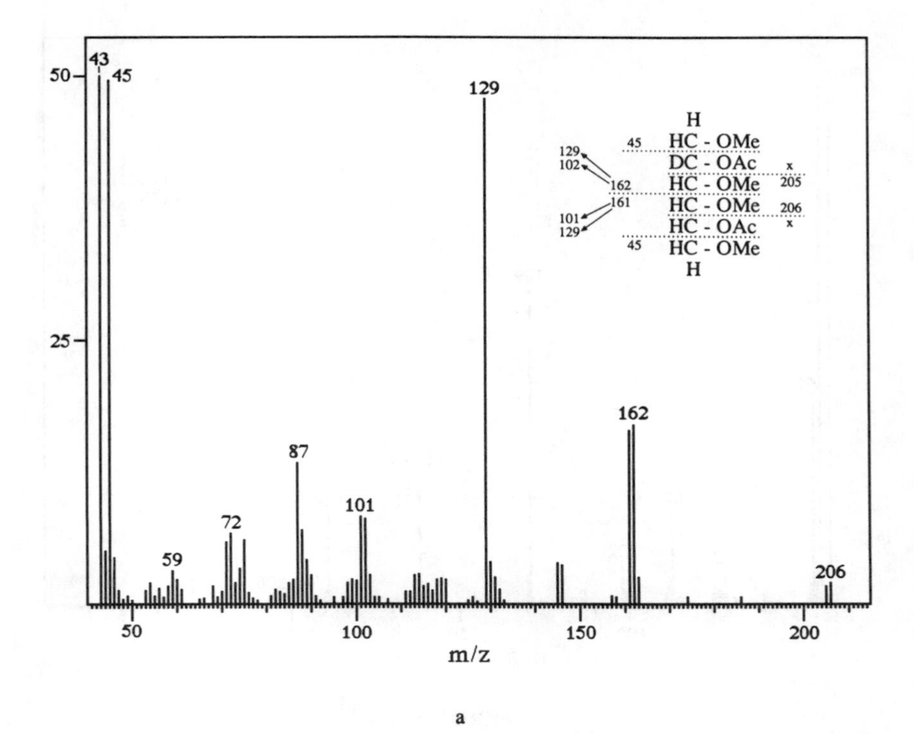

a

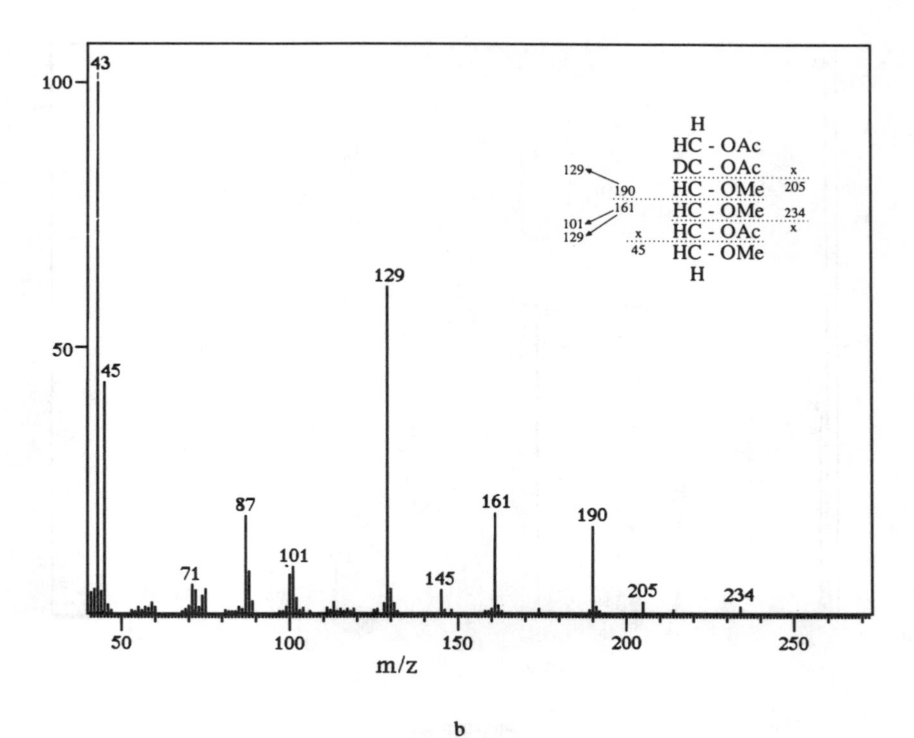

b

FIGURE 8. Spectra of derivatives of hexofuranoses: a, 2,5-di-*O*-acetyl-(2-deuterio)-1,3,4,6-tetra-*O*-methylhexitol; b, 1,2,5-tri-*O*-acetyl-(2-deuterio)-3,4,6-tri-*O*-methyl hexitol; c, 2,5,6-tri-*O*-acetyl-(2-deuterio)-1,3,4-tri-*O*-methyl hexitol; d, Mixture of (b) and (c); e, 1,2,5,6-tetra-*O*-acetyl-(2-deuterio)-3,4-di-*O*-methyl hexitol.

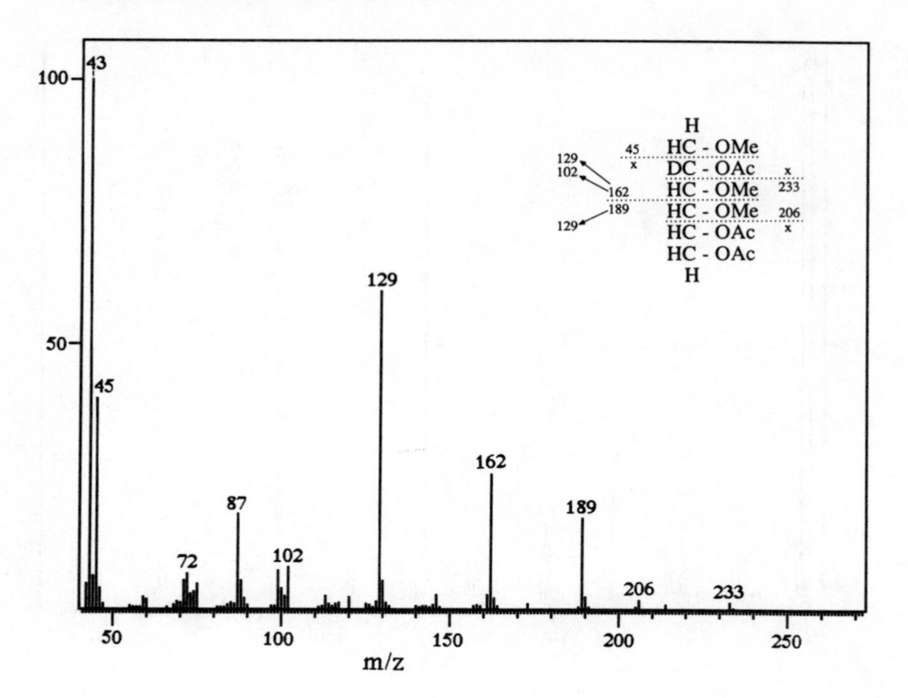

FIGURE 8c

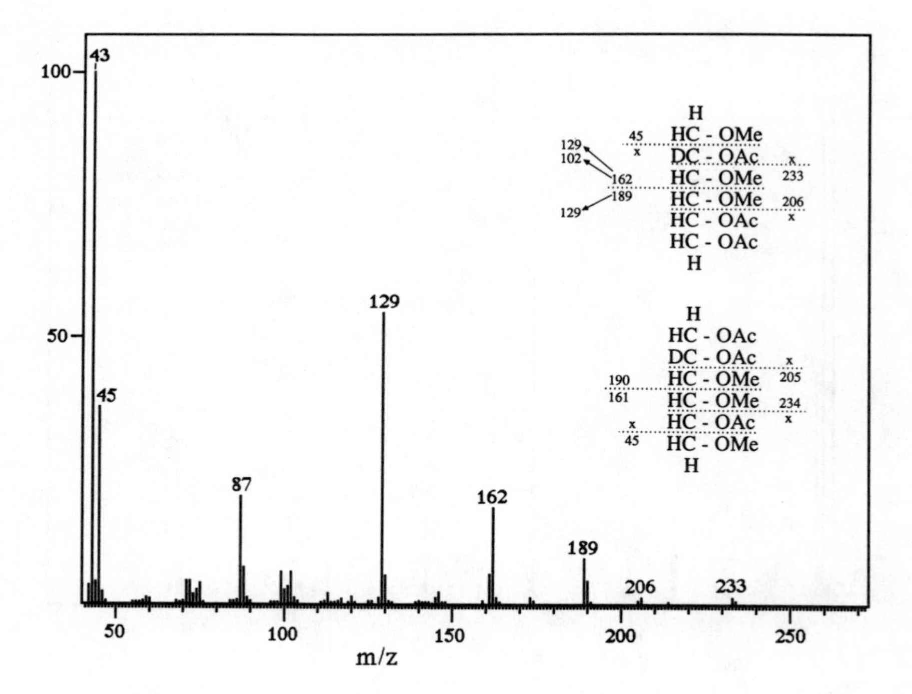

FIGURE 8d

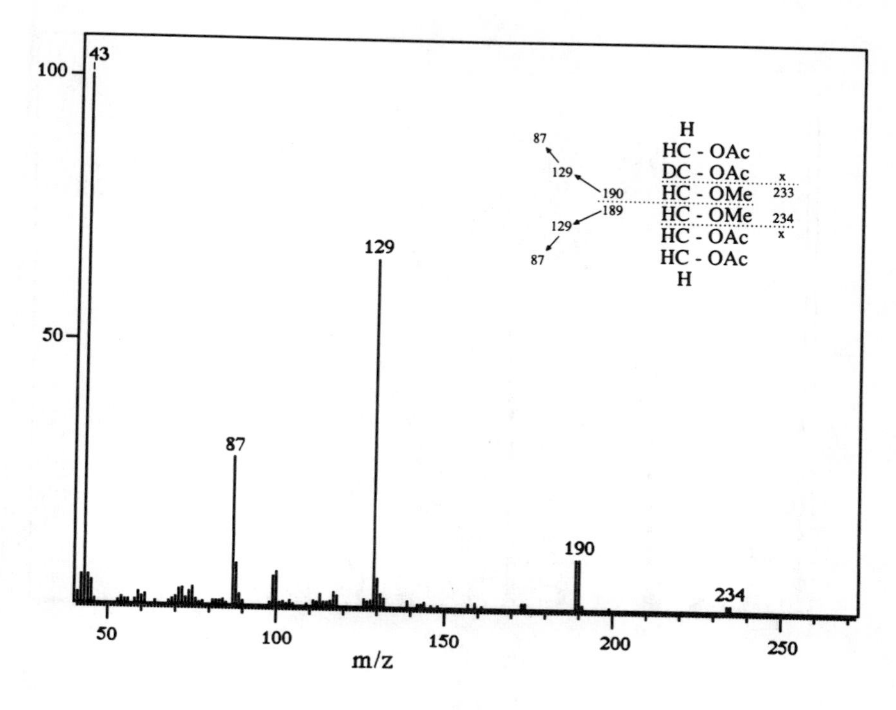

FIGURE 8e

lipopolysaccharides (rhamnose and dideoxygalactose). Their fragmentation is straightforward by virtue of the unique diagnostic methyl or methylene group. The derivative from 2-linked 6-deoxyhexose units contains contiguous methoxylated carbons that yield m/z 131, diagnostic of both the terminal methyl group and *O*-4 methylation, and m/z 190 which, for reasons described earlier, indicates that the acetylation site is the *O*-2 position (Figure 9a). Less frequent primary fragmentation between C-2 and C-3 yields significant amounts of m/z 175, and, upon loss of acetic acid by β-elimination, m/z 115; these fragments confirm linkage at *O*-2. Acetylation of *O*-3 still results in m/z 131, but m/z 118 becomes the major primary fragment (Figure 9b). One possible diagnostic primary fragment is m/z 247 which may or may not be detected. It will lose methanol by β-elimination (m/z 215) or acetic acid by γ-elimination (m/z 187) which should be detected (Figure 9b). Linkage at the *O*-4 position eliminates the diagnostic m/z 131, but cleavage of the contiguous methoxylated carbons yields the unique m/z 203, and m/z 143 results from loss of acetic acid (Figure 9c). The m/z 118 fragment confirms that the *O*-2 was methylated.

A common branched deoxysugar unit from plant pectins is 2,4-linked rhamnose, and its fragmentation is quite easily deduced. Primary fragmentation is on either side of the single methoxylated carbon yielding nearly equal amounts of m/z 203, indicating *O*-4 acetylation, and m/z 190, indicating *O*-2 acetylation. The major secondary fragments m/z 143 and 130 are from the usual loss of acetic acid by β-elimination (Figure 9d). The significant m/z 101 detected is from loss of methanol from the less stable m/z 143, whereas m/z 88 is from loss of ketene from m/z 130.

G. Alditol Acetates

Completely acetylated alditols give quite characteristic spectra, but because there is little complexity in the fragmentation, pentitol and hexitol acetates can appear similar (Figure 10a, b). Further, because cleavage between contiguous acetoxylated carbons is rare, compared to fission of the acetylium ion (m/z 43), the signals from higher masses are quite

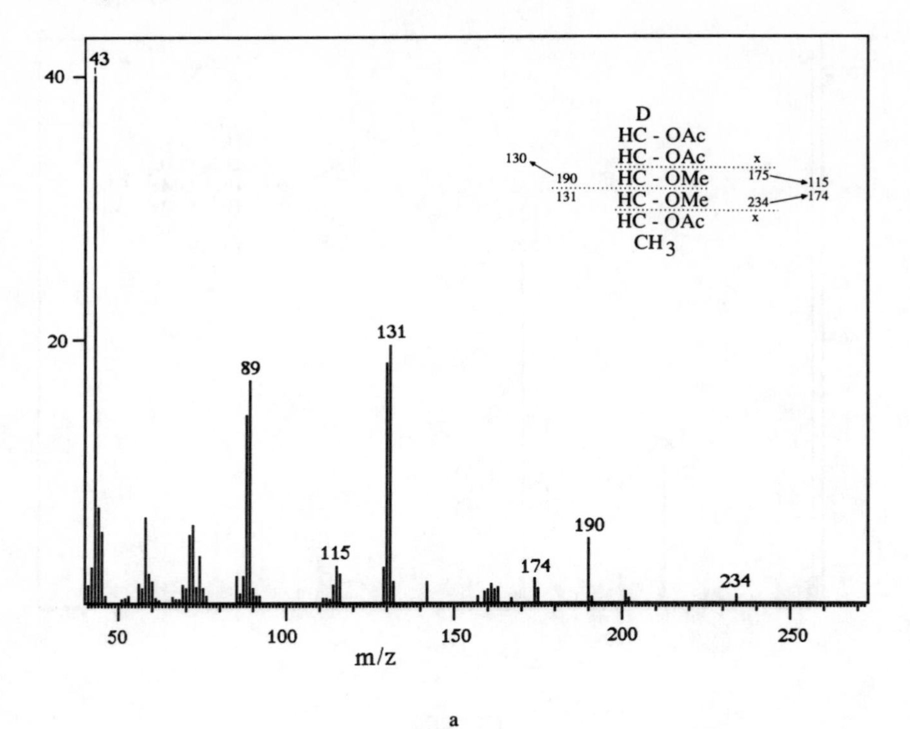

a

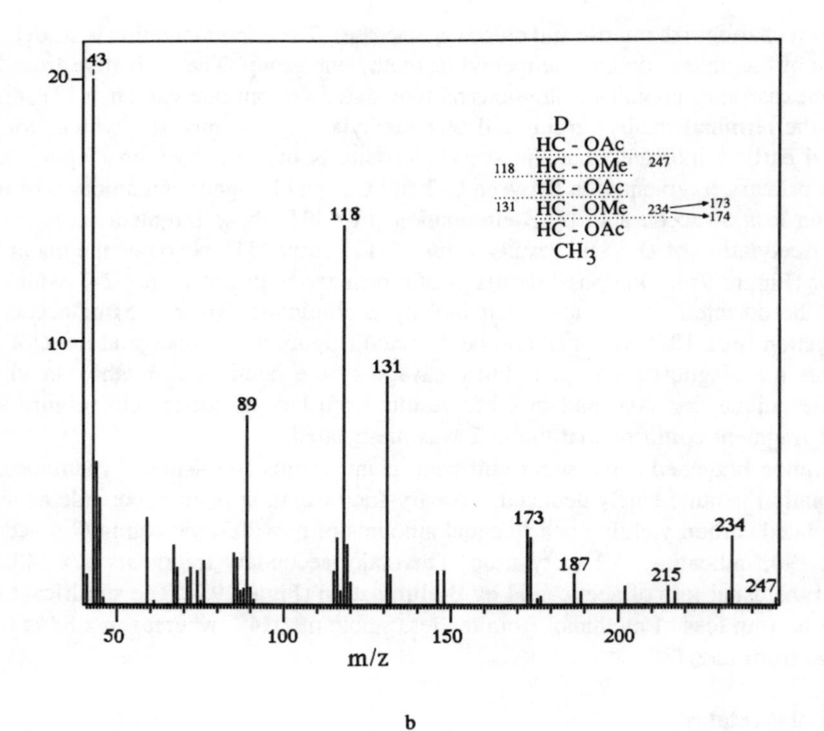

b

FIGURE 9. Spectra of derivatives of deoxyhexitols: a, 1,2,5-tri-*O*-acetyl-(1-deuterio)-3,4-di-*O*-methyl-6-deoxy-hexitol; b, 1,3,5-tri-*O*-acetyl-(1-deuterio)-2,4-di-*O*-methyl-6-deoxy-hexitol; c, 1,4,5-tri-*O*-acetyl-(1-deuterio)-2,3-di-*O*-methyl-6-deoxy-hexitol; d, 1,2,4,5- tetra-*O*-acetyl-(1-deuterio)-3-*O*-methyl-6-deoxy-hexitol.

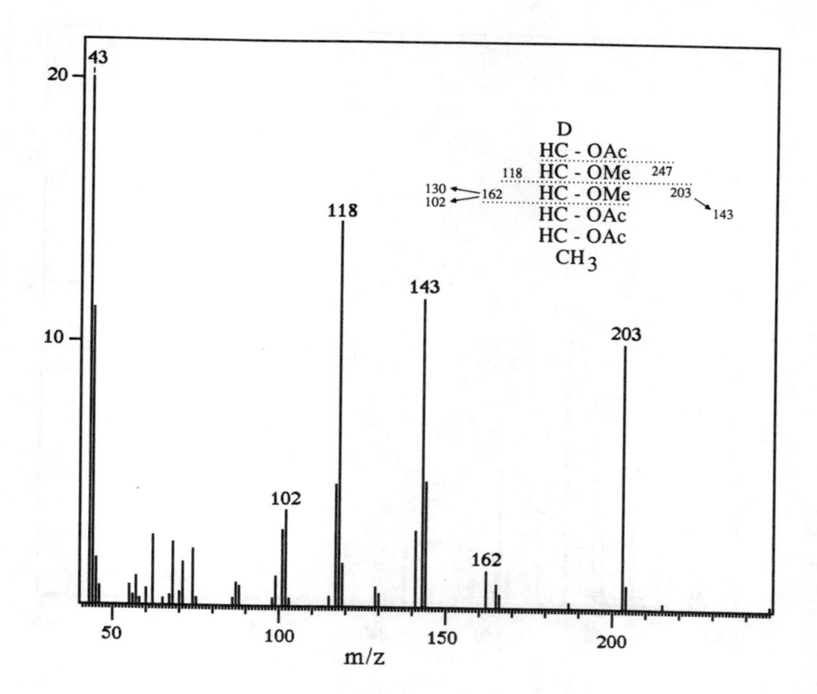

FIGURE 9c

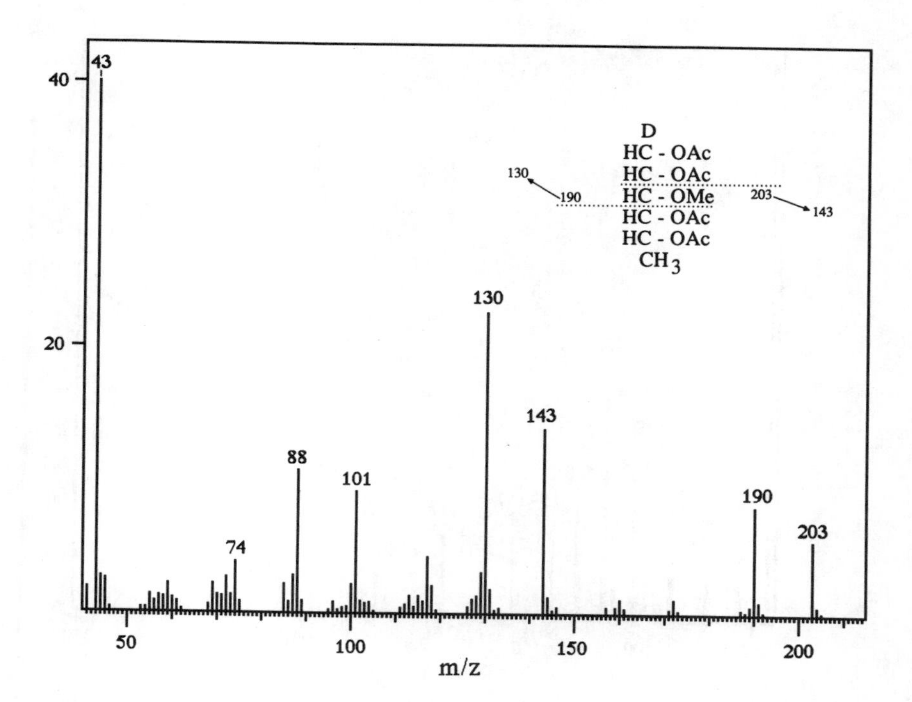

FIGURE 9d

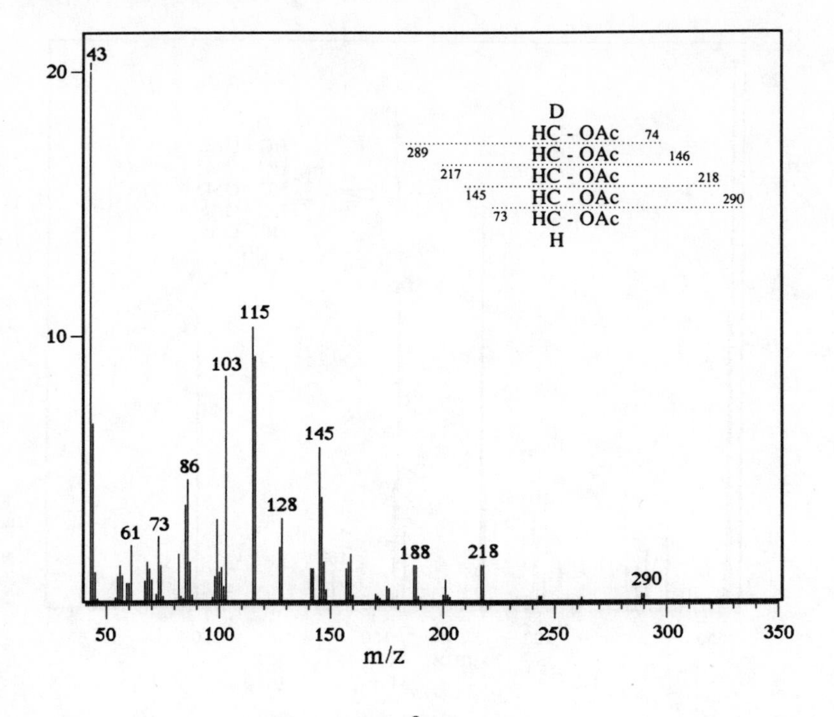

a

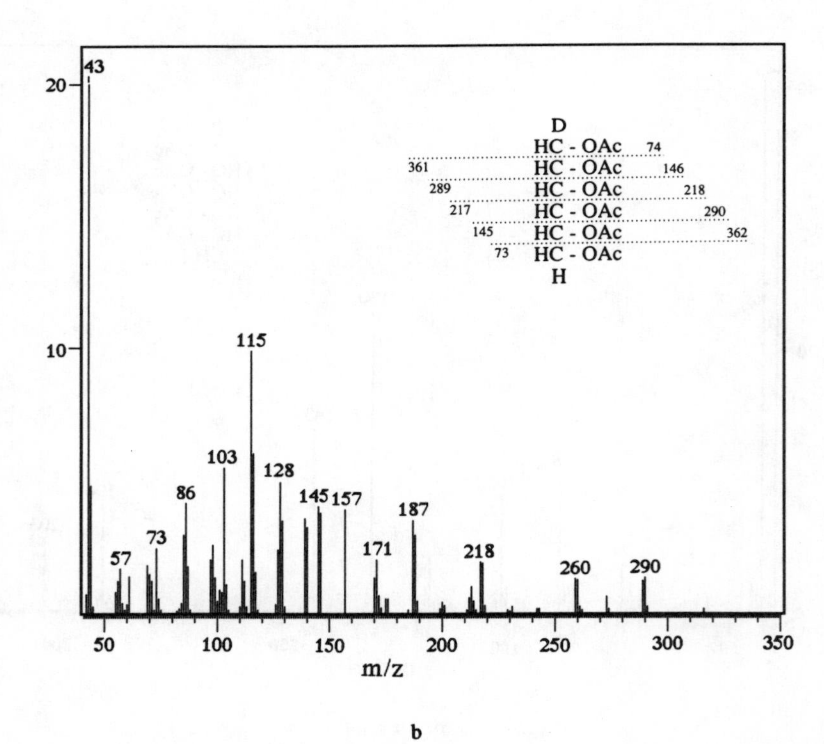

b

FIGURE 10. Spectra of unmethylated alditol acetate derivatives: a, 1-deuterio pentitol pentaacetate; b, 1-deuterio hexitol hexaacetate; c, 1-deuterio-6-deoxy-hexitol pentaacetate; d, inositol hexaacetate.

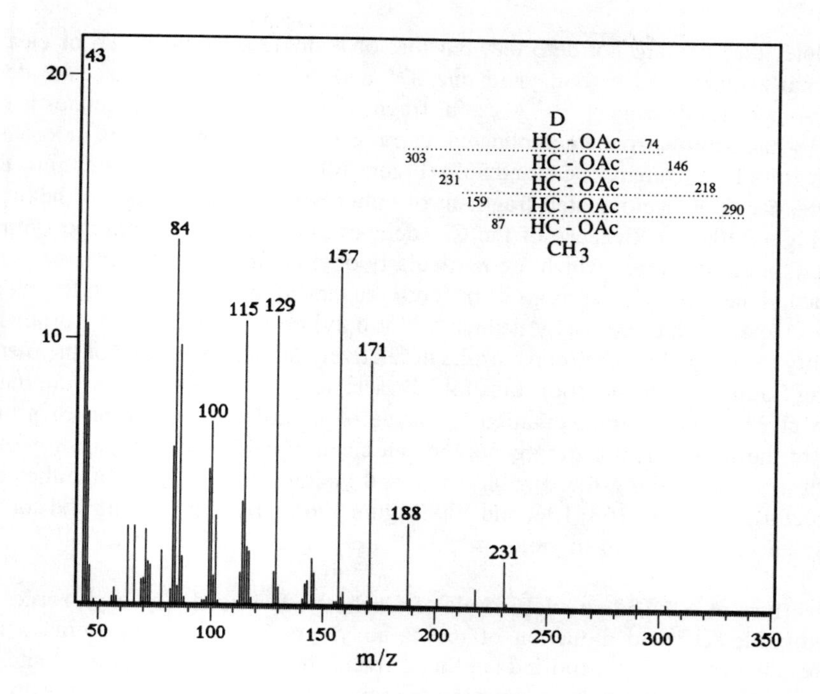

FIGURE 10c

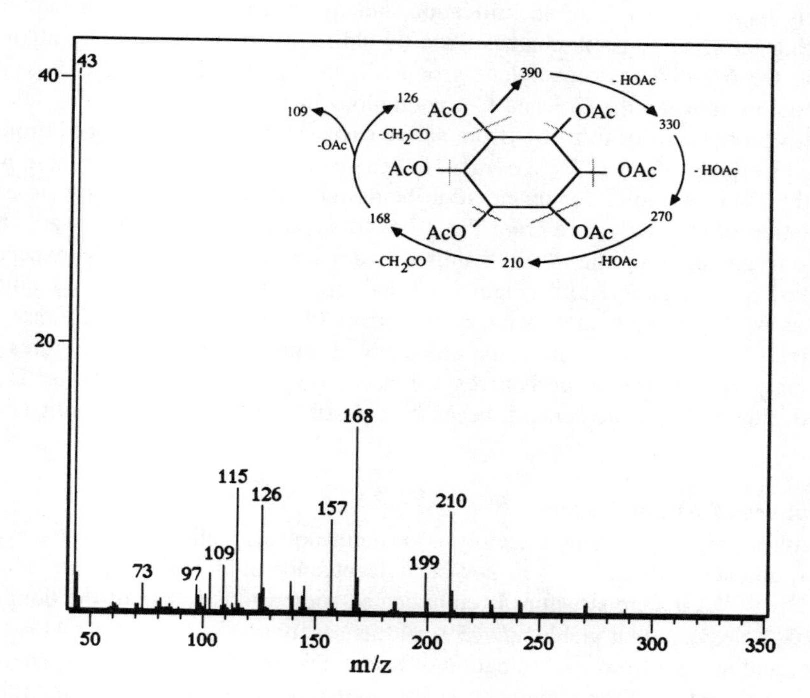

FIGURE 10d

weak. Molecular ions are not detected, but loss of a single acetoxyl group or cleavage of the terminal acetoxylated carbon yield m/z 375 and 361 from the hexaacetate derivative which is sometimes detected.[106,113] A useful fragment diagnostic of hexaacetates is m/z 259 (and 260) which results from a complicated rearrangement reaction resulting in loss of acetic anhydride (m/z 102) from m/z 361 (and 362) (Figure 10b).[113] Often, the highest mass detected is from the 4-carbon acetoxylated fragment of either pentitol pentaacetate or hexitol hexaacetate (Figure 10a, b). Because of the C-1 deuterium, "doublets" from the symmetrical fragmentation are detected, which are particularly obvious in the higher masses.

The methyl and methylene groups of the deoxysugars are quite resistant to fragmentation, resulting in spectra that are easily deduced. Not only are unique fragments produced, but the stability of the methylene groups results in relatively higher abundance of the methylene-containing fragments (Figure 10c). Inositol hexaacetate, typically used as the internal standard, also gives a unique fragmentation by virtue of its homocyclic carbon ring structure. Because of the ring structure, fragmentation can occur either by consecutive loss of ketene and acetic acid, preserving the ring structure and yielding m/z 390, 330 (neither detected in our spectra), 270, 210, 168, 126, and 109 (Figure 10d) or by ring splitting and consecutive loss of ketene or acetic acid to yield, m/z 241, 199, 157, 139, 115, and 97.[114]

H. Derivatives of 6,6-Dideuteriohexitols of Carboxyl-Reduced Glycosyluronic Acids

Carbodiimide-activated reduction of the carboxyl groups of glycosyluronic acids with $NaBD_4$ results in an easily identified tag that provides not only diagnostic analysis, but also quantitative determination of the proportion of uronic acid to its corresponding neutral sugar. The sugar derivatives are obviously not resolved chromatographically, so mass spectral analysis is the only means of identification and quantitation. The fragmentation of the nonreducing terminal sugar is characterized by shifts of two mass units by all fragments containing the 6,6-dideuteriomethylene group (Figure 11a). Because the C-1 is also deuterated, fission between the C-3 and C-4 contiguous methoxylated carbons yields m/z 162 and various proportions of m/z 161 (from native sugar) and 163 (from reduced uronic acid). Significant spillover of ^{13}C (7.77% of m/z 162) can confound quantitation of these particular masses, the obvious choice for quantitation being m/z 207 and 205 (Figures 4c and 11a). Fragmentation of the derivative from the 4-linked sugar is also straightforward, the most diagnostic fragment being the m/z 235 shifted two mass units from m/z 233 expected from the 4-linked hexopyranosyl unit (Figures 6c and 11b). The alditol acetate is quite easily deduced as well. Fragmentation between C-3 and C-4 yields m/z 218 and either m/z 217 from native sugar or m/z 219 from the dideuterated reduced uronic acid (Figures 10b and 11c). If only alditol acetate derivatives are being prepared, the quantitation is actually simplified if the C-1 is undeuterated, because of possible problems with quantitation by ^{13}C spillover.

I. Derivatives of Amino Sugars

Derivatives prepared by either acetolysis or methanolysis, followed by ester hydrolysis, reduction, and acetylation, result in 2-*N*-methylacetamide alditols that can be identified by EI-MS.[12,81,115,116] Linkage structure is complicated somewhat because of the dominance of the C-2, C-3 cleavage that yields m/z 159, and its secondary fragments m/z 117, from loss of ketene, and m/z 99 from α-elimination of acetic acid from the C-1 acetoxylated carbon. Linkage is deduced from the remainder of the fragments which follow the same rules as the neutral hexitols. For example, the derivative from nonreducing terminal amino sugar yields m/z 161 and 203 upon cleavage of the contiguous methoxylated carbons; the m/z 205 demonstrates that no acetylation, other than upon the *O*-5 ring carbon, is present (Figure 12a). The derivative from 4-linked amino sugar yields a strong m/z 233, and the lack of m/z 189 is a clear demonstration that the *O*-4, not the *O*-6, bears the additional acetyl group

(Figures 6c and 12b). The derivative from 3-linked amino sugar can be misleading for lack of abundant fragments. The m/z 275 and 161, and their principal secondary fragments m/z 215 and 101 from loss of acetic acid, are diagnostic of the linkage (Figure 12c). The derivative of 4,6-linked amino sugar, on the other hand, gives a strong m/z 261 fragment and the expected secondary fragments (Figure 12d).

J. Miscellaneous Sugars

Unusual sugars common to certain bacterial lipopolysaccharides are heptitols,[6] 3,6-dideoxysugars,[12,79,117] and 3-deoxy-octulosonic acid, and KDO,[118] the last found in an unusual plant pectic polysaccharide as well.[119] The epimers tyvelose (3,6-dideoxy-D-*arabino*-hexose) and abequose (3,6-dideoxy-D-*xylo*-hexose) are resolved chromatographically on high-polarity phases, preceding even 2,3,4-Me_3-rhamnose.[89-90] The internal methylene group is resistant to cleavage, yielding easily deducible diagnostic fragments (Figure 13a).[12,79] With KDO, the acid-labile octulosonic acid can be methylated, its acid group carboxyl-reduced to the respective octitol, and the acetylated permethylated derivatives separated by GLC.[118,119] Reduction of this ketosugar produces the two possible epimers which may or may not be resolved chromatographically.[116] The principal fragmentation of some of these derivatives is given in Figure 13b. Galactofuranose, another relatively uncommon hexofuranose, is a unique aldose, as opposed to the more common ketose, fructofuranose. Its fragmentation is equally unique, yielding m/z 89 from the terminal methoxylated carbons (Figure 13c).[80] Neuraminic acid and its respective endogenously acetylated derivative are common modifier groups on some glycoproteins. They are extremely acid-labile, however, and are best determined as permethylated and methyl glycosides, rather than as acetates.[120,121]

K. Fine Structure

1. Mild-Acid Hydrolysis

With polysaccharides, methylation analysis provides general linkage structure, but little sequence information. Oligosaccharides comprising more than three unit linear, or four unit branched, chains present ambiguity that sometimes can be eliminated by taking advantage of some of the different properties of the component sugars. Carbohydrates containing acid-labile constituents, such as deoxyhexose, dideoxyhexose, and furanose, offer such possibilities.[8,9,17] When done in tandem with material processed normally, permethylated polymers are subjected to mild acid hydrolysis, cleaving only the labile constituent and unmasking a hydroxyl group. Upon second methylation, this hydroxyl is tagged specifically with CD_3I[122] or CH_3CH_2I,[123] and the position of the former attachment identified by EI-MS of this derivative. The analysis is supported by comparison of samples from single and double methylation by loss of a derivative of a linear or branched sugar residue and the acid-labile constituent, and the gain of a derivative of a terminal or linear sugar, respectively.

2. β-Elimination of 4-O-Substituted Glycosyluronic Acids

Although the general recommendation is to reduce the carboxyl groups on glycosyluronic acid units of polymers before permethylation, there are special cases where information can be gained by not doing so. Pectic polysaccharides, rich in (1-4) α-D-galacturonan, are not amenable for these reactions, but for those which are not 4-*O*-substituted, a "double methylation" reaction is useful. Permethylation of carbohydrate containing uronic acid introduces a 4-*O*-methyl group which makes the glycosyl uronic acid susceptible to β-elimination in alkali.[8,9,17] The initial methylation neutralizes the methanide preparation so rapidly that such elimination is prevented in a single methylation.[16] Subjecting the permethylated uronic acid to a second methanide preparation induces β-elimination and unmasks a hydroxyl group denoting the position on the sugar to which it was attached.[123,124] For example, glucuronoarabinoxylans from many plant species contain a xylan backbone with arabinosyl, and

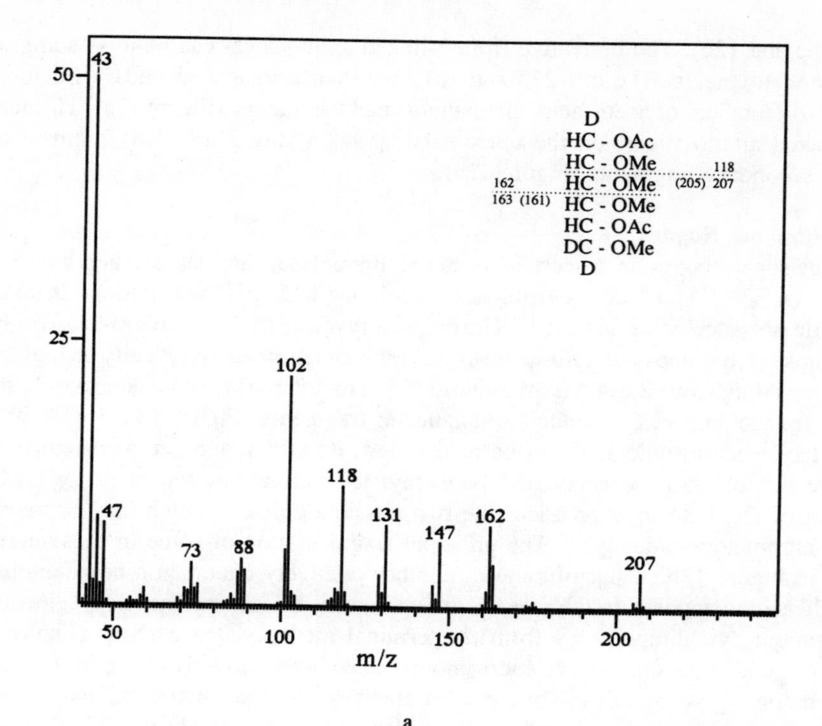

D
HC - OAc
118
HC - OMe
235 (233)
HC - OMe
162
HC - OAc
HC - OAc
47
DC - OMe
D
43
47
87
102
118
144
162
175
235
50
25
50
100
150
200
m/z
b

FIGURE 11. Spectra of 6,6-dideuterio carboxyl-reduced glycosyl uronic acids: a, 1,5-di-*O*-acetyl-(1,6,6-trideuterio)-2,3,4,6-tetra-*O*-methyl hexitol; b, 1,4,5-tri-*O*-acetyl-(1,6,6-trideuterio)-2,3,6-tri-*O*-methyl hexitol; c, 1,6,6-trideuterio hexitol hexaacetate.

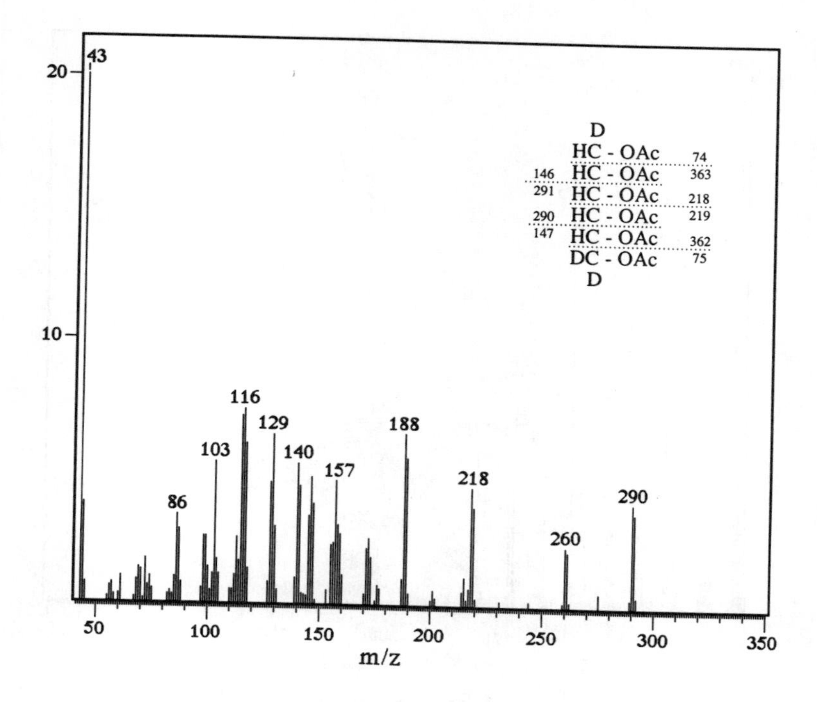

FIGURE 11c

glucuronosyl side chains. The glucosyluronic acid may only be present at 10% of the abundance of arabinose and its attachment site can be masked. The arabinofuranosyl units can be hydrolyzed before the first methylation by mild acid to yield a linear permethylated xylan and its uronic acid constituents. A second introduction of methanide induces β-elimination, and these sites are ethylated with CH_3CH_2I, rather than CD_3I, to resolve the unique linkage derivatives from the abundant 2,3-Me_2 xylose (Figure 14a).[125]

3. *Identification of Endogenously Methylated Sugars*

Some sugar units bear endogenous methyl groups, such as 2-*O*-methylfucose, 2-*O*-methylxylose, and 4-*O*-methylglucosyluronic acid. These methyl positions are revealed simply by permethylation with CD_3I rather than CH_3I. For example, the carboxyl on 4-*O*-methyl glucosyluronic acid is reduced to yield a mixture of free and methyl-substituted derivatives. Recalling the fragmentation of the C-1 deuterated derivative of terminal hexose, cleavage between the C-3 and C-4 methoxylated carbons yields nearly equal amounts of m/z 161 and 162 (Figure 3b). Two deuteriomethyl groups on each of these fragments adds 6 amu to each fragment, yielding m/z 167 and 168 (Figure 14b).[125] Endogenous methylation becomes apparent as these derivatives yield m/z 164, instead, and the amounts can be quantified by ratio analysis of m/z 164 and 167 (Figure 14b).

4. *Detection of Endogenous Acetylation*

Another significant modification of polysaccharides that affects their solubility is endogenous acetylation. These groups are quite labile in alkali and are lost upon contact with the methanide or OH^- preparations. A method described by DeBelder and Norrman[126] reveals the position of these acetyl groups indirectly by first protecting the free hydroxyls with methyl vinyl ether, which forms mixed acetals. The acetyl groups are eliminated upon introduction of the methanide ion, and these hydroxyl groups are methylated specifically. Acid hydrolysis of the sugars also cleaves the acetals from the protected hydroxyls, and the

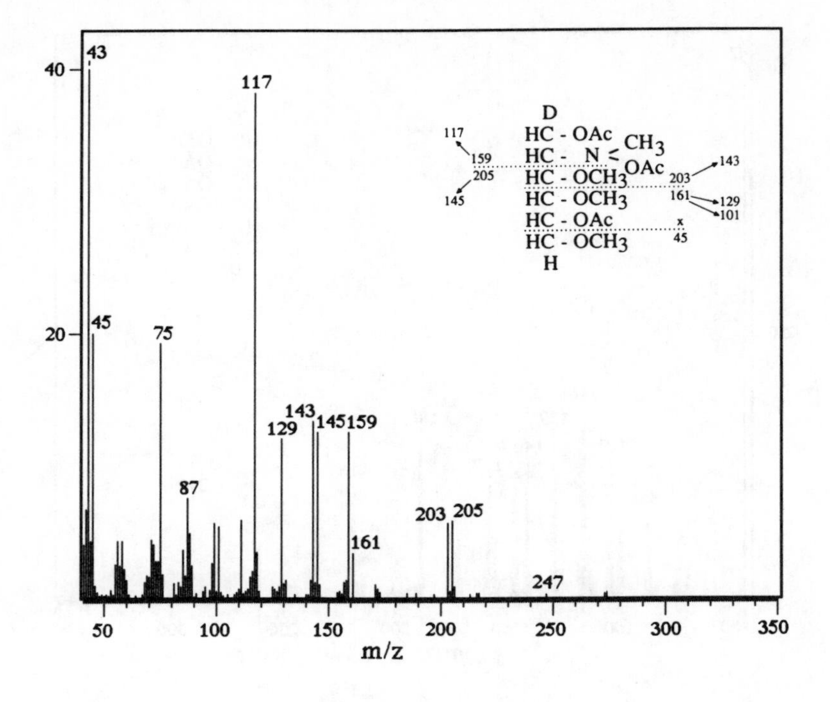

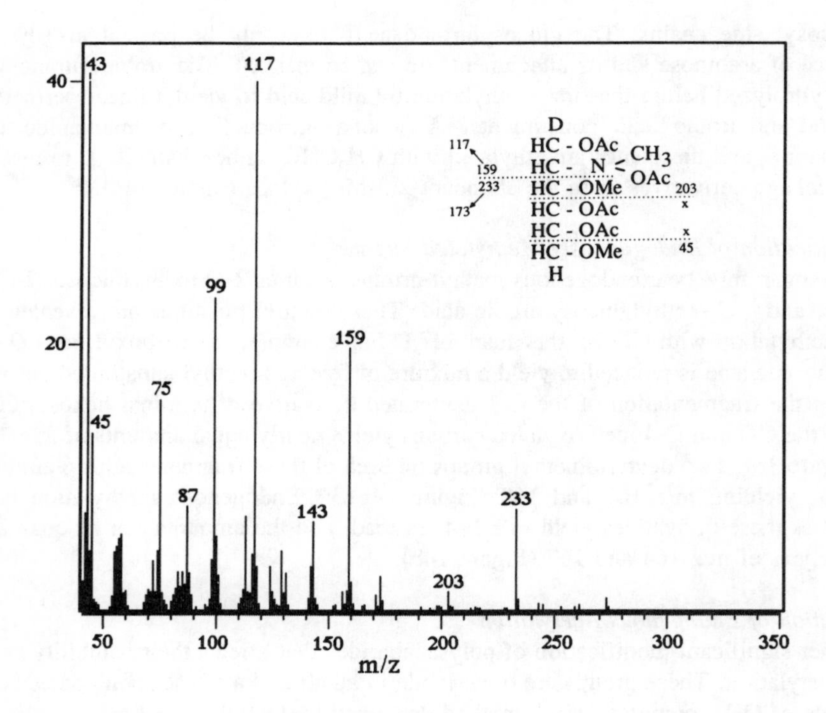

FIGURE 12. Spectra of derivatives of amino sugars: a, 1,5-di-*O*-acetyl-(1-deuterio)-(2-*N*-methylacetamide)-3,4,6-tri-*O*-methyl hexitol; b, 1,4,5-tri-*O*-acetyl-(1-deuterio)-(2-*N*-methylacetamide)-3,6-di-*O*-methyl hexitol; c, 1,3,5-tri-*O*-acetyl-(1-deuterio)-(2-*N*-methylacetamide)-4,6-di-*O*-methyl hexitol; d, 1,4,5,6-tetra-*O*-acetyl-(1-deuterio)-(2-*N*-methylacetamide)-3-*O*-methyl hexitol. (Redrawn from Schwarzmann, G. O. H., and Jeanloz, R. W., *Carbohydr. Res.*, 34, 161, 1974. With permission.)

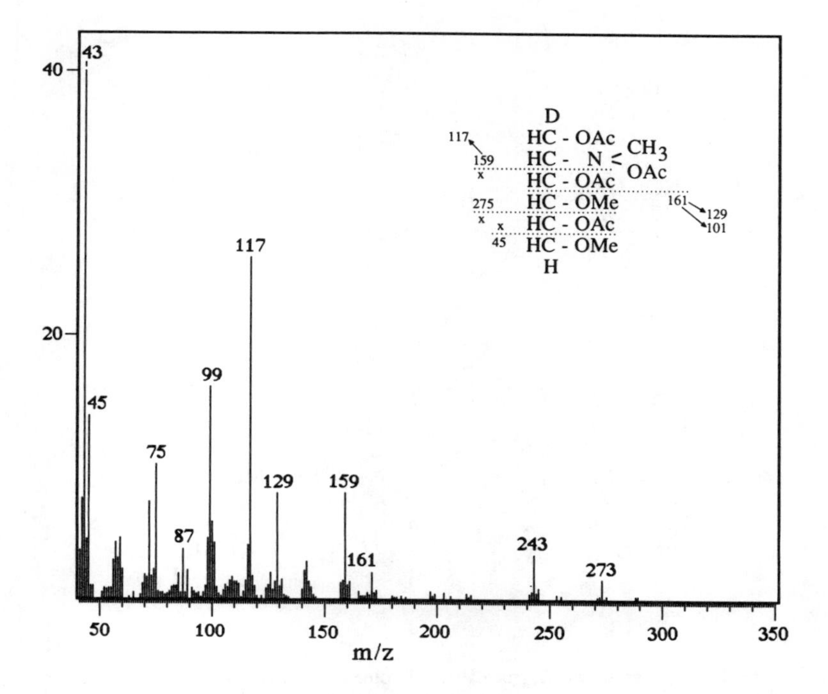

FIGURE 12c

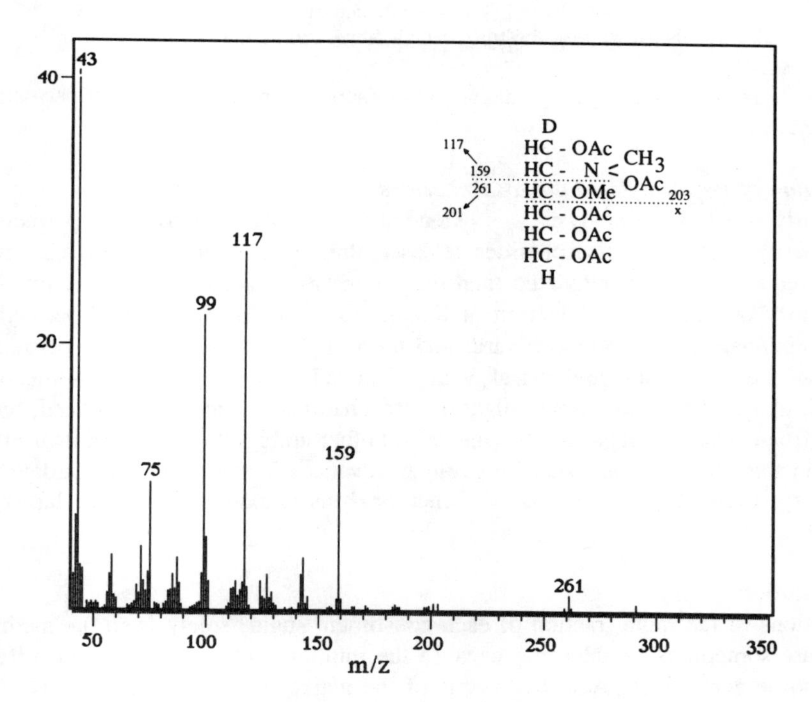

FIGURE 12d

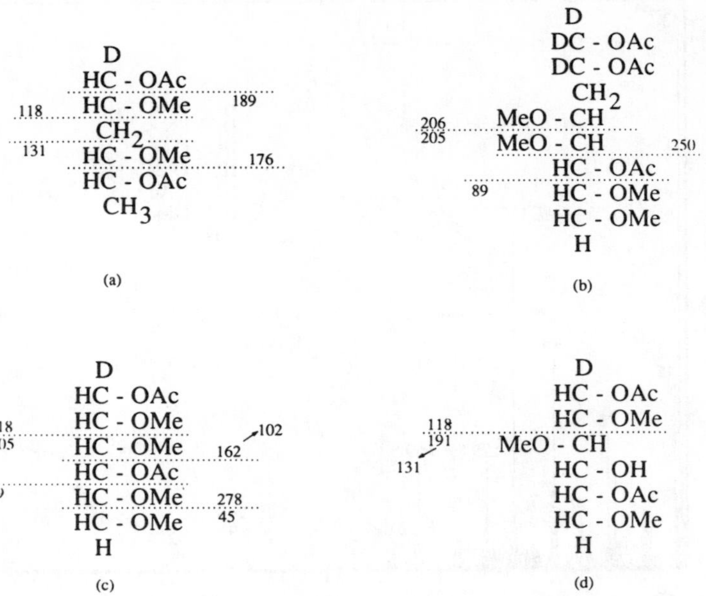

D
HC - OAc
118 HC - OMe 189
CH_2
131 HC - OMe 176
HC - OAc
CH_3

(a)

D
DC - OAc
DC - OAc
CH_2
206 MeO - CH
205 MeO - CH 250
HC - OAc
89 HC - OMe
HC - OMe
H

(b)

D
HC - OAc
118 HC - OMe 102
205 HC - OMe 162
HC - OAc
89 HC - OMe 278
HC - OMe 45
H

(c)

D
HC - OAc
118 HC - OMe
191 MeO - CH
131 HC - OH
HC - OAc
HC - OMe
H

(d)

FIGURE 13. Predicted fragmentation of some rare sugars and an "underacetylated" glycosyl derivative: a, 1,5-di-*O*-acetyl-(1-deuterio)-2,4-di-*O*-methyl 3,6-dideoxyhexitol; b, 1,2,6-tri-*O*-acetyl-(1,1,2-trideuterio)-4,5,7,8-tetra-*O*-methyl 3-deoxyoctitol; c, 1,4-di-*O*-acetyl-(1-deuterio)-2,3,5,6-tetra-*O*-methyl hexitol; d, 1,5-di-*O*-acetyl-(1-deuterio)-2,3,6-tri-*O*-methyl-(4-hydroxy)-hexitol.

former positions of the acetyl groups deduced after separation of the family of permethylated alditol acetates.

5. *Separation of Permethylated Oligosaccharides*

Very early on, linkage analysis was extended from monosaccharides to permethylated oligosaccharides. Di- and trisaccharides released through chemical or enzymic digestion were end reduced and permethylated, and these derivatives separated directly by GLC on lower-polarity phases.[100-102] Deduction of linkage to the reducing terminal permethylated alditol or aminoalditol is straightforward, and the C-1 deuterium introduced permits identification of the otherwise symmetrical 3- and 4-linked hexitol and 2- and 4-linked pentitol.[100-102] Linkage of the penultimate sugar in trisaccharides is sometimes deduced, but here, the lower frequencies of diagnostic fragments and other ambiguities diminish its usefulness. Advance in this area is superseded by stronger advances in fast atom bombardment-mass spectrometry (FAB-MS) where most sequence analysis is extended to much larger oligosaccharides.

L. Quantitation

Calculations of the mole fraction of each constituent sugar, solely from the methylation analysis, are sometimes troublesome because the multitude of steps results in differential losses of some derivatives. Acid hydrolysis of the native polymer, or acetolysis or methanolysis of glycoproteins, and direct preparation of alditol acetates is quantitative, based on response factors determined with pure standard monosaccharides or related polymers. Hence, mole fraction of each sugar expected can be deduced reasonably well. This is particularly important for amino sugars where substantial losses are to be expected.

Similar direct response factors for partially methylated alditol acetates do not exist for lack of adequate standards. Response factors of partially methylated and ethylated alditol

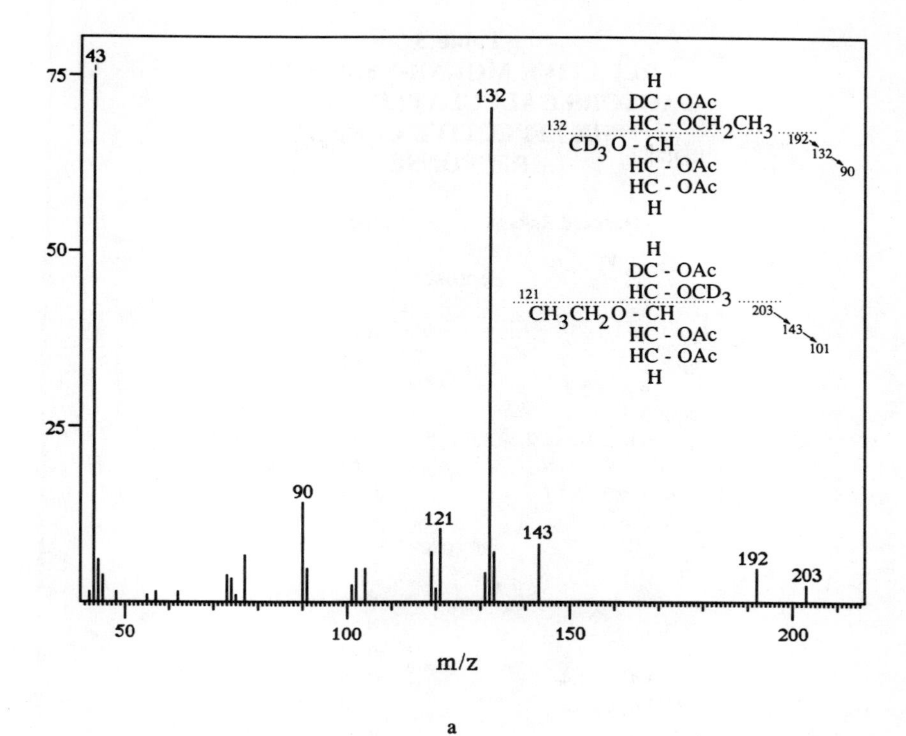

a

b

FIGURE 14. Spectra of partially ethylated and partially deuteriomethylated derivatives: a, Mixture of 1,5-di-*O*-acetyl-(1-deuterio)-2,3,4,6-tetra-*O*-deuteriomethyl hexitol and 1,5-di-*O*-acetyl-(1 deuterio)-2,3,6-tetra-*O*-deuteriomethyl-4-*O*-methyl hexitol; b, Mixture of 1,4,5-tri-*O*-acetyl-(1-deuterio)-2-*O*-ethyl-3-*O*-deuteriomethyl pentitol and 1,4,5-tri-*O*-acetyl-(1-deuterio)-2-*O*-deuteriomethyl-3-*O*-ethyl pentitol.

Table 3
RELATIVE MOLAR-RESPONSE FACTORS CALCULATED BASED ON THE EFFECTIVE CARBON RESPONSE

Deduced linkage	Methyl	Ethyl
Pentoses		
t-*f*	0.60	0.90
t-*p*	0.61	0.91
2- 3- 4-*p* or 5-*f*	0.66	0.86
2- or 3-*f*	0.65	0.85
2,3- 3,4- 2,4-*p* or 2,5- 3,5-*f*	0.70	0.80
2,3-*f*	0.69	0.79
2,3,4-*p* or 2,3,5-*f*	0.75	0.75
Hexose		
t-	0.70	1.10
2- 3- or 4-	0.74	1.03
6-	0.75	1.04
2,3- 3,4- or 2,4-	0.79	0.99
2,6- 3,6- or 4,6-	0.80	1.00
2,3,4-	0.84	0.94
2,3,6- 2,4,6- or 3,4,6-	0.84	0.95
2,3,4,6-	0.89	0.89
6-Deoxyhexose		
t-	0.70	1.00
2- 3- or 4-	0.75	0.95
2,3- 2,4- or 3,4-	0.79	0.89
2,3,4-	0.84	0.84

From Sweet, D. P., Shapiro, R. H., and Albersheim, P., *Carbohydr. Res.*, 40, 217, 1975. With permission.

acetates are calculated by the "effective carbon response"[127] (Table 3), originally based on the predicted ionization potential of organic constituents in a flame ionization detector (FID).[128,129] Using these calculations, one can obtain a reasonable estimation of the proportion of each derivative, but not the relative losses. These effective carbon response factors apply only to (FID) and not total ion detection by MS, as the latter values reflect counts independent of fragment size or potential. Total ion values are used either with or without correction, except perhaps relative to the number of carbons (or number of methyl and acetyl groups) per derivative, since this will reflect the relative number of ions generated. Mole fractions estimated directly from MS should be reasonably close to those calculated from FID, and mole fractions based on summation of all derivatives of each sugar should be consistent with the monosaccharide analysis.

"Undermethylation" can be a persistent problem with certain types of polysaccharides. With experience, a quick glance at the chromatogram will immediately reveal the extent of undermethylation. Certainly, the glycosyl units of most carbohydrates are more often simply linked rather than branched, and each branch point will terminate in an additional nonreducing terminal unit. By nature of a "failed" methylation, partial permethylation induces precipitation of the remainder of the polymer, and hence, unmethylated alditol acetates will predominate in the mixture of derivatives, rather than a wide range of derivatives of apparent branch points. Small, usually equivalent, amounts of virtually all possibilities of derivatives from apparently "doubly" and "triply" branched units are observed, but only rarely does one observe this kind of complexity in nature. The derivatives from the true branch points will predominate, and derivatives representative of the nonreducing terminal sugars will be present in stoichiometric amounts. There is an additional factor to appreciate, however. Handa and Montgomery[130] demonstrated that the methylation of hydroxyl groups of methyl mannoside was not random and proceeded at different rates depending on the method employed; the 3–OH is apparently the most difficult to methylate in the methylsulfinylmethonide ion preparation. Steric hinderance, particularly likely in more highly branched polymers, could also result in undermethylation of specific hydroxyl groups. It is obviously important to add a slightly excess stoichiometric amount of CH_3I. One may sometimes observe that methylation may be reasonably complete, yet amounts of the derivatives from nonreducing terminal sugars are not stoichiometric. There are two possible problems in this case. First, terminal substitutions are generally from more acid-labile groups, such as arabinofuranose, fucose, or rhamnose, and preferential decomposition during hydrolysis may have occurred. Second, these same derivatives are more volatile and can be lost during the evaporations of acid or solvents, particularly if one insists on removal of borate by repeated methanol evaporation after reduction,[77] rather than by direct partitioning.[88]

"Underacetylation" can also be a problem, and just as methylation of certain hydroxyls can be difficult, acetylation can be equally nonrandom. In some preparations we find significant amounts of C4–OH from 2,3,6-methylglucitol (4-linked glucosyl units) remaining unacetylated. The derivative is quite volatile, nevertheless, eluting just before the fully acetylated derivative. The diagnostic m/z 191 and 131 help identify this artifact (Figure 13d).

ACKNOWLEDGMENTS

We thank Anna Olek for the artwork. We also thank Prof. B. A. Stone, Department of Biochemistry, LaTrobe University, Bundoora, Australia, and Prof. J.-P. Joseleau, CNRS, Grenoble, France, for their critical review of the manuscript and many helpful suggestions. Journal paper No. 11,678 of the Purdue University Agricultural Experiment Station.

REFERENCES

1. **Geyer, R., Geyer, H., Kuhnhardt, S., Mink, W., and Stirm, S.,** Methylation analysis of complex carbohydrates in small amounts: capillary gas chromatography-mass fragmentography of methylalditol acetates obtained from *N*-glycosidically linked glycoprotein oligosaccharides, *Anal. Biochem.*, 133, 197, 1983.
2. **Harris, P. J., Henry, R. J., Blakeney, A. B., and Stone, B. A.,** An improved procedure for the methylation analysis of oligosaccharides and polysaccharides, *Carbohydr. Res.*, 127, 59, 1984.
3. **Waege, T. J., Darvill, A. G., McNeil, M., and Albersheim, P.,** Determination, by methylation analysis, of the glycosyl-linkage compositions of microgram quantities of complex carbohydrates, *Carbohydr. Res.*, 123, 281, 1983.
4. **Dutton, G. G. S.,** Applications of gas-liquid chromatography to carbohydrates: Part I., *Adv. Carbohydr. Chem. Biochem.*, 28, 11, 1973.
5. **Dutton, G. G. S.,** Applications of gas-liquid chromatography to carbohydrates: Part II., *Adv. Carbohydr. Chem. Biochem.*, 30, 9, 1974.
6. **Jansson, P.-E., Kenne, L., Liedgren, H., Lindberg, B., and Lonngren, J.,** A practical guide to the methylation analysis of carbohydrates, *Chemical Communication,* No. 8, University of Stockholm, Sweden, 1976.

7. **Lindberg, B.,** Methylation analysis of polysaccharides, *Methods Enzymol.*, 28, 178, 1972.
8. **Lindberg, B. and Lonngren, J.,** Methylation analysis of complex carbohydrates: General procedure and application for sequence analysis, *Methods Enzymol.*, 50, 3, 1978.
9. **Rauvala, H., Finne, J., Krusius, T., Karkainen, J., and Jarnefelt, J.,** Methylation techniques in the structural analysis of glycoproteins and glycolipids, *Adv. Carbohydr. Chem. Biochem.*, 38, 389, 1981.
10. **York, W. S., Darvill, A. G., McNeil, M., Stevenson, T. T., and Albersheim, P.,** Isolation and characterization of plant cell walls and cell wall components, *Methods Enzymol.*, 118, 3, 1986.
11. **Kochetkov, N. K. and Chizhov, O. S.,** Mass spectrometry of carbohydrate derivatives, *Adv. Carbohydr. Chem.*, 21, 39, 1966.
12. **Bjorndal, H., Hellerqvist, C. G., Lindberg, B., and Svensson, S.,** Gas-liquid chromatography and mass spectrometry in methylation analysis of polysaccharides, *Angew. Chem. Int. Ed. Engl.*, 9, 610, 1970.
13. **Lönngren, J. and Svensson, S.,** Mass spectrometry in structural analysis of natural carbohydrates, *Adv. Carbohydr. Chem. Biochem.*, 29, 41, 1974.
14. **Ballou, C. E.,** Alkali-sensitive glycosides, *Adv. Carbohydr. Chem.*, 9, 59, 1954.
15. **Aspinall, G. O., Greenwood, C. T., and Sturgeon, R. J.,** The degradation of xylans by alkali, *J. Chem. Soc.*, 3667, 1962.
16. **Lindberg, B., Lonngren, J., and Thompson, J. L.,** Degradation of polysaccharides containing uronic acid residues, *Carbohydr. Res.*, 28, 351, 1973.
17. **Lindberg, B.,** Structural studies of polysaccharides, *Chem. Soc. Rev.*, 10, 409, 1981.
18. **Aspinall, G. O., Gestetner, B., Molloy, J. A., and Uddin, M.,** Pectic substances from lucerne *(Medicago sativa)*. II. Acidic oligosaccharides from partial hydrolysis of leaf and stem pectic acids, *J. Chem. Soc.*, 2554, 1968.
19. **Axelsson, K. and Bjorndal, H.,** Polysaccharides elaborated by *Fomes annosus* Fr. Cooke. I. A water-soluble acidic polysaccharide from the fruit bodies, *Acta Chem. Scand.*, 24, 713, 1970.
20. **Larm, O., Lindberg, B., Svensson, S., and Kabat, E. A.,** Structural studies on *Pneumococcus* Type II capsular polysaccharide, *Carbohydr. Res.*, 22, 391, 1972.
21. **Taylor, R. L. and Conrad, H. E.,** Stoichiometric depolymerization of polyuronides and glycosaminoglycuronans to monosaccharides following reduction of their carbodiimide-activated carboxyl groups, *Biochemistry*, 11, 1383, 1972.
22. **Anderson, M. A. and Stone, B. A.,** A radiochemical approach to the determination of carboxylic acid groups in polysaccharides, *Carbohydr. Polym.*, 5, 115, 1985.
23. **Blumenkrantz, N. and Asboe-Hansen, G.,** New method for quantitative determination of uronic acids, *Anal. Biochem.*, 54, 484, 1973.
24. **Knutson, C. A. and Jeanes, A.,** A new modification of the carbazole analysis: application to heteropolysaccharides, *Anal. Biochem.*, 24, 470, 1968.
25. **Galambos, J. T.,** The reaction of carbazole with carbohydrates. I. Effect of borate and sulfamate on the carbazole color of sugars, *Anal. Biochem.*, 19, 119, 1967.
26. **Chaudari, A. S., Bishop, C. T., and Dudman, W. F.,** Structural studies on the specific capsular polysaccharide from *Rhizobium trifolii*, TA-1, *Carbohydr. Res.*, 28, 221, 1973.
27. **Kabat, E. A., Bassett, E. W., Pryzwansky, K., Lloyd, K. O., Kaplan, M. E., and Layug, E. J.,** Immunochemical studies on blood groups. XXXIII. The effects of alkaline borohydride and of alkali on blood group A, B, and H substances, *Biochemistry*, 4, 1632, 1965.
28. **Carlson, D. M.,** Oligosaccharides isolated from pig submaxillary mucin, *J. Biol. Chem.*, 241, 2984, 1966.
29. **Iyer, R. N. and Carlson, D. M.,** Alkaline borohydride degradation of blood group H substance, *Arch. Biochem. Biophys.*, 142, 101, 1971.
30. **Tarentino, A. L. and Maley, F.,** Purification and properties of an endo-β-*N*-acetylglucosaminidase from *Streptomyces griseus*, *J. Biol. Chem.*, 249, 811, 1974.
31. **Elder, J. H. and Alexander, S.,** Endo-β-*N*-acetylglucosaminidase F: endoglycosidase from *Flavobacterium meningosepticum* that cleaves both high mannose and complex glycoproteins, *Proc. Natl. Acad. Sci. U.S.A.*, 79, 4540, 1982.
32. **Trimble, R. B. and Maley, F.,** Optimizing hydrolysis of *N*-linked high mannose oligosaccharides by endo-β-*N*-acetylglucosaminidase H, *Anal. Biochem.*, 141, 515, 1984.
33. **Tarentino, A. L. and Maley, F.,** A comparison of the substrate specificities of endo-β-*N*-acetylglucosaminidases from *Streptomyces griseus* and *Diplococcus pneumoniae*, *Biochem. Biophys. Res. Commun.*, 67, 455, 1975.
34. **Plummer, T. H., Jr., Elder, J. H., Alexander, S., Phelan, A. W., and Tarentino, A. L.,** Demonstration of peptide: *N*-glycosidase F activity in endo-β-*N*-acetylglucosaminidase F preparations, *J. Biol. Chem.*, 259, 10700, 1984.
35. **Umemoto, J., Bhavanandan, V. P., and Davidson, E. A.,** Purification and properties of an endo-α-*N*-acetyl-D-galactosaminidase from *Diplococcus pneumoniae*, *J. Biol. Chem.*, 252, 8609, 1977.

36. **Lamblin, G., Lhermitte, M., Klein, A., Roussel, P., van Halbeek, H., and Vliegenthart, J. F. G.,** Carbohydrate chains from human bronchial mucus glycoproteins: a wide spectrum of oligosaccharide structures, *Biochem. Soc. Trans.*, 12, 599, 1984.
37. **Purdie, T. and Irvine, J. C.,** The alkylation of sugars, *J. Chem. Soc.*, 83, 1021, 1903.
38. **Haworth, W. N.,** A new method of preparing alkylated sugars, *J. Chem. Soc.*, 107, 8, 1915.
39. **Kuhn, R., Trischmann, H., and Low, I.,** Zur permethylierung von zuchern und glykosiden, *Angew. Chem.*, 67, 32, 1955.
40. **Walker, H. G., Jr., Gee, M., and McCready, R. M.,** Complete methylation of reducing carbohydrates, *J. Org. Chem.*, 27, 2100, 1962.
41. **Corey, E. J. and Chaykovsky, M.,** Methylsulfinylcarbanion, *J. Am. Chem. Soc.*, 84, 866, 1962.
42. **Price, G. G. and Whiting, M. C.,** Some analytical and preparative uses of the sodium derivative of dimethylsulfoxide, ''dimsylsodium'', *Chem. Ind. (Lond.)*, 775, 1963.
43. **Hakomori, S.-I.,** A rapid permethylation of glycolipid, and polysaccharide catalyzed by methylsulfinyl carbanion in dimethyl sulfoxide, *J. Biochem. (Tokyo)*, 55, 205, 1964.
44. **Anderson, D. M. W. and Cree, G. M.,** Studies on uronic acid materials. XIV. Methylation with the sodium hydride-methyl iodide-dimethyl sulfoxide system, *Carbohydr. Res.*, 2, 162, 1966.
45. **Sandford, P. A. and Conrad, H. E.,** Structure of the *Aerobacter aerogenes* A3(S1) polysaccharide. I. A re-examination using improved procedures for methylation analysis, *Biochemistry*, 5, 1508, 1966.
46. **Steiner, E. C. and Gilbert, J. M.,** The acidities of weak acids in dimethylsulfoxide, *J. Am. Chem. Soc.*, 85, 3054, 1963.
47. **Hiller, L. K., Jr.,** Conductometric titration of weak acids in dimethylsulfoxide using dimsylsodium reagent, *Anal. Chem.*, 42, 30, 1970.
48. **Rauvala, H.,** Use of triphenylmethane as an indicator of complete methylation of glycolipids and glycopeptides, *Carbohydr. Res.*, 72, 257, 1979.
49. **Phillips, L. R. and Fraser, B. A.,** Methylation of carbohydrates with dimsyl potassium in dimethyl sulfoxide, *Carbohydr. Res.*, 90, 149, 1981.
50. **Valent, B. S., Darvill, A. G., McNeil, M., Robertsen, B. K., and Albersheim, P.,** A general and sensitive chemical method for sequencing the glycosyl residues of complex carbohydrates, *Carbohydr. Res.*, 79, 165, 1980.
51. **Exner, J. H. and Steiner, E. C.,** Solvation and ion pairing of alkali-metal alkoxides in dimethyl sulfoxide, *J. Am. Chem. Soc.*, 96, 1782, 1974.
52. **Rose, K., Simona, M. G., and Offord, R. E.,** Amino acid sequence determination by g.l.c. - mass spectrometry of permethylated peptides, *Biochem. J.*, 215, 261, 1983.
53. **Ciucanu, I. and Kerek, F.,** A simple and rapid method for the permethylation of carbohydrates, *Carbohydr. Res.*, 131, 209, 1984.
54. **Finne, J., Krusius, T., and Rauvala, H.,** Use of potassium *tert*-butoxide in the methylation of carbohydrates, *Carbohydr. Res.*, 80, 336, 1980.
55. **Brauman, J. I., Bryson, J. A., Kahl, D. C., and Nelson, N. J.,** Equilibrium acidities in dimethyl sulfoxide, *J. Am. Chem. Soc.*, 92, 6679, 1970.
56. **Blakeney, A. B. and Stone, B. A.,** Methylation of carbohydrates with lithium methylsulphinyl carbanion, *Carbohydr. Res.*, 140, 319, 1985.
57. **Parente, J. P., Cardon, P., Leroy, Y., Montreuil, J., Fournet, B., and Ricart, G.,** A convenient method for methylation of glycoprotein glycans in small amounts by using lithium methylsulfinyl carbanion, *Carbohydr. Res.*, 141, 41, 1985.
58. **Kvernheim, A. L.,** Methylation analysis of polysaccharides with butyllithium in dimethyl-sulfoxide, *Acta Chem. Scand. Ser. B*, 41, 150, 1987.
59. **Itakura, Y. and Komori, T.,** Structure elucidation of two new oligosaccharide sulfates, versicoside B and versicoside C, *Liebigs Ann. Chem.*, 359, 1986.
60. **Mansson, J.-E., Mo, H., Egge, H., and Svennerholm, L.,** Trisialosyllactosylceramide (GT3) is a ganglioside of human lung, *FEBS Lett.*, 196, 259, 1986.
61. **Uchikawa, K., Sekikawa, I., and Azuma, A.,** Structural studies on teichoic acids in cell walls of several serotypes of *Listeria monocytogenes*, *J. Biochem. (Tokyo)*, 99, 315, 1986.
62. **Lomax, J. A., Gordon, A. H., and Chesson, A.,** Methylation of unfractionated, primary and secondary cell-walls of plants, and the location of alkali-labile substituents, *Carbohydr. Res.*, 122, 11, 1983.
63. **Chanzy, H., Dube, M., and Marchessault, R. H.,** Crystallization of cellulose with *N*-methylmorpholine *N*-oxide: a new method for texturing cellulose, *J. Polym. Sci., Polym. Lett. Ed.*, 17, 219, 1979.
64. **El-Kafrawy, A.,** Investigations of the cellulose/LiCl/dimethylacetamide and cellulose/LiCl/*N*-methyl-2-pyrrolidinone solutions by ^{13}C NMR spectroscopy, *J. Appl. Polym. Sci.*, 27, 2435, 1982.
65. **Hudson, S. M. and Cuculo, J. A.,** The solubility of unmodified cellulose: a critique of the literature, *J. Macromol. Sci., Rev. Macromol. Chem.*, 18, 1, 1980.
66. **Isogai, A., Ishizu, A., Nakano, J., Eda, S., and Kato, K.,** A new facile methylation method for cell-wall polysaccharides, *Carbohydr. Res.*, 138, 99, 1985.

67. **Isogai, A., Ishizu, A., and Nakano, J.,** Preparation of tri-*O*-benzylcellulose by the use of nonaqueous cellulose solvents, *J. Appl. Polym. Sci.*, 29, 2097, 1984.
68. **Joseleau, J.-P., Chambat, G., and Chumpitazi-Hermoza, B.,** Solubilization of cellulose and other plant structural polysaccharides in 4-methylmorpholine *N*-oxide: an improved method for the study of cell wall constituents, *Carbohydr. Res.*, 90, 339, 1981.
69. **Narui, T., Takahashi, K., Kobayashi, M., and Shibata, S.,** Permethylation of polysaccharides by a modified Hakomori method, *Carbohydr. Res.*, 103, 293, 1982.
70. **Hellerqvist, C. G., Ruden, U., and Makela, P. H.,** The group C-type modification of the B-type lipopolysaccharide in a hybrid between *Salmonella* groups B and C., *Eur. J. Biochem.*, 25, 96, 1972.
71. **Saier, M. H., Jr. and Ballou, C. E.,** The 6-*O*-methylglucose-containing lipopolysaccharide of *Mycobacterium phlei*, *J. Biol. Chem.*, 243, 4319, 1968.
72. **Talmadge, K. W., Keegstra, K., Bauer, W. D., and Albersheim, P.,** The structure of plant cell walls. I. The macromolecular components of the walls of suspension-cultured sycamore cells with a detailed analysis of pectic polysaccharides, *Plant Physiol.*, 51, 158, 1973.
73. **Mort, A. J., Parker, S., and Kuo, M.-S.,** Recovery of methylated saccharides from methylation reaction mixtures using Sep-Pak® C_{18} cartridges, *Anal. Biochem.*, 133, 380, 1983.
74. **Dubois, M., Gilles, K. A., Hamilton, J. K., Rebers, P. A., and Smith, F.,** Colorimetric method for determination of sugars and related substances, *Anal. Chem.*, 28, 350, 1956.
75. **Conrad, H. E., Bamburg, J. R., Epley, J. D., and Kindt, T. J.,** The structure of the *Aerobacter aerogenes* A3(S1)polysaccharide. II. Sequence analysis and hydrolysis studies, *Biochemistry*, 5, 2808, 1966.
76. **Croon, I., Herrstrom, G., Kull, G., and Lindberg, B.,** Demethylation and degradation of sugars in acid solution, *Acta Chem. Scand.*, 14, 1338, 1960.
77. **Albersheim, P., Nevins, D. J., English, P. D., and Karr, A.,** A method for the analysis of sugars in plant cell-wall polysaccharides by gas-liquid chromatography, *Carbohydr. Res.*, 5, 340, 1967.
78. **Berst, M., Hellerqvist, C. G., Lindberg, B., Luderitz, O., Svensson, S., and Westphal, O.,** Structural investigations on T1 lipopolysaccharides, *Eur. J. Biochem.*, 11, 353, 1969.
79. **Hellerqvist, C. G., Lindberg, B., Svensson, S., Holme, T., and Lindberg, A. A.,** Structural studies on the *O*-specific side-chains of the cell-wall lipopolysaccharide from *Salmonella typhimurium* 395 MS, *Carbohydr. Res.*, 8, 43, 1968.
80. **Hellerqvist, C. G., Lindberg, B., Svensson, S., Holme, T., and Lindberg, A. A.,** Structural studies of the *O*-specific side-chains of the cell-wall lipopolysaccharides from *Salmonella newport* and *Salmonella kentucky*, *Carbohydr. Res.*, 14, 17, 1970.
81. **Stellner, K., Saito, H., and Hakomori, S.-I.,** Determination of aminosugar linkages in glycolipids by methylation. Aminosugar linkages of ceramide pentasaccharides of rabbit erythrocytes and of Forssman antigen, *Arch. Biochem. Biophys.*, 155, 464, 1973.
82. **Caroff, M. and Szabo, L.,** *O*-Demethylation of per-*O*-methylated derivatives of 2-amino-2-deoxy-hexitols during acid hydrolysis and acetolysis, *Carbohydr. Res.*, 84, 43, 1980.
83. **Finne, J. and Rauvala, H.,** Determination (by methylation analysis) of the substitution pattern of 2-amino-2-deoxyhexitols obtained from *O*-glycosidic carbohydrate units of glycoproteins, *Carbohydr. Res.*, 58, 57, 1977.
84. **Hase, S. and Rietschel, E. T.,** Methylation analysis of glucosaminitol and glucosamino-glucosaminitol disaccharides. Formation of 2-deoxy-2-(*N*-acetylacetamido)-glucitol derivatives, *Eur. J. Biochem.*, 63, 93, 1976.
85. **Chambers, R. E. and Clamp, J. R.,** An assessment of methanolysis and other factors used in the analysis of carbohydrate-containing materials, *Biochem. J.*, 125, 1009, 1971.
86. **Chaplin, M. F.,** A rapid and sensitive method for the analysis of carbohydrate components in glycoproteins using gas-liquid chromatography, *Anal. Biochem.*, 123, 336, 1982.
87. **Maltby, D., Carpita, N. C., Montezinos, D., Kulow, C., and Delmer, D. P.,** β-1,3-Glucan in developing cotton fibers: structure, localization, and relationship of synthesis to that of secondary wall cellulose, *Plant Physiol.*, 63, 1158, 1979.
88. **Blakeney, A. B., Harris, P. J., Henry, R. J., and Stone, B. A.,** A simple and rapid preparation of alditol acetates for monosaccharide analysis, *Carbohydr. Res.*, 113, 291, 1983.
89. **Bjorndal, H., Lindberg, B., and Svensson, S.,** Gas-liquid chromatography of partially methylated alditols as their acetates, *Acta Chem. Scand.*, 21, 1801, 1967.
90. **Lönngren, J. and Pilotti, A.,** Gas-liquid chromatography of partially methylated alditols as their acetates II, *Acta Chem. Scand.*, 25, 1144, 1971.
91. **Harris, P. J., Bacic, A., and Clarke, A. E.,** Capillary gas chromatography of partially methylated alditol acetates on a SP-2100 wall-coated open-tubular column, *J. Chromatogr.*, 350, 304, 1985.
92. **Klok, J., Cox, H. C., De Leeuw, J. W., and Schenck, P. A.,** Analysis of synthetic mixtures of partially methylated alditol acetates by capillary gas chromatography, gas chromatography-electron impact mass spectrometry and gas-chemical ionization mass spectrometry, *J. Chromatogr.*, 253, 55, 1982.

93. **Lomax, J. A. and Conchie, J.,** Separation of methylated alditol acetates by glass capillary gas chromatography and their identification by computer, *J. Chromatogr.*, 236, 385, 1982.
94. **Lomax, J. A., Gordon, A. H., and Chesson, A.,** A multiple-column approach to the methylation analysis of plant cell-walls, *Carbohydr. Res.*, 138, 177, 1985.
95. **Shibuya, N.,** Gas-liquid chromatographic analysis of partially methylated alditol acetates on a glass capillary column, *J. Chromatogr.*, 208, 96, 1981.
96. **Grob, K. and Grob, G.,** Capillary columns with immobilized stationary phases. II. Practical advantages and details of procedure, *J. Chromatogr.*, 213, 211, 1981.
97. **Grob, K., Grob, G., and Grob, K., Jr.,** Capillary columns with immobilized stationary phases. I. A new simple preparation procedure, *J. Chromatogr.*, 211, 243, 1981.
98. **Bacic, A., Harris, P. J., Hak, E. W., and Clarke, A. E.,** Capillary gas chromatography of partially methylated alditol acetates on a high-polarity bonded-phase vitreous-silica column, *J. Chromatogr.*, 315, 373, 1984.
99. **Geyer, R., Geyer, H., Kuhnhardt, S., Mink, W., and Stirm, S.,** Capillary gas chromatography of methylhexitol acetates obtained upon methylation of *N*-glycosidically linked glycoprotein oligosaccharides, *Anal. Biochem.*, 121, 263, 1982.
100. **Karkkainen, J.,** Analysis of disaccharides as permethylated disaccharide alditols by gas-liquid chromatography-mass spectrometry, *Carbohydr. Res.*, 14, 27, 1970.
101. **Karkkainen, J.,** Structural analysis of trisaccharides as permethylated trisaccharide alditols by gas-liquid chromatography-mass spectrometry, *Carbohydr. Res.*, 17, 11, 1971.
102. **Mononen, I., Finne, J., and Karkkainen, J.,** Analysis of permethylated hexopyranosyl-2-acetamido-2-deoxyhexitols by g.l.c.-m.s., *Carbohydr. Res.*, 60, 371, 1978.
103. **Henry, R. J., Harris, P. J., Blakeney, A. B., and Stone, B. A.,** Separation of alditol acetates from plasticizers and other contaminants by capillary gas chromatography, *J. Chromatogr.*, 262, 249, 1983.
104. **Laine, R. A.,** Enhancement of detection for partially methylated alditol acetates by chemical ionization mass spectrometry, *Anal. Biochem.*, 116, 383, 1981.
105. **McNeil, M. and Albersheim, P.,** Chemical-ionization mass spectrometry of methylated hexitol acetates, *Carbohydr. Res.*, 56, 239, 1977.
106. **Golovkina, L. S., Chizhov, O. S., and Wulfson, N. S.,** Mass-spektrometritcheskoe issledowanie uglewodow soobshenie 9. Acetaty polilow, *Izv. Akad. Nauk S.S.S.R., Ser. Khim.*, 1915, 1966.
107. **Bjorndal, H., Lindberg, B., and Svensson, S.,** Mass spectrometry of partially methylated alditol acetates, *Carbohydr. Res.*, 5, 433, 1967.
108. **Bjorndal, H., Lindberg, B., Pilotti, A., and Svensson, S.,** Mass spectra of partially methylated alditol acetates. II. Deuterium labelling experiments, *Carbohydr. Res.*, 15, 339, 1970.
109. **Weast, R. C. and Astle, M. J., Eds.,** *Handbook of Chemistry and Physics,* CRC Press, Boca Raton, Fla., 1982, B-257.
110. **Lindberg, B., Lonngren, J., and Thompson, J. L.,** Methylation studies on levans, *Acta Chem. Scand.*, 27, 1819, 1973.
111. **Hancock, R. A., Marshall, K., and Weigel, H.,** Structure of the levan elaborated by *Streptococcus salivarius* Strain 51: an application of chemical ionisation mass-spectrometry, *Carbohydr. Res.*, 49, 351, 1976.
112. **Carpita, N. C., Housley, T. L., Hendrix, J. E., and Sillerud, L. O.,** New features of plant fructans by methylation analysis and carbon-13 n.m.r. spectroscopy, *Carbohyd. Res.*, in press.
113. **DeJongh, D. C.,** Mass spectrometry in carbohydrate chemistry. Acyclic peracetates of pentoses and hexoses, *J. Org. Chem.*, 30, 453, 1965.
114. **Sherman, W. R., Eilers, N. C., and Goodwin, S. L.,** Combined gas chromatography-mass spectrometry of the inositol trimethylsilyl ethers and acetate esters, *Org. Mass Spectrom.*, 3, 829, 1970.
115. **Schwarzmann, G. O. H. and Jeanloz, R. W.,** Separation by gas-liquid chromatography, and identification by mass spectrometry, of the methyl ethers of 2-deoxy-2-(*N*-methylacetamido)-D-glucose, *Carbohydr. Res.*, 34, 161, 1974.
116. **Tai, T., Yamashita, K., and Kobata, A.,** Synthesis and mass fragmentographic analysis of partially *O*-methylated 2-*N*-methylglucosamines, *J. Biochem. (Tokyo)*, 78, 679, 1975.
117. **Hellerqvist, C. G., Lindberg, B., Svensson, S., Holme, T., and Lindberg, A. A.,** Structural studies on the *O*-specific side chains of the cell wall lipopolysaccharides from *Salmonella typhi* and *S. enteritidis*, *Acta Chem. Scand.*, 23, 1588, 1969.
118. **Tacken, A., Brade, H., Unger, F. M., and Charon, D.,** G.l.c.-m.s. of partially methylated and acetylated derivatives of 3-deoxyoctitols, *Carbohydr. Res.*, 149, 263, 1986.
119. **York, W. S., Darvill, A. G., McNeil, M., and Albersheim, P.,** 3-Deoxy-D-*manno*-2-octulosonic acid (KDO) is a component of rhamnogalacturonan II, a pectic polysaccharide in the primary cell walls of plants, *Carbohydr. Res.*, 138, 109, 1985.
120. **Rauvala, H. and Karkkainen, J.,** Methylation analysis of neuraminic acids by gas chromatography-mass spectrometry, *Carbohydr. Res.*, 56, 1, 1977.

121. **Tai, T., Yamashita, K., Ogata-Arakawa, M., Koide, N., Muramatsu, T., Iwashita, S., Inoue, Y., and Kobata, A.,** Structural studies of two ovalbumin glycopeptides in relation to the endo-β-*N*-acetylglucosaminidase specificity, *J. Biol. Chem.*, 250, 8569, 1975.
122. **Lindberg, B., Lonngren, J., and Nimmich, W.,** Structural studies of the *Klebsiella* O Group 9 lipopolysaccharide, *Carbohydr. Res.*, 23, 47, 1972.
123. **Lindberg, B., Lonngren, J., and Thompson, J. L.,** Degradation of polysaccharides containing uronic acid residues, *Carbohydr. Res.*, 28, 351, 1973.
124. **Darvill, J. E., McNeil, M., Darvill, A. G., and Albersheim, P.,** Structure of plant cell walls. XI. Glucuronoarabinoxylan, a second hemicellulose in the primary cell walls of suspension-cultured sycamore cells, *Plant Physiol.*, 66, 1135, 1980.
125. **Carpita, N. C. and Whittern, D.,** A highly substituted glucuronoarabinoxylan from developing maize coleoptiles, *Carbohydr. Res.*, 146, 129, 1986.
126. **DeBelder, A. N. and Norrman, B.,** The distribution of substituents in partially acetylated dextran, *Carbohydr. Res.*, 8, 1, 1968.
127. **Sweet, D. P., Shapiro, R. H., and Albersheim, P.,** Quantitative analysis by various g.l.c. response-factor theories for partially methylated and partially ethylated alditol acetates, *Carbohydr. Res.*, 40, 217, 1975.
128. **Ackman, R. G.,** The flame ionization detector: further comments on molecular breakdown and fundamental group responses, *J. Gas Chromatogr.*, 6, 497, 1968.
129. **Addison, R. F. and Ackman, R. G.,** Flame ionization detector molar responses for methyl esters of some polyfunctional metabolic acids, *J. Gas Chromatogr.*, 6, 135, 1968.
130. **Handa, N. and Montgomery, R.,** Partial methylation of methyl α-D-mannopyranoside. Preparation and distribution of mono-, di-, and tri-methyl ethers of D-mannose, *Carbohydr. Res.*, 11, 467, 1969.
131. **Shea, E. M. and Carpita, N. C.,** Separation of partially methylated alditolacetates on SP-2330 and HP-1 vitreous silica capillary columns, *J. Chromatogr.*, in press.

Chapter 10

FAST ATOM BOMBARDMENT-MASS SPECTROMETRY (FAB-MS): SAMPLE PREPARATION AND ANALYTICAL STRATEGIES

Anne Dell and Jane E. Thomas-Oates

TABLE OF CONTENTS

I. INTRODUCTION

Fast Atom Bombardment (FAB) is a mass spectrometric (MS) ionization technique[1] which is particularly suited for the examination of biomolecules such as carbohydrates. It differs from the more traditional techniques of electron impact (EI) and chemical ionization (CI) in that the sample is ionized directly from solution in a matrix by bombardment with accelerated atoms or ions; thus, it is now possible to obtain good data from polar, nonvolatile, thermally labile compounds which were traditionally difficult to analyze by MS. In the FAB process, samples are ionized by the addition of a proton or cation (positive ion mode) or loss of a proton (negative ion mode) giving pseudomolecular ions as major signals in the spectrum. These define the composition of the sample in terms of the number of hexosyl, deoxyhexosyl, etc. residues. During ionization, some internal energy is imparted to the molecule resulting in fragmentation at some of the glycosidic linkages. The spectrum of fragment ions produced reflects the sequence of the residues in the molecule. The amount of unambiguous sequence information which can be derived in a FAB experiment varies according to the type of sample examined, and in this chapter, we discuss the most appropriate derivatization strategies for optimizing sequence data.

II. FAST ATOM BOMBARDMENT-MASS SPECTROMETRY (FAB-MS) PRINCIPLES AND PRACTICE

In the FAB experiment, an accelerated beam of xenon atoms is fired from an atom gun towards a small metal target which has previously been loaded with a viscous liquid, referred to as the matrix, containing the sample to be analyzed. When the atom beam collides with the matrix, kinetic energy is transferred to the surface molecules, many of which are sputtered out of the liquid into the high vacuum of the ion source. A significant number of these molecules is ionized during the sputtering process. Thus, gas-phase ions are generated without prior volatilization of the sample. The sputtered ions, which include both molecular and fragment ions, are accelerated from the ion source and analyzed in the same manner as for EI- or CI-MS.

The signal received at the collector may be recorded either by use of an oscillographic recorder on UV or heat-sensitive paper, or through a data system. Each method has its advantages and drawbacks and the individual FAB-MS user normally prefers one or the other system. The oscillograph records exactly the electrical signals received at the collector during any individual scan, so that peak shapes, spikes, and background noise are all faithfully recorded for that scan. In contrast, the data system is set up to reject any signals falling below a preset intensity and it presents the spectrum as a series of lines, so that the original peak shapes are lost. With computer acquisition, several scans are acquired and these may be averaged. Random noise or electrical spikes are not seen in the spectrum. Using computer averaging techniques, better data can usually be obtained from low-abundance, high-mass ions than can be achieved by recording a single slow scan on chart paper.

Mass assignment of peaks on an oscillographic recording is accomplished by manually counting up through the spectrum from the low mass end. This is made possible by the fact that FAB spectra are characterized by a background in which there is a peak at every mass. This background is generated by side reactions occurring during the FAB-MS experiment in which matrix and sample molecules break down and ionize randomly. This method gives a mass assignment which constitutes the nominal mass of the sample (i.e., based on C = 12, H = 1, O = 16, etc.) rather than its accurate mass (based on C = 12.00, H = 1.01, O = 15.99); the nominal mass obtained for the "average" carbohydrate works out at about half an amu less than its accurate mass for every 1000 mass units. In contrast, mass assignments obtained through the data system are accurate masses, which the computer

Table 1
USEFUL MATRICES AND MATRIX ADDITIVES FOR FAB-MS

Matrix/additive	Applications	Comments
Glycerol	Broad applications; especially good for polar compounds	Normally try glycerol alone before trying in a mixture with thioglycerol for unknowns
Glycerol: thioglycerol	Very effective for hydrophobic samples; e.g., some native, most permethylated and peracetylated derivatives	Thioglycerol can be used on its own, but glycerol enhances spectral lifetime because of its lower volatility
Thioglycerol	Glycolipids	Ref. 11
Dilute HCl	Broad applications	In positive, improves carbohydrates with basic groups; in negative, protonates additional carboxyl groups in molecules with multiple negative charges
Sodium acetate	Glycosphingolipids	0.1% solution in methanol added to target before matrix gives strong $(M + Na^+)^+$ molecular ions and fragment ions usually weaker than from $(M + H)^+$ ions
NH_4SCN	Permethylated oligosaccharides	1 $\mu\ell$ 100 m*M* NH_4^+ gives good cationization when no amino sugars present; the SCN^- helps negative spectra and probably assists desorption as a chaotropic agent

assigns by comparison with the spectrum and known accurate masses of a calibration standard, such as CsI, which gives cluster ions spaced every 260 amu up to at least mass 15,000.

A typical procedure for running a sample is as follows. The sample (1 to 50 μg depending on the quantity available) is dissolved in 10 $\mu\ell$ of 5% aqueous acetic acid (polar samples) or methanol (nonpolar samples), and a 1 $\mu\ell$ aliquot is removed using a graduated 5 $\mu\ell$ micropipette (e.g., Supracaps) and gently blown into 1 to 2 $\mu\ell$ of the matrix (glycerol, thioglycerol, etc. see Table 1) previously placed on the target. The probe carrying the loaded target is then introduced into the ion source through a vacuum lock system. The solvent evaporates in the vacuum and sometimes causes the matrix to coevaporate, so that it is useful to withdraw the probe again after going through the vacuum lock to check that sufficient matrix is still present after evaporation. If necessary, extra matrix should be added using a micropipette prior to reinserting the probe.

III. WHAT A FAB SPECTRUM LOOKS LIKE

Signals originating from the sample are observed as clusters superimposed on the background. Clusters rather than single peaks, e.g.,

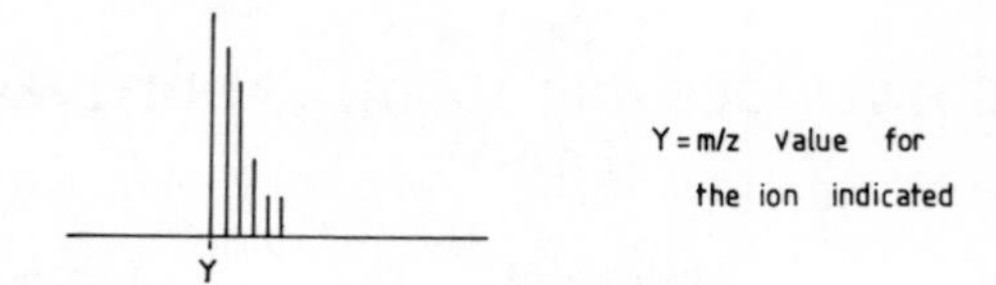

are present for two reasons. First, the isotope contribution of ^{13}C (1.1% natural abundance) means that for every carbon atom in the molecule there is a 1.1% chance that it will be a ^{13}C atom and so have an atomic mass of 13 instead of 12. The relative intensity of the different peaks in the cluster reflects this probability distribution, i.e., the height of the signal at m/z Y + 1 reflects the probability of finding one ^{13}C atom in the molecule, while that of the signal at m/z Y + 2 reflects the probability of there being two ^{13}C atoms, etc. As the number of carbon atoms in the molecule increases, so does the chance of including at least one ^{13}C atom, until by the time the molecule contains more than 90 carbon atoms it is "certain" (90 × 1.1% = 99%) that there will be one ^{13}C atom and so the isotope peak at m/z Y + 1 will be the most intense peak in the cluster:

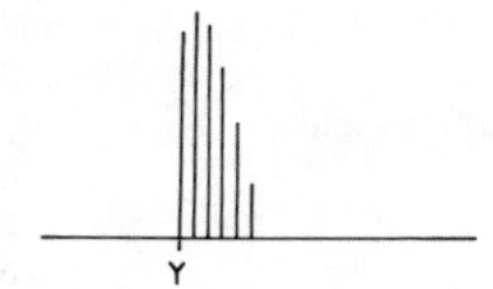

Thus, the shape of a signal cluster is characteristic of the molecular mass (i.e., the number of carbon atoms) of the ion from which it arises — the higher the mass of the cluster, the more intense the isotope peak may be expected to be. The experienced operator can glean useful information from an isotope pattern which is not that normally found within a certain mass range.

The second reason for observing a cluster of peaks is the generation of minor byproducts at the same time as generating the major ion. These byproducts are always less intense than the major ion, but contribute to the shape of the cluster, e.g., underivatized carbohydrates frequently oxidize to yield an ion two mass units lower than the major signal.

The individual signals in a cluster will be observed when the instrumental resolution is set to resolve nominal masses. At high mass (e.g., greater than 4000) it may be necessary to optimize sensitivity at the expense of resolution and unresolved signals are then observed as envelopes. The centroid of the envelope corresponds to the chemical mass of the ion. Its value is ascertained either by mass marking against a calibrated mass marker on UV chart paper or by computer calibration. In both cases the alkali halides (CsI, RbI, KI) are normally chosen as reference compounds.

With some experience, a quick glance at the spectrum, even before it has been counted, can give useful information on how well the sample is working, e.g., a weak or irregular background above about m/z 900 suggests that it is improbable that molecular ions will be observed above this mass, while an intense and regular series of background ions well above m/z 2000 indicates either high mass sample ions, or that impurities are present which are giving rise to the high background. The matrix does not normally give significant background above m/z 900.

Strategies for improving or changing the appearance of a spectrum include the use of different matrices and/or matrix additives. The choice of matrix depends on the type of sample or derivative being run (Table 1). For example, permethylated samples give very poor data when glycerol alone is used as the matrix, while affording excellent spectra from thioglycerol or a 1:1 mixture of glycerol and thioglycerol. It is often advantageous to use the mixture even for samples which work well in glycerol alone in order to simplify the spectrum of matrix signals. Glycerol alone gives strong $[xM + H]^+$ cluster ions at 92 amu intervals, sometimes up to about m/z 1500, depending on the sample. The addition of thioglycerol suppresses all cluster ions above about m/z 400. This can be an advantage for the rate cases when a sample ion corresponds in mass to one of the $[xM + H]^+$ ions.

The use of matrix additives can be a very powerful means of altering the appearance of the data, e.g., the addition of 1 $\mu\ell$ of 0.1 *M* HCl solution directly to the matrix while running a polyanionic species can be valuable in displacing cations with protons and so reduce the number of molecular species present, simplifying the spectrum. A summary of the most useful matrices and matrix additives which can be employed in various applications is given in Table 1.

IV. OBTAINING FAB DATA

FAB-MS data can be obtained on native carbohydrate samples or on their derivatives. Native samples will give good data as long as they are free from salts and other surface active contaminants, such as detergents (surfactants occupy the matrix surface from which desorption takes place, displacing the sample and preventing it from ionizing).

The derivatives commonly used in FAB-MS of carbohydrates tend to be those which are used in classical carbohydrate chemical analysis, such as the permethyl and the peracetyl derivatives.

FAB-MS analysis can, therefore, fit very conveniently into a preexisting analytical scheme, extending and complementing the classical data. Having said this, it should be pointed out that new or adapted protocols have been introduced for preparing derivatives specifically for FAB-MS. These are faster and/or cleaner and/or more convenient to carry out on a practical level, and produce derivatives which are also suitable for the classical procedures.

A good example of where a modification constitutes an improvement over traditional protocols is the peracetylation reaction. Instead of using the usual pyridine/acetic anhydride recipe, a reaction which gives byproducts which suppress FAB spectra, we have found it much more convenient to use a modification[2] of the method of Bourne et al.[3] This involves adding 200 $\mu\ell$ of a 2:1 (v/v) mixture of trifluoroacetic anhydride/glacial acetic acid to the native carbohydrate, letting this stand at room temperature for 10 min and then removing the volatile reagents under a stream of nitrogen. This protocol, by avoiding the use of pyridine and/or heat, yields a derivative which is clean enough to run immediately, without purification unless the native sample was contaminated with salt. Peracetylation can be carried out successfully even in the presence of salt and converts the aqueous soluble carbohydrate into an organic-soluble derivative. The derivative can be very conveniently separated from any salt by partition between chloroform and water: the derivative is dissolved in 1 mℓ of chloroform, which is washed three or four times in 2 mℓ of water to remove the salt, before being dried down and analyzed by FAB-MS.

It should be noted that this acetylation protocol does not remove base labile groups,[4] such as *O*-acetyl esters, from the molecule. It has, however, been shown to hydrolyze a very acid labile linkage[27] and so caution should be exercised in choosing this acetylation method if the molecule is thought to contain any particularly acid labile linkages.

Permethylation techniques are ubiquitous in carbohydrate analysis and are almost always based on Hakomori's classical method[5] with a few notable exceptions.[6] Each lab has its own variations on the basic method, but all such variations are matters of individual convenience and preference, and generally produce comparable results. Purification techniques vary much more widely than do the reaction conditions. A particularly nice cleanup procedure, using Sep-Pak® cartridges was developed by Waeghe et al,[7] for producing derivatives pure enough to obtain data on low microgram quantities of sample. This protocol is a convenient way to yield permethylated carbohydrates clean enough to give good FAB data.[2] The permethylation and purification protocols used in our laboratory are based on Hakomori's and Waeghe's methods and are routinely used to permethylate carbohydrates and glycopeptides for FAB-MS analysis.

Permethylation is achieved by a double methylation modification of the Hakomori method. The dimsyl sodium base is prepared by heating sodium hydride in dimethyl sufoxide (DMSO) (50 μg/μℓ) at 90°C for 20 min, followed by cooling and centrifugation (20 min at 3000 rpm). The sample is dissolved in 100 μℓ of dry DMSO in a stoppered tube and 300 μℓ of base added. After 75 sec, 50 μℓ of methyl iodide is added and the reaction allowed to proceed at room temperature for 10 min. An additional 500 μℓ of base is then added, followed by an additional 500 μℓ of methyl iodide and the reaction continued for an additional 20 min at room temperature. The reaction is terminated by the addition of about 2 mℓ of water — the resulting mixture is concentrated under a stream of nitrogen and then dried thoroughly by lyophilization. Purification of the products is then achieved on a C18 Sep-Pak® (Waters, Millford, Mass.) cartridge after the method of Waeghe.[7] The Sep-Pak® is conditioned using 5 mℓ of water, 5 mℓ of ethanol, 5 mℓ of acetonitrile, and a final 10 mℓ of water. The permethylated sample is dissolved in 200 μℓ of 1:1 (v/v) methanol/water, loaded onto the Sep-Pak® and the Sep-Pak® is then washed with 3 mℓ of water to elute salts. The sample is eluted using 1.5 mℓ each of 15% aqueous acetonitrile, 50% aqueous acetonitrile, 75% aqueous acetonitrile, and pure acetonitrile, each fraction being collected separately and dried down under vacuum. The permethylated products are generally recovered in the 75% wash. Details and safety precautions of methylation reactions may be found in Chapter 9.

For GC-MS studies, oligosaccharides are normally reduced prior to preparation of the peracetyl and permethyl derivatives. Although FAB-MS will give perfectly good information on the reduced derivatized sample, it works equally well on unreduced derivatives. Thus, if derivatization is being performed solely for the purposes of obtaining good FAB-MS data, rather than as part of a concerted GC-MS and FAB-MS strategy, then reduction is an unnecessary step.

Less commonly used derivatives which have been applied to specific problems are summarized, with appropriate experimental details, in Table 2.

Because FAB-MS data can be obtained rapidly from samples in solution, it is an ideal technique for monitoring the progress of chemical and enzymatic degradation reactions while they are being carried out. A variety of hydrolyses are amenable to monitoring by FAB-MS, including aqueous acid hydrolysis, methanolysis, and acetolysis. A suitable procedure for aqueous acid hydrolysis or methanolysis is as follows: the sample (1 to 10 μg) is dried down in a small Reacti-vial. Approximately 20 μℓ of the reagent (typically 0.1 *M* aqueous H_2SO_4, or 0.1 *M* HCl in methanol) is added and the mixture is incubated at 40 to 60°C for a time course. At suitable intervals (e.g., 5 min, 30 min, 2 hr, etc.), a 1 μℓ sample is removed from the reaction mixture and loaded directly into the appropriate matrix for FAB-MS analysis. With the use of a small heating block, these reactions can be conveniently carried out on the machine console and the data obtained immediately. Sample aliquots from acetolysis do require a very simple workup before the FAB-MS data can be obtained, but this can be completed in a matter of minutes. The acetolysis reaction is quenched in water and then the products are extracted into chloroform or dichloromethane, the organic layer is washed two or three times with water, and the organic solvent is removed under a stream of nitrogen. The sample is then taken up and loaded into the matrix for immediate analysis. The acetolysis reaction is quenched in water and then the products are extracted into chloroform or dichloromethane, the organic layer is washed two or three times with water, and the organic solvent is removed under a stream of nitrogen. The sample is then taken up and loaded into the matrix for immediate analysis.

Table 2
LESS COMMONLY USED DERIVATIVES FOR FAB-MS OF CARBOHYDRATES

Derivative	Application	Recipe	Data
Methyl ester	Confirmation of presence of carboxyls; introduction of ^2H label	Solid sample is mixed with 200 μℓ 1 *M* methanolic HCl incubated at room temp. 15 min	Positive FAB obtained directly on an aliquot of the reaction mixture; add mass increment of 14 amu (17 for ^2H) for every Me-esterified COOH, 1:1 CH_3OH: C^2H_3OH may help in interpretation of spectra
Pentafluorobenzyloxime	Labeling of reducing end; enhancement of sensitivity in negative	Solid sample is mixed with 200 μℓ 30 mg/mℓ PFB hydroxylamine ·HCl in anhydrous pyridine; incubated 2 hr at 100°C	Negative FAB directly on an aliquot of reaction mixture, or after drying down and redissolving; reducing end residue incorporates 195 amu; allows reducing end to be distinguished
N-Acetylation	Labeling of free primary amino groups; incorporation of ^2H label; especially useful in glycopeptide studies	Solid sample dissolved in 100 μℓ water, 500 μℓ 4:1 MeOH: acetic anhydride added; incubated room temp. 10 to 60 min	FAB after drying and redissolving sample; 42 amu incorporated for every primary amino group; 1:1 H:^2H acetic anhydride may assist spectral interpretation
Reductive amination	Incorporation UV absorbing group; labels reducing end; increases sensitivity in positive	Up to 1 mg of solid sample dissolved in 10 μℓ water + 40 μℓ reagent (1 mmol aniline or analogue, 35 mg $NaBH_4CN$, 41 μℓ glacial acetic acid + 350 μℓ MeOH) incubated 80°C, 30 min; products isolated in organic solvent[21]	Incorporation of relevant mass increment (77, 119, 150 amu) at reducing end
N-(2-pyridinyl)-glycosylamine	Fluorescent labeling of reducing end; improved sensitivity in positive FAB; improved HPLC chromatographic and detection properties	Solid sample dissolved in water with 100-fold excess of 2-aminopyridine (pH 8.0); incubated 65°C, 3 hr or more[22]	FAB of HPLC fractions; 78 amu increment incorporated at reducing end

suitable for enzyme activity, then the digest may be carried out in a salt buffer and aliquots from the reaction are dried down, acetylated using the recipe given earlier, and dried down under nitrogen. The acetylated sample is taken up into chloroform and the water-soluble impurities are removed by washing with water. FAB-MS analyses are then carried out on the acetylated enzyme-released carbohydrates.

V. STRATEGIES FOR SAMPLE HANDLING

A. Information from Native Samples

The way in which a sample is analyzed depends on the nature of the compound, how it has been prepared, and what sort of data is required from it. Almost always the first approach is to obtain, or attempt to obtain, information on the native sample, using both positive and negative FAB-MS.

The spectrum obtained from a native sample primarily contains molecular weight information in the form of one or more pseudomolecular ions — the most commonly observed of these are $(M + H)^+$, $(M + NH_4)^+$, $(M + Na)^+$, and $(M + K)^+$, with masses corresponding to M + 1, M+ 18, M + 23, and M + 39, respectively, in positive and $(M - H)^-$ and $(M + Cl)^-$, with masses for M − 1 and M + 35 and 37 in negative. Other ions, which may be observed depending on the nature and quantity of the sample, correspond to adducts formed between sample molecules, or between sample and matrix molecules, e.g., $(2M + H)^+$, the ionized dimer, and $(M + G + H)^+$, an ionized sample/matrix adduct.

The molecular weight data derived from these molecular ions provides information on the composition of the molecule (for residue masses see Table 4) and usually reveals the presence of any additional noncarbohydrate components such as amino acid residues, phosphoryl, acetyl, formyl, or other groups, which would not necessarily be identified in classical procedures. The native FAB-MS data will also reveal any heterogeneity in a sample, such as contamination of a pentasaccharide preparation with tetra- or hexasaccharides, or a variation in the degree of phosphorylation or acetylation between components of the sample, etc.

B. Information from Derivatives

If there is sufficient sample, fragment ions may also be seen in the native spectrum along with the pseudomolecular ions. These fragments are generated by breakdown of the molecules in the ion source, and they reflect the structure of the sample. A summary of the fragmentation pathways, common to all classes of saccharide-containing compounds, is given in Table 3.

Although fragment ions from a native sample reflect its structure, they may not allow unambiguous determination of sequence. This is because

1. The structures of reducing and nonreducing-end fragments are analogous, i.e., pathways B and C in Table 3 cannot be distinguished, unless the oligosaccharide is reduced or is a lyase-type product.
2. More than one glycosidic linkage can be cleaved (“double cleavage”), resulting in fragment ions which contain neither the original reducing end nor the original nonreducing end of the oligosaccharide, while having the same masses as reducing or nonreducing end ions.

In order to determine the sequence of residues unambiguously, the sample must be derivatized prior to FAB-MS analysis. Derivatization solves the problem of double cleavage ions since internal cleavages will give nonreducing ends with free hydroxyl groups. Derivatization also enhances, and in many cases directs, cleavage so that a specific type of fragment ion is reliably produced, permitting unambiguous sequencing. The residue masses for the permethyl and peracetyl derivatives are given in Table 4. These derivatives are the ones recommended for sequencing purposes. The choice of a derivative for a particular application, and the type of data which may be obtained using it, are discussed later.

The abundance of fragment ions can sometimes be enhanced by the use of collision gases. This type of MS is referred to as MS/MS and the interested reader should consult reviews for further information.[8] In our experience, sufficient fragmentation data is normally present in the conventional FAB spectra of suitable derivatives to permit reliable sequencing. There is, as yet, no published data which indicates that MS/MS can provide information not available from simple experiments of the type described in the applications below. However, if MS/MS instrumentation is available, it should be used to provide confirmatory sequence data; it seems apparent that more valuable data would be obtained from MS/MS of permethylated and peracetylated rather than native samples, given the characteristic and unambiguous fragmentation data known to be generated from these derivatives.

Table 3
PATHWAYS OF FRAGMENTATION OF OLIGOSACCHARIDES IN FAB-MS

	COMMENTS
SCHEME A 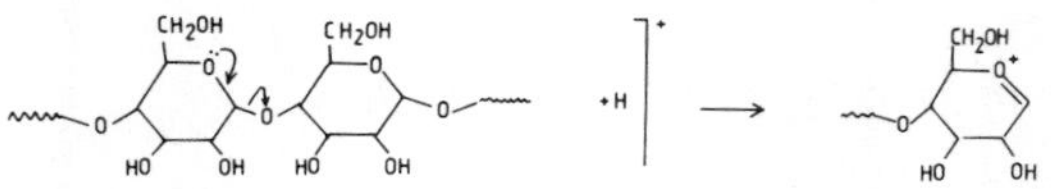	Charge retained on nonreducing fragment; positive mode only; "A-type" cleavage
SCHEME B 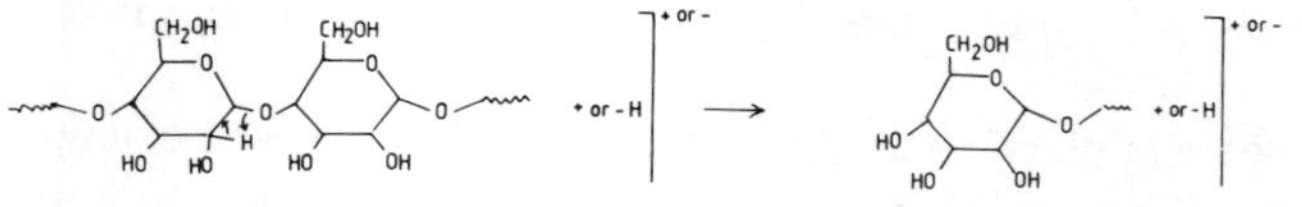	Charge retained on reducing end fragment; positive and negative modes; "β-cleavage"
SCHEME C 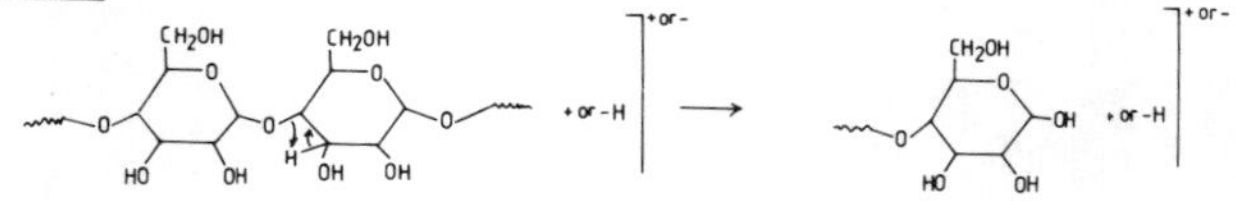	Charge retained on nonreducing end; positive and negative modes
SCHEME D 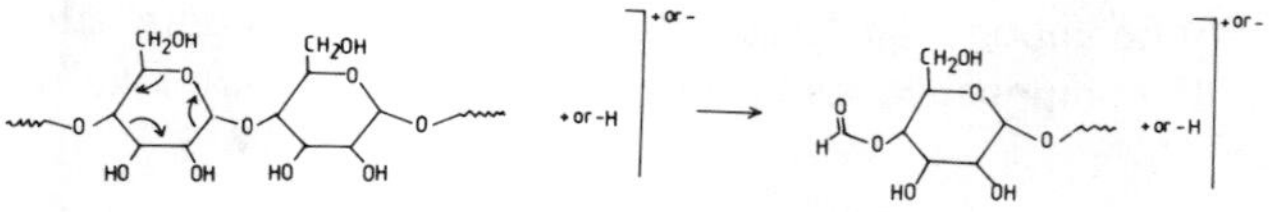	Charge retained on reducing fragment; 28 amu heavier than analogous Pathway B ion; positive and negative modes
SCHEME E 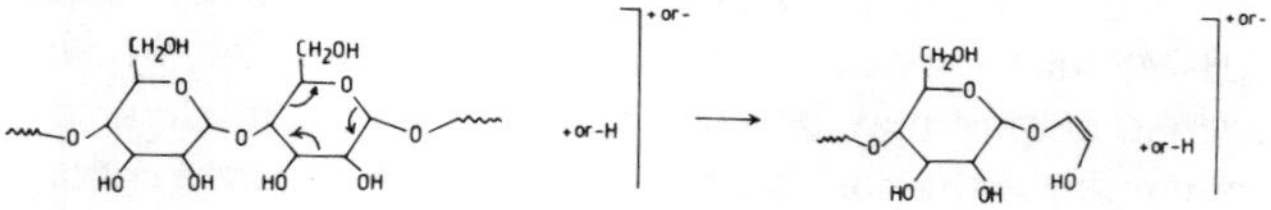	Charge retained on nonreducing fragment; 42 amu heavier than Pathway C ions; major pathway in negative; rarely in positive

After Dell, A., *Adv. Carbohydr. Chem. Biochem.*, 45, 19, 1987.

Table 4
INCREMENTAL RESIDUE MASSES FOR THE COMMONLY FOUND RESIDUES AND THEIR ROUTINE DERIVATIVES

Residue	Native	Permethyl	Peracetyl	^{2}H-acetyl
Hex	162	204	288	297
Deoxyhex	146	174	230	236
HexNAc	203	245	287	293
HexU	176	218	260	266
NANA	291	361	417	426
Pent	132	160	216	222

Note: Figures are in atomic mass units (amu).

VI. LIMITATIONS OF FAB-MS

It is appropriate that FAB-MS and traditional gas chromatography-mass spectrometry (GC-MS) techniques appear side by side in this book since the two methods of analysis give complementary data and involve the use of the same types of derivatives. FAB-MS does not differentiate carbohydrate isomers while GC-MS does; linkage data are not usually provided by FAB-MS, while methylation analysis solves this. FAB-MS does not give anomeric information directly, but enzymic experiments for defining anomeric configuration can be assisted by both FAB-MS and GC-MS.

The remainder of this chapter illustrates the strengths of FAB-MS in carbohydrate analysis, including its sensitivity, speed, ease of execution, and its ability to yield unique data which no other technique can provide.

VII. STRATEGIES FOR SOLVING PROBLEMS

The key to solving a particular structural problem lies in choosing the appropriate strategy for its analysis, i.e., knowing which experiments to perform to obtain the most information from the sample available.

In order to best present a description of handling strategies, we have chosen examples of five general structural problems, and we discuss the most appropriate ways to approach them.

A. Small Oligosaccharides

1. Background Information

A sample is believed to contain an oligosaccharide(s) of molecular weight about 2000 daltons. The oligosaccharide may be endogenously acetylated, contaminated with salt, and may contain other oligosaccharide components. Only microgram quantities are available.

2. Approach

The initial approach is to dissolve the sample in 10 μℓ aqueous solvent, usually 5% acetic acid, and remove 1 μℓ for FAB-MS in the positive ion mode. Depending on whether or not the sample works well, a second aliquot may be removed for negative mode analysis.

If there is not too much salt contamination, a positive spectrum will be obtained containing one or more pseudomolecular ions and possibly also some fragment ions. Typically, a pair of pseudomolecular ions will be observed for each component — one for $(M + H)^+$ and the other either for $(M + NH_4)^+$ or $(M + Na)^+$. Heterogeneity in the preparation may be seen at this point, with pseudomolecular ions showing up for more than one component. The molecular weights of the components allow their compositions to be assigned, so that the carbohydrate residues and any other groups present are identified. It should be noted here that FAB-MS can only give information in terms of Hex, HexNAc, HexU, etc. and cannot distinguish isomers. Assignments of glucose, galactose, and mannose, for example, must be made from other data such as GC-MS.

Fragment ions observed in the spectrum are generally distinguishable from molecular ions since they are of lower intensity and are also less likely to be cationized. They give some information, including an indication of how much sample is present, since it is rare to see significant fragmentation of a native sample when less than 1 to 5 μg is loaded. As discussed above, unambiguous sequence data are frequently not gleaned from any fragmentation seen in the native spectrum. Thus, regardless of whatever happens during the experiment on the native sample, the next step is to make the peracetyl derivative of a further 1 to 2 μℓ dried down from the original solution of the native carbohydrate. The purpose of this experiment is to provide sequence data, and, if necessary, to remove salt from the sample. If the FAB-MS data on the native sample or prior experiments suggested that the oligosaccharide(s) contains endogenous acetyl groups, then, instead of acetylating using glacial acetic acid, the deuterio form should be substituted. The endogenous acetyl group will then be distinguishable by mass from introduced deuterioacetyl groups.

Both peracetyl and perdeuterioacetyl derivatives give intense molecular ions and also yield abundant fragment ions via pathway A (Table 3), giving sequence data from the nonreducing terminus right up to the molecular ion. These A-type fragment ions, in addition to allowing the sequence of the residues in the oligosaccharide to be identified, also permit the location of branch points and endogenous substituents to be determined. A residue containing a branch point or an endogenous acetyl group among deuterioacetyl groups has a distinct mass which allows it to be identified. When *N*-acetylhexosamine residues are present, ions formed by cleavage of the glycoside linkage on the reducing side of the amino sugar are highly favored. Thus, when an oligosaccharide contains *N*-acetylhexosamine residues, fragmentation occurs predominantly at these residues.

It is fortunate that, despite the large increase in residue mass resulting from peracetylation (peracetyl Hex = 288 amu, *cf.* native Hex = 162 amu) there is such a very great improvement in sensitivity (about 50-fold over that of the native oligosaccharide) that good data should be obtainable on derivatizing 1 to 2 μg and loading about 1/10 of the derivative. This improvement in sensitivity can frequently allow additional components to be identified in a mixture when they are present in such low abundance as to go unobserved in the spectrum of the native carbohydrate.

One possible ambiguity can arise in the interpretation of spectra of peracetylated derivatives and this originates from the residue masses of peracetyl Hex (288 amu) and peracetyl HexNAc (287 amu) being separated by only one mass unit. In a heterogeneous mixture of oligosaccharides containing both Hex and HexNAc residues, it is not always possible to make a firm identification of all components of the mixture, for example, $Hex_4HexNAc$ (m/z 1542) in the presence of an abundant $Hex_3HexNAc_2$ signal (m/z 1541). At this mass, the ^{13}C isotope signal for the lower mass ion is significant, so that a small contribution from the higher mass $Hex_4HexNAc$, superimposed on the isotope peak, may go undetected. In such a case, permethylation of the remainder of the acetylated sample resolves the ambiguity. Permethylation using standard techniques, as given above, will displace all the *O*-acetyl groups and replace them with *O*-methyl ethers, while leaving *N*-acetyl groups intact and mono-*N*-methylating. The mass difference between permethyl $Hex_4HexNAc$ (m/z 1108) and permethyl $Hex_3HexNAc_2$ (m/z 1149) is now 41 amu, so that the presence of the di-HexNAc component can be confirmed positively. The fragmentation which occurs in permethylated samples also follows pathway A, and the same selectivitiy for cleavage after HexNAc residues is observed. The HexNAc cleavage is often exclusive in the case of permethylated samples, especially of larger oligosaccharides, while low abundance fragmentation at other sites may be found with the peracetylated derivative.

If sequence ambiguities remain after interpretation of the spectra of peracetylated and permethylated derivatives (these ambiguities normally arise because of the limited cleavages at Hex residues), then these can normally be resolved by FAB-MS analysis of degradation products of the permethylated sample. These products are best prepared by acid hydrolysis of the permethylated sample. Methanolysis is a mild means of accomplishing this while aqueous acid provides a more forceful approach. Whichever method is chosen, the progress of the hydrolysis can be monitored directly by FAB-MS during the reaction (see Section IV) so that a picture can be built up of the products at every time point. This is a fast and convenient way to carry out a degradation, and has the advantage of being economical on sample since the reaction is being monitored as it proceeds, rather than having to guess at how long to hydrolyze and make a suitable number of time points "blindly". Hydrolysis or methanolysis of each glycosidic linkage yields a free hydroxyl group on the newly generated nonreducing end. Thus, the number of free hydroxyl groups in each hydrolytic fragment defines the extent of branching or substitution.

3. Summary of Data Obtained

This complete procedure involving analysis of the native sample, followed by acetylation, and then permethylation and hydrolysis, can be carried out on as little as 20 μg of native carbohydrate to give full sequence information, including the location of substituents and branch points. The methods are applicable to mixtures, but all components may not be sequenced if the mixture is complex or if relative amounts of each component vary significantly. High-performance liquid chromatography (HPLC) should be used in these cases to separate the permethylated or peracetylated components (see later).

B. Large Glycopeptides or Polysaccharides

1. Background Information

A sample contains a very large (5,000 to 50,000 amu) glycopeptide or polysaccharide, typically of the lactosaminoglycan type. The structures are likely to be branched and to be internally heterogeneous.

2. Approach

The compound is so large that FAB-MS on the native material is unlikely to be successful. Therefore, a suitable strategy begins with the permethyl derivative, which, after purification (see above for protocol) is subjected to FAB-MS. Despite the very high mass of the native compound, now made even higher on permethylation, significant data may be obtained from the spectra of fragment ions. The molecular ions are well outside the mass range of most instruments and no attempt is made to record them. As shown by our studies on the Band 3 glycoprotein of red cell surfaces,[9] this type of permethylated molecule gives intense fragment ions in the 100 to 3000 mass range corresponding to cleavage after every HexNAc residue (pathway A, Table 3) present towards the nonreducing ends of branches. The pattern of fragment ions allows a map of all nonreducing terminal structures to be built up. This means that a branched structure containing a variety of different antigenic structures will give fragments from every branch, so that all the major antigenic structures present may be identified.

At the low mass end of the spectrum, diagnostic HexNAc cleavage signals are observed for each nonreducing moiety. These are at m/z 260 ($HexNAc^+$), m/z 376 ($NeuNAc^+$), m/z 406 ($NeuNG^+$), and m/z 464 ($Hex\text{-}HexNAc^+$). Signals at higher mass are built up from these by the addition of mass increments for the extending residues of 174 (Fuc), 204 (Hex), 245 (HexNAc), 361 (NeuNAc), 391 (NeuNG), and 449 (Hex-HexNAc) amu. The pattern of branching can be deduced in a particular chain from the ions generated, e.g.,

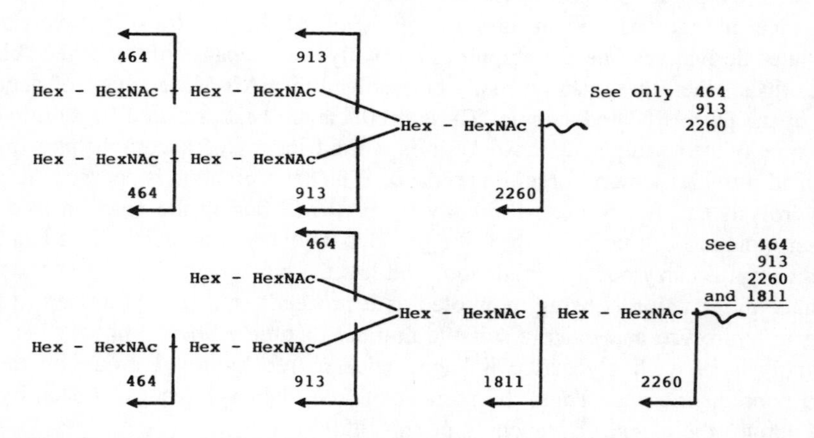

The oxonium ions below mass 1000 may lose methanol via beta-elimination to form an ion of structure:

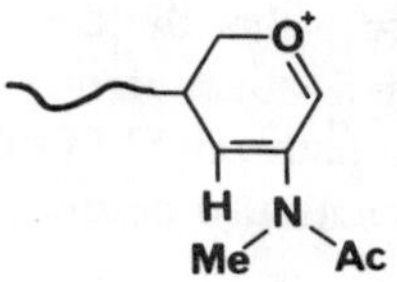

This results in prominent signals 32 amu below each sequence signal and occurs when the 3 position of the HexNAc residue contains a methoxyl group. If the oxonium ion is accompanied by a signal 206 amu, instead of 32 amu, below it, then the substituent on position 3 was terminal fucose rather than a methoxyl group. A similar reasoning can be applied to other 3-linked substituents. If the lactosaminoglycan chain is extended through position 3 of the GlcNAc residues, as in the type 1 backbone (Galβ1-3GlcNAc) then −32 signals are never seen, allowing chain types 1 and 2 to be distinguished.

This protocol using permethyl derivatives is appropriate for the rapid screening of cell surface glycoproteins for particular antigenic structures and is applicable to any glycopeptide or polysaccharide which contains HexNAc residues and is available in amounts greater than about 20 μg.

Acetolysis is an associated technique which yields similar, albeit partial, nonreducing terminal information. It has the advantage that HexNAc residues are not required for successful results; large glycoproteins, glycopeptides, and polysaccharides all give data whether or not HexNAc is present. Acetolysis is discussed in more detail in the glycoprotein section below.

3. *Summary of Data Obtained*

About 30μg of a high-molecular-weight glycopeptide is sufficient to provide information on the structures and branching pattern of all nonreducing terminal chains. It allows type 1 and type 2 chains to be differentiated and can give information on the linkage position of substituents on the HexNAc residues.

C. Sulfated Oligosaccharides

1. *Background Information*

An oligosaccharide is released from a glycosaminoglycan by chemical or enzymic degradation. The size, extent of sulfation, and residue sequence are required from 20 μg of sample.

2. *Approach*

The sample is dissolved in 10 μℓ aqueous solvent, and the negative FAB spectrum is obtained on a 1 μℓ aliquot of the native compound. The native FAB spectra of such oligosaccharides are characterized[10] by two types of ions: first, $(M - H)^-$ pseudomolecular ions are observed for the components present, with each sulfate group normally existing as its sodium salt, and second, fragment ions are produced which represent loss of successive sodium sulfite groups and so are separated by mass increments of 102 amu. The pseudomolecular ions allow residue composition to be assigned, and also give information on the number of sulfate groups in the oligosaccharide. In contrast, the fragment ions unfortunately do not provide any additional information.

In order to determine residue sequence information and to locate the sites of sulfation, it is necessary to make derivatives. The sulfate groups are acid-labile, but are stable under the permethylation conditions given above. Thus, the first step is to make the permethyl derivative and purify it using the Sep-Pak® procedure already given: the permethylated sulfated product usually elutes earlier than uncharged species and is generally recovered in the 50% acetonitrile wash. A negative FAB spectrum of this product will contain $(M - H)^-$ molecular ions for the permethylated components with the sulfate groups usually in their free acid form. The masses of these ions confirm that the components seen in the native spectrum have not lost any sulfates during the derivatization procedure. In addition to molecular ions, fragment ions are also generated, via pathway C (Table 3), which allow determination of partial or complete residue sequence and also define the residues on which sulfate groups are located.

Complementary data may be obtained in the positive mode after acetylating the permethylated sample, using the acetylation method given above. During this reaction, the sulfate groups are lost and replaced by acetyl groups. The positive FAB spectrum of this derivative, with its oxonium fragment ions (pathway A, Table 3), confirms the assignments made from the negative spectrum.

This protocol provides a long overdue method for analysis of glycosaminoglycan oligosaccharides, which, until now, have proved somewhat disappointing as mass spectral analytes.

3. Summary of Data Obtained

Sample heterogeneity, molecular composition, residue sequence, and location of the residues on which sulfates are attached can be determined with 10 μg of sample.

D. Glycolipids

1. Background Information

A carbohydrate-containing sample is isolated by chloroform/methanol extraction procedures and so is thought to be glycolipid in nature.

2. Approach

Glycolipids, especially gangliosides, have been shown to work very well in the negative ion mode when suitable matrices, such as thioglycerol[11] or triethanolamine/1,1,3,3-tetramethyl urea[12] are employed. Thus, the first step in the analysis is to dissolve the sample in about 10 μℓ methanol and use 1 μℓ for negative ion FAB-MS. Abundant $(M - H)^-$ pseudomolecular ions for the different molecular components in the fraction will be observed; dosing the matrix with acid often encourages ionization. Fragment ions which allow sequence assignment are also observed in the negative mode,[13] and are generated by pathways B and C (Table 3). In the case of glycolipids, the ions produced by these two pathways are distinct in mass since the lipid moiety is a built-in marker of the reducing end of the molecule. Ions generated by pathway B thus contain the mass increment corresponding to the base and fatty acids; their heterogeneity reveals variations in this part of the molecule. Pathway C ions are always accompanied by signals 2 amu below them, which arise from oxidative reactions. Far from being a disadvantage, such signals provide a very characteristic doublet pattern which stands out from the background, allowing nonreducing sequence ions to be instantly identified.

Although native glycolipids run so well in FAB-MS because of their lipid portion which allows the molecules to orientate themselves at the matrix surface, their permethyl derivatives often work even better. Permethylation should thus be carried out if insufficient quantities are available to obtain complete data on the native sample, or in order to improve the sensitivity to detect minor components in a mixture.

Positive FAB spectra of permethylated glycolipids are characterized by $(M + H)^+$ and $(M + Na)^+$ pseudomolecular ions, and by two types of fragment ions. The first type of fragment ions are generated from the nonreducing end of the molecule by pathway A (Table 3) and give the sequence of residues in the glycan part of the molecule. As discussed above, preferential cleavage of HexNAc residues occurs if these are present and such cleavages will be accompanied by ions generated by β-elimination reactions which provide further information on the linkage of the backbone and substituents (see Example 2, above). The less-favored cleavage after Hex residues is more likely to be observed from $(M + H)^+$, rather than $(M + Na)^+$, molecular ions. The second set of fragment ions originates from cleavage of and charge retention on the ceramide moiety. These ions give information on the nature of the base and fatty acid substituent(s). Further information on the fatty acid is given by a cleavage ion arising only from the $(M + H)^+$ molecular ion which loses the acyl group, retaining charge on the sphingosine base. The mass difference between this fragment and the molecular ion reflects the number of carbon atoms and the degree of saturation of the acyl group which has been lost.

3. Summary of Data Obtained

The sequence of the carbohydrate residues in the glycan part and characterization of the base and fatty acid components can be determined using 5 μg of sample.

E. Glycoproteins

1. Background Information

An intact protein has been isolated and information is required on whether or not it is glycosylated. If glycosylation is detected, then further data is required, such as whether the chains are *N*- or *O*-linked, the location of the glycosylation site(s) in the protein backbone, and the structures of the carbohydrates linked in these positions.

2. Approach

a. Detection of Glycosylation

In order to determine whether a protein contains any sugar chains, it is subjected to a modification of the classical acetolysis reaction with the progress of the reaction being monitored by FAB-MS.[14] Intact native protein (10 to 50 μg) is mixed with 1 mℓ of the acetolysis reagent, which comprises a 10:10:1 (v/v/v) mixture of acetic anhydride/glacial acetic acid/concentrated sulfuric acid, and incubated at 60°C. A convenient time point for removal of the first 200 μℓ aliquot is 30 min. This aliquot is quenched with about 1 mℓ of water, extracted into chloroform, and the organic layer washed three times with 1 mℓ of water, prior to drying down for FAB-MS analysis. The timing of subsequent aliquots depends on the data obtained from previous time points but is usually on the order of 2 hr, 6 hr, and 18 hr. During the acetolysis reaction, acetylation and hydrolysis of sugar chains occur to produce fully acetylated carbohydrate fragments. The positive FAB spectrum of the extracted aliquot contains pseudomolecular ions for the released carbohydrate chains: $(M + H)^+$ molecular ions are observed for amino sugar containing fragments, while $(M + NH_4)^+$ ions are generated from carbohydrate fragments with no amino sugar residues. These molecular ions allow compositions to be assigned for the released fragments. Fragment ions formed by pathway A (Table 3) may also be observed and serve to confirm the conclusions drawn from the molecular ion data.

If the protein is glycosylated, the presence of pseudomolecular and fragment ions not only answers the question "is there glycosylation?", but also provides information on the nature of the sugar chains. For example, very short fragments, especially in the earliest time points, suggest *O*-linked chains, while early time points giving large hydrolysis fragments suggest that *N*-linked chains are present. Moreover, the compositions of the large fragments allow the type of *N*-linked chain to be identified: a composition of $Hex_{3\text{-}6}$ indicates high mannose, while compositions of Hex(x)HexNAc(y) originate from complex or lactosaminoglycan chains. Mixtures of both types may also be detected. Surprisingly, sialic acids are fairly resistant to acetolysis and sialylated fragments will be observed allowing identification of NeuNAc and NeuNG.

Having determined that the protein is glycosylated, and having received an indication of the types of chains present, the next step is to examine these sugar chains in more detail.

Separate protocols are used to obtain data efficiently on the *O*-linked and *N*-linked chains, and these are discussed in the two following sections.

b. O-Linked Chains

O-Linked chains are released using the traditional base hydrolysis and reduction procedure,[15] but the sugar chains are then isolated and analyzed using a protocol developed in this laboratory,[16] which allows rapid and convenient cleanup of the sample for FAB-MS. The dry, intact glycoprotein or tryptic glycopeptide is incubated at 45°C with 200 $\mu\ell$ of 1 *M* $NaBH_4$ in 0.05 *M* NaOH solution for 16 hr. The mixture is then cooled, mixed with 100 $\mu\ell$ of methanol, concentrated under a stream of nitrogen, and then freeze-dried. The product contains a mixture of reduced oligosaccharides, protein fragments, and salts and as such cannot be directly analyzed by FAB-MS. The next step in the procedure is, therefore, acetylation. The reason for this lies in the ease with which the sugar chains may then be isolated and cleaned up for FAB-MS. Sufficient TFAA/acetic acid acetylating reagent is added to the salt/sugar/protein mixture to be in excess after neutralizing any residual base present. The mixture is allowed to react at room temperature for 10 min, and the reagents are removed under a stream of nitrogen. The acetylated oligosaccharides are dissolved in chloroform and the salts and other aqueous soluble impurities removed by washing the organic phase several times with water. The mixture of reduced acetylated oligosaccharides generated in this way is then analyzed by positive ion FAB-MS. The spectrum obtained will contain both molecular ions, allowing composition to be defined, and fragment ions, giving sequence data, for the reduced acetylated oligosaccharide chains released.

It is quite normal for several different *O*-linked chains to be present on any glycoprotein, and this technique reveals any such heterogeneity. However, purification may be necessary to reveal any minor components which may not be sufficiently abundant to be seen in the unpurified sample, due to suppression or masking by more abundant components, or organic soluble contaminants. Furthermore, purification of components allows any structural ambiguities, such as the presence in the mixture of two components with different branching patterns, to be resolved. The fact that the reduction reaction can be carried out using a tritiated label, and that the chains are acetylated, makes the sample particularly amenable to reverse-phase HPLC in order to separate individual components. Typical HPLC conditions for such separations are elution of the acetylated carbohydrates from a C_{18} column using a linear gradient of 10% aqueous acetonitrile to 100% acetonitrile over 2 hr. Separation of radiolabeled components also permits quantitation of the different species in the mixture.

FAB-MS analysis of the radioactively labeled components may also be assisted in the case of high-molecular-weight species by either de-*O*-acetylation or "back permethylation". Both procedures significantly reduce the molecular weight of the component, since each acetyl group is removed completely (mass reduction of 42 amu) or replaced by a methyl group (mass reduction of 28 amu). Back permethylation is achieved by carrying out a normal permethylation reaction, as given above, on the reduced acetylated carbohydrate. The advantages of back permethylation are that purification is efficient and the sample is still derivatized, so all the advantages of increased sensitivity, generation of unambiguous sequence data, and suitability for subsequent time course hydrolysis experiments are still available, with some decrease in mass. De-*O*-acetylation gives a greater reduction in mass, but with the sacrifice of the sensitivity and fragmentation advantages of derivatives.

This procedure of reductive elimination of *O*-linked chains, followed by acetylation and extraction of released chains can, in our experience, release some *N*-linked chains in some examples, as already reported.[17] This does not, however, cause any confusion or problems, since *N*-linked chains are generally distinguishable from *O*-linked chains by their structures. This protocol has the advantage of yielding detailed information from very little sample. For example, data may be obtained on all the known *O*-linked chains of glycophorin using only 20 μg of intact glycoprotein.[28]

Alternative procedures for preparing *O*-linked oligosaccharides for structural analysis are available: a very successful protocol was employed by Fukuda's group[18] in studies of leukosialin. The *O*-linked chains were reductively eliminated using standard procedures incorporating a tritiated label and were purified on Sephadex® G-50 and G-25 gel filtration columns, followed by chromatography on a QAE-Sephadex® A-25 column, and subsequent desalting steps. The oligosaccharides generated in this way were then permethylated and analyzed by FAB-MS and methylation analysis. The major difference between the two strategies lies in the purification steps prior to the MS experiments.

c. *N*-Linked Chains

The strategy for analyzing N-linked chains is based on the use of *N*-glycanase. The protocol relies on the fact that the enzyme has a broad sugar specificity, and releases the entire carbohydrate portion from the peptide by cleavage of the GlcNAc-Asn linkage, leaving the peptide intact. This leaves both the sugar and peptide free for further analysis and so allowed a strategy to be developed (detailed by Carr[19] and based on the original concept of Morris et al.[20] for determining the site(s) of glycosylation in the protein and for analyzing the types of carbohydrate chains present.[16]

The intact glycoprotein is digested extensively with trypsin: the glycoprotein is dissolved in 250 μℓ 50 m*M* ammonium bicarbonate solution, buffered to pH 8.3, and trypsin is added to bring the solution to a concentration of 1:50 (w/w) enzyme/substrate, with the digestion being carried out at 37°C for 6 to 8 hr. The digest produces a mixture of tryptic peptides and glycopeptides. The digestion is stopped by freeze-drying which also removes the volatile buffer. The sample is then dissolved in 20 μℓ 5% acetic acid and 1 μℓ is added to a matrix of glycerol with some added thioglycerol and a positive FAB spectrum obtained. The spectrum or FAB "map" will contain molecular ions derived from most of the peptides which are within the mass range scanned at full sensitivity. Glycopeptide signals are not usually seen since unglycosylated peptides ionize in preference to glycosylated ones. If the protein sequence is known, the molecular ions can be attributed to the expected tryptic peptides by comparing the observed masses with the calculated values for each peptide.

The remainder of the trypsin digested glycoprotein is then digested with *N*-glycanase: 5 units of enzyme are used for 100 μg of protein in 200 μℓ of 50 m*M* ammonium bicarbonate buffer (pH 8.5), and the digest carried out at 37°C for 18 hr. The volatile buffer allows FAB spectra to be obtained directly from the freeze-dried digest without further purification. A positive FAB map is obtained, as before, of the newly digested sample. Most of the signals in the new map will be the same as those in the original map. Any new signals appearing after the second digestion are derived from peptides originally linked via Asn residues to a carbohydrate chain. These peptides contain Asp at the glycosylation site. As before, provided the protein sequence is known, the newly observed tryptic peptides can be located within the protein. Signals from released carbohydrate are generally not seen in these maps since peptides ionize more readily than sugars.

It has been shown[21] that *N*-glycanase is ineffective in removing *N*-linked chains which are attached to either the *N*-terminal or *C*-terminal residues of peptides. Thus, when *N*-glycosylation occurs on the residue immediately preceding or following a tryptic cleavage site, the carbohydrate will not be removed by *N*-glycanase and so no new peptide signal will be generated and a site of glycosylation will be missing from the map. If this situation is suspected, alternative proteolytic enzymes, e.g., chymotrypsin, may be used instead of trypsin.

In order to study the released carbohydrates, the doubly digested mixture is acetylated with excess TFAA/acetic acid as described in Section IV. After normal extraction procedures, the acetylated carbohydrates are analyzed as outlined above for the acetylated *O*-linked chains, i.e., first as the peracetyl derivatives and then, after back permethylation, as the permethyl derivatives. The FAB spectra of the permethylated samples are particularly valuable for defining the structures of complex-type chains. The discussion given earlier for lactosaminoglycans is relevant to these molecules, since the same types of fragment ions will be produced. The only major difference is the fact that molecular ions will be observed as well as HexNAc fragment ions.

The disadvantages of using *N*-glycanase to release *N*-linked chains are two-fold: its failure to release terminal linked carbohydrate, and glycanase is very expensive. If carbohydrate structure alone is required, without a requirement to assign sites of glycosylation, it may be more convenient to release the glycans using hydrazine[22] or to digest with pronase.[23] In either case, the recommended strategy for further analysis is to permethylate the oligosaccharides or glycopeptides for FAB-MS and methylation analysis.

3. *Summary of Data Obtained*

About 200 μg of intact glycoprotein is normally sufficient to determine whether or not there is glycosylation, the heterogeneity and type of that glycosylation, the composition and sequence of the *O*-linked chains, the points of attachment of the *N*-linked chains, the compositions of high mannose chains, and the sequences of complex-type chains. Obviously the percentage of carbohydrate present in the glycoprotein will influence the amount of data obtained on the carbohydrate structures.

REFERENCES

1. **Barber, M., Bordoli, R. S., Sedgwick, R. D., and Tyler, A. N.,** Fast atom bombardment of solids (F.A.B.): a new ion source for mass spectrometry, *J. Chem. Soc. Chem. Commun.*, 325, 1981.
2. **Abraham, D., Blakemore, W. F., Dell, A., Herrtage, M. E., Jones, J., Littlewood, J. T., Oates, J., Palmer, A. C., Sidebotham, R., and Winchester, B.,** The enzymic defect and storage products in canine fucosidosis, *Biochem. J.*, 221, 25, 1984.
3. **Bourne, E. J., Stacey, M., Tatlow, J. C., and Tedder, J. M.,** Studies on trifluoroacetic acid. I. Trifluoroacetic anhydride as a promoter of ester formation between hydroxy-compounds and carboxylic acids, *J. Chem. Soc.*, 2976, 1949.
4. **Dell, A. and Tiller, P. R.,** A novel mass spectrometric procedure to rapidly determine the position of *O*-aceylated residues in the sequence of naturally occurring oligosaccharides, *Biochem. Biophys. Res. Commun.*, 135, 1126, 1986.
5. **Hakomori, S.-I.,** A rapid permethylation of glycolipid, and polysaccharide catalyzed by methylsulfinyl carbanion in dimethylsulfoxide, *J. Biochem. (Tokyo)*, 55, 205, 1964.
6. **Ciucanu, I. and Kerek, F.,** A simple and rapid method for the permethylation of carbohydrates, *Carbohydr. Res.*, 131, 209, 1984.
7. **Waeghe, T. J., Darvill, A. G., McNeil, M., and Albersheim, P.,** Determination, by methylation analysis, of the glycosyl-linkage compoositions of microgram quantities of complex carbohydrates, *Carbohydr. Res.*, 123, 281, 1983.
8. **McLafferty, F. W., Ed.,** *Tandem Mass Spectrometry*, John Wiley & Sons, New York, 1983.
9. **Fukuda, M., Dell, A., and Fukuda, M. N.,** Structure of foetal lactosaminoglycan, the carbohydrate moiety of Band 3 isolated from human umbilical cord erythrocytes, *J. Biol. Chem.*, 259, 4782, 1984.
10. **Carr, S. A. and Reinhold, V. N.,** Structural characterization of sulfated glycosaminoglycans by fast atom bombardment mass spectrometry: application to chondroitin sulfate, *J. Carbohydr. Chem.*, 3, 381, 1984.
11. **Egge, H., Dabrowski, J., and Hanfland, P.,** Structural analysis of blood-group ABH, I, i, Lewis and related glycosphingolipids, application of FAB mass spectrometry and high resolution proton NMR, *Pure Appl. Chem.*, 56, 807, 1984.
12. **Arita, M., Iwamori, M., Higuchi, T., and Nagai, Y.,** 1,1,3,3-tetramethylurea and triethanolamine as a new useful matrix for Fast Atom Bombardment mass spectrometry of gangliosides and neutral glycosphingolipids, *J. Biochem. (Tokyo)*, 93, 319, 1983.
13. **Arita, M., Iwamori, M., Higuchi, T., and Nagai, Y.,** Negative ion fast atom bombardment mass spectrometry of gangliosides and asialogangliosides: a useful method for the structural elucidation of gangliosides and related neutral glycosphingolipids, *J. Biochem. (Tokyo)*, 94, 249, 1983.

14. **Naik, S., Oates, J. E., Dell, A., Taylor, G. W., Dey, P. M., and Pridham, J. B.,** A novel mass spectrometric procedure for the rapid determination of the types of carbohydrate chains present in glycoproteins: application to alpha-galactosidase I from *Vicia faba* seeds, *Biochem. Biophys. Res. Commun.*, 132, 1, 1985.
15. **Blanchard, D., Cartron, J.-P., Fournet, B., Montreuil, J., van Halbeek, H., and Vliegenthart, J. F. G.,** Primary structure of the oligosaccharide determinant of blood group Cad specificity, *J. Biol. Chem.*, 258, 7691, 1983.
16. **Dell, A. and Rogers, M. E.,** Novel FAB-MS procedures for glycoprotein analysis, in Proc. 13th Int. Carbohydrate Symp., Ithaca, New York, 1986, 274.
17. **Rasilo, M-L. and Renkonen, O.,** Mild alkaline borohydride treatment liberates *N*-acetylglucosamine-linked oligosaccharide chains of glycoproteins, *FEBS Lett.*, 135, 38, 1981.
18. **Fukuda, M., Carlsson, S. R., Klock, J. C., and Dell, A.,** Structures of *O*-linked oligosaccharides isolated from normal granulocytes, chronic myelogenous leukemia cells, and acute myelogenous leukemia cells, *J. Biol. Chem.*, 261, 12796, 1986.
19. **Carr, S. A. and Roberts, G. D.,** Carbohydrate mapping by mass spectrometry: a novel method for identifying attachment sites of Asn-linked sugars in glycoproteins, *Anal. Biochem.*, 157, 396, 1986.
20. **Morris, H. R., Panico, M., Barber, M., Bordoli, R. S., Sedgwick, R. D., and Tyler, A. N.,** Fast atom bombardment: a new mass spectrometric method for peptide sequence analysis, *Biochem. Biophys. Res. Commun.*, 101, 623, 1981.
21. **Tarentino, A. L., Gomez, C. M., and Plummer, T. H., Jr.,** Deglycosylation of asparagine-linked glycans by peptide: N-glycosidase F, *Biochemistry,* 24, 4665, 1985.
22. **Hase, S., Ibuki, T., and Ikenaka, T.,** Reexamination of the pyridylamination used for fluorescence labeling of oligosaccharides and its application to glycoproteins, *J. Biochem.*, 95, 197, 1984.
23. **Spiro, R. G.,** Characterization of carbohydrate units of glycoproteins, *Methods in Enzymol.*, 8, 26, 1966.
24. **Wang, W. T., LeDonne, N. C., Jr., Ackerman, B., and Sweeley, C. C.,** Structural characterization of oligosaccharides by high-performance liquid chromatography, fast-atom bombardment-mass spectrometry, and exoglycosidase digestion, *Anal. Biochem.*, 141, 366, 1984.
25. **Her, G. R., Santikarn, S., Reinhold, V. N., and Williams, J. C.,** Simplified approach to HPLC precolumn fluorescent labeling of carbohydrates: *N*-(2-pyridinyl)-glycosylamines, *J. Carbohyd. Chem.*, 6, 129, 1987.
26. **Dell, A.,** FAB-mass spectrometry of carbohydrates, *Adv. Carbohydr. Chem. Biochem.*, 45, 19, 1987.
27. **Ballou, C. E. and Dell, A.,** Unpublished observations on methylglucose polysaccharide from *Mycobacterium.*
28. **Dell, A., Thomas-Oates, J. E., Rogers, M. E., and Tiller, P. R.,** Novel fast atom bombardment mass spectrometric procedures for glycoprotein analysis, *Biochimie,* in press.

Chapter 11

LOCATION OF CELLULOSE SUBSTITUENTS

Karl-Gunnar Rosell

TABLE OF CONTENTS

I. INTRODUCTION

Cellulose and its derivatives are extensively used in both consumer and industrial applications. Many review articles dealing with synthesis, properties, manufacture and composition, economics, and uses of cellulose derivatives have been published, but methods for determination of the distribution of substituents in these derivatives has not been reviewed. Cellulose consists of a linear backbone of β-(1 → 4)-linked D-glucopyranosyl residues as shown in Figure 1 and is not soluble in water. Its commercial derivatives, however, are mostly water soluble and function as thickeners, bulking agents, stabilizers, protective colloids, suspending agents, binders, moisture retainers, and emulsifiers to identify just a few of the more important functions. The solubility of cellulose derivatives depends upon the nature of the substituents, the degree of substitution (DS), and the distribution of substituted sites.

Modified cellulosics are prepared by replacing the hydrogen atom of the hydroxyl groups in the glucose residues with different chemical groups, or by partial hydrolysis of fully substituted polymer. Each anhydroglucose residue has three hydroxyl groups, one primary at the 6 position and two secondary at the C-2 and C-3 positions, available for substitution. The polymer can, therefore, have a maximum DS of three per glucose residue. In most cases, the desired properties are achieved by making a partially substituted derivative — that is, by substituting, on the average, only one or two of the three hydroxyl hydrogens. Thus, the original homopolymer is transformed into a copolymer of eight monomers randomly distributed along the polymer chain. They are shown in Figure 2. When alkylating reagents (e.g., ethylene oxide), with an alternate reaction site capable of chain extension are used, the hydroxyl group of the substituent can be further substituted, forming side chains. The extent of substitution of these derivatives is measured as the molar substitution, the number of moles of substituent groups per mole of glucose. The ratio of molar substitution to DS is a measure of the average length of the derivative chain. Cellulose derivatives with mixed substituents also exist, adding further complexity to the analysis of the distribution of these substituents.

The two most important analytical criteria concerning the molecular structure of cellulose derivatives are the distribution of the substituents and the distribution of the molecular weight. The distribution of substituents actually consists of two independent parts — distribution of substituents along the polymer chain and within the polymer unit. Distribution along the polymer chain has been suggested to influence solubility, stability, shearing behavior, rheology, and degradation stability, while distribution within the polymer unit will influence solubility, stability and dissolving properties.[1] Thus, the amount, DS, and the distribution of the substituents play a major role in the properties of these derivatives and, therefore, accurate methods for determining these parameters are of great importance. The DS can be determined by gas-liquid chromatography (GLC), conductometric and potentiometric titration, colorimetry, enzymatic degradation, infrared (IR) spectrophotometry and, in recent years, by nuclear magnetic resonance (NMR) spectroscopy.

This chapter deals mainly with the use of chemical methods, combined with gas-liquid chromatography-mass spectrometry (GLC-MS), for determination of different substituents, the degree of substitution, and the distribution of the substituents in cellulose. References may also be made to the use of other methods. The techniques described here are also valid for other modified polysaccharides, especially for starch derivatives. The cellulose derivatives have been divided into three groups depending on their chemical structures: esters, ethers, and other compounds.

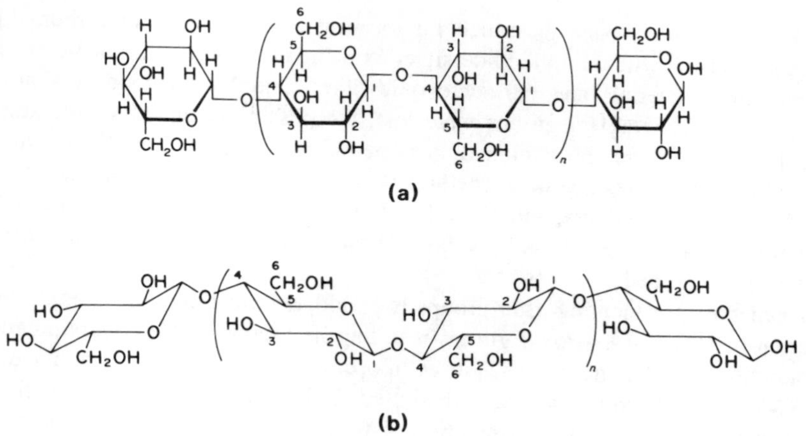

(a)

(b)

FIGURE 1. Structure of cellulose: (a) Haworth projection; (b) conformational representation.

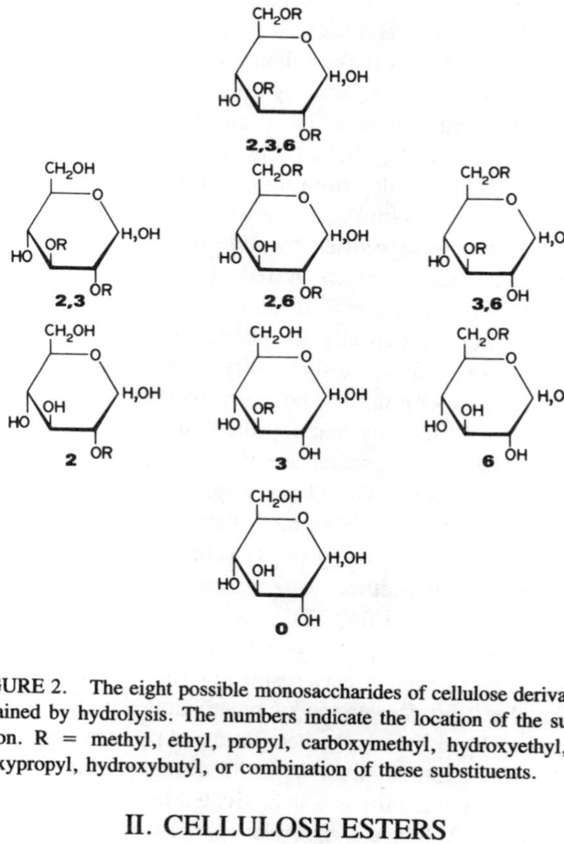

FIGURE 2. The eight possible monosaccharides of cellulose derivatives obtained by hydrolysis. The numbers indicate the location of the substitution. R = methyl, ethyl, propyl, carboxymethyl, hydroxyethyl, hydroxypropyl, hydroxybutyl, or combination of these substituents.

II. CELLULOSE ESTERS

A. General

The preparation, properties, and uses of cellulose esters have recently been reviewed.[2] In contrast to cellulose ethers, esters are not stable to acid or base treatment and, therefore, cleavage of the glycosidic linkages are not possible while retaining the ester groups.

Several methods have been described for locating *O*-acyl groups in carbohydrates, for example, methylation with methyl iodide-silver oxide in dimethylformamide of the acetylated and deacetylated material[3] and periodate oxidation.[4] The first method was not entirely satisfactory since some acetyl groups were lost, the possibility of acetyl migration during the methylation could not be excluded, and periodate oxidation and NMR studies have limited application. A more general method is one in which acyl groups are replaced by *O*-methyl groups in accordance with the following sequence: (1) protection of free hydroxyl groups, (2) deacylation, (3) *O*-methylation, (4) removal of protecting groups. Steps 2 and 3 can often be combined.

Another method for locating ester groups is periodate oxidation. This technique is based on oxidation of adjacent hydroxyl groups resulting in breakage of the carbon-carbon bond and introduction of aldehyde groups. For cellulose, oxidation will only take place when the C-2 and C-3 hydroxyl groups are not substituted. This method, therefore, has limited use, but has been employed in the determination of sulfate groups in the 6 position of cellulose (see Section II.C.3).

B. Organic Esters

1. Cellulose Acetate

Cellulose acetate is the most important organic ester of cellulose due to its extensive applications in fibers, plastics, and coatings. It is manufactured by reacting cellulose with acetic anhydride, utilizing acetic acid as the solvent, and sulfuric acid as a catalyst. Cellulose acetate with varying DS is obtained by acid hydrolysis of the cellulose triacetate.

O-Acetyl groups are generally labile to base, and the liberated acid can be determined colorimetrically, titrimetrically, or by GLC. NMR has also been employed to determine the acetyl content. A micromethod for determining *O*-acetyl groups in saccharides and wood[5] has been described as follows: the acetyl groups are split off by reaction with sodium ethoxide and the resulting ethyl acetate is saponified by addition of water. After cation exchange to remove sodium ions, the acetic acid is converted to its benzyl ester via its tetrabutyl ammonium salt. The benzyl ester is determined quantitatively by GLC.

The location of acetyl groups in partially acetylated polysaccharides is attended by several difficulties. These acetyl groups are not only readily hydrolyzed under basic conditions and, to a lesser extent, under acidic conditions, but may also migrate.[6,7]

Phenylcarbamoyl groups, formed by reacting the hydroxyl groups with phenylisocyanate under conditions where the acetyl groups are stable and do not migrate, were the first protective groups to be employed.[8] The *O*-acetyl groups were then removed by acid hydrolysis, the modified polysaccharide was methylated with methyl iodide-silver oxide in dimethylformamide, and the protective groups were finally removed by reduction with lithium aluminum hydride. This method was not entirely satisfactory, since some of the *O*-acetyl groups were not accounted for. The reason for this is presumably that during the time-consuming acidic deacetylation, phenylcarbamoyl groups may be lost, and also that the reductive removal of these groups may result in demethylation.[9]

Another method involves the conversion of unsubstituted hydroxyl groups to methoxyethylacetals on reaction with methyl vinyl ether, followed by base-catalyzed de-*O*-acetylation and methylation with methyl iodide-silver oxide in dimethylformamide.[10] The distribution of substituents in two *O*-acetyldextrans was investigated by this method and no loss of acetyl groups could be detected.[10] When the methylation was performed with dimethylsulfinyl anion and methyl iodide in dimethylsulfoxide (Hakomori methylation),[11] *O*-acetyl groups were cleaved and the de-*O*-acetylation step could be omitted.[12] The reaction sequence is depicted in Figure 3. After acid hydrolysis, partially methylated sugars, in which the methoxyl groups marked the position of the original acetyl groups, were obtained. Separation of partially methylated alditol acetates is shown in Figure 6. This method has been used to

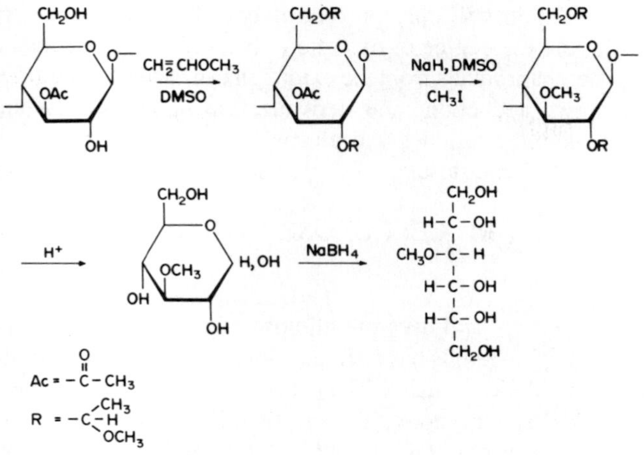

FIGURE 3. Scheme for determination of the position of *O*-acetyl groups in polysaccharides.

locate *O*-acetyl groups in secondary cellulose acetate,[12,13] birch xylan,[14] and pine glucomannan.[15,16] A more elaborate method is a nitration step for protection of the hydroxyl groups.[17] Characterization of cellulose acetate has also been performed by NMR.[18]

C. Inorganic Esters

1. Cellulose Nitrate

Cellulose nitrate is the most important inorganic ester of cellulose. It finds application in plastics, lacquers, coatings, and explosives. It is manufactured by treating cellulose with nitric acid in the presence of sulfuric acid and water. The amount of water will determine the DS.

Hydrazine hydrate in ethanol is effective in achieving denitration of carbohydrate nitrates without effecting methyl groups.[19,20] Thus, methylation of cellulose nitrate, followed by denitration, hydrolysis, reduction, derivatization, and analysis by GLC, should result in alditols with methyl groups in the positions (2, 3, or 6), which were not occupied by nitrate groups in the native polysaccharide. The comparison of the methylated sugars obtained from nitrated parent compound and the denitrated polysaccharides would also give the location of the nitrate groups.

Use of both ^{1}H- and ^{13}C-NMR, as well as solid state ^{13}C-NMR of cellulose nitrate, has been found to be a useful technique for qualitative and quantitative analyses of the distribution of substituents.[21,22]

2. Cellulose Phosphate

Dephosphorylation without cleavage of the glycosidic linkages can be achieved by treatment with HF at 0°C[23] or by using enzymes.[24] Since phosphate groups have been shown to be stable during methylation,[25] methylation analysis of native and dephosphorylated polysaccharides would, therefore, indicate the distribution of the phosphate ester groups.

3. Cellulose Sulfate

Since polysaccharides can be desulfated without depolymerization, comparison of the methylated sugars formed from the sulfated parent compound and the corresponding desulfated polysaccharides can give information on the location of sulfate esters. Desulfation of cellulose sulfate can be performed by treatment with cold methanolic hydrogen chloride.[26]

Periodate oxidation has been used to determine the percentage of sulfate substitution at

C-6 of cellulose.[27] Oxidation will only proceed if both C-2 and C-3 hydroxyl groups are free, and from the DS and the degree of oxidation, the percentage of substitution at C-6 can be calculated. The different products from the oxidation can easily be identified and quantified by GLC after hydrolysis, reduction, and derivatization. The products are erythritol and ethylene glycol from oxidized residues and glucose from nonoxidized residues. ^{13}C-NMR-spectral analysis of sulfated and desulfated polysaccharides has been reported.[28]

III. CELLULOSE ETHERS

A. General

The preparation, properties, and uses of cellulose ethers have recently been reviewed.[29] A wide variety of analytical methods has been developed over the years for the determination of molar substitution in cellulose ethers. They include titrimetry, gravimetry, colorimetry, and, more recently, NMR spectroscopy. The classical Zeisel distillation method,[30] in combination with gas chromatography (GC) to obtain the selectivity needed for the analysis of mixed cellulose ethers, has been commonly employed. An improved method has been developed by Hodges et al.[31] using adipic acid to catalyze the hydroiodic acid cleavage of the substituted alkoxyl groups quantitatively to their corresponding alkyliodides. An *in situ* xylene extraction of the alkyl iodides in a sealed vial allows for determination using GLC of methoxy, ethoxy, hydroxyethoxy, or hydroxypropoxy substitution in mixed or homogeneous cellulosic ethers. The chart of a separation is shown in Figure 4. Also the determination of carboxymethyl groups in cellulose ethers, using essentially the same method, has been reported.[32]

B. Hydrolysis

In contrast to cellulose esters, ethers are stable to both acid and alkaline treatment. The routes for preparing volatile derivatives of neutral polysaccharides and their ethers are shown in Figure 5. For acidic polymers, other routes have to be used (see Section III.F). Determination of the chemical composition of polysaccharides starts with an initial hydrolysis into monosaccharides. Thus, the conditions of hydrolysis must be carefully chosen and controlled. The goal is to obtain complete hydrolysis of the glycosidic linkages without affecting the sugar residues or the substituents (Figure 5.**i**). Mineral acids of different concentrations, such as hydrochloric acid and sulfuric acid, have been employed. Trifluoroacetic acid, which is volatile and thus easily removed, has also been used.[33]

Fully or highly alkylated polysaccharides are often not soluble in aqueous solutions. For this reason methanolysis, formolysis, or treatment with 72% sulfuric acid is usually performed before hydrolysis with hot dilute acid (Figure 5.**ii**). In this manner, degradation and dealkylation are kept to a minimum. Treatment of these polymers with 90% formic acid at 100°C for 2 hr, followed by treatment with 0.67 *M* trifluoroacetic acid at 100°C for 16 hr, has been shown to result in complete hydrolysis without affecting the substituents in several cellulose ethers.[34] The carboxymethyl groups have also been shown not to be affected by hydrolysis and remain attached to the glucose, as has been established with the aid of radiolabeled substituents in carboxymethyl cellulose (CMC).[35]

An alternative to hydrolysis is methanolysis (Figure 5.**iii**). The modified cellulose is treated with hydrogen chloride (usually 1 to 3%) in methanol for 16 hr at reflux temperature. The result is a mixture of methyl glycosides which can be analyzed directly as their acetates or trimethylsilyl (TMS) derivatives (Figure 5.**iv**). Unfortunately glycosidation affords several peaks from each sugar and thus problems with separation and identification of the partially alkylated monosaccharides can occur. This can be avoided by acid hydrolysis (Figure 5.**v**) and reduction of the resulting monosaccharides to alditols (Figure 5.**vi**). An advantage of methanolysis is that the carboxyl groups of uronic acids and carboxymethyl groups of CMC

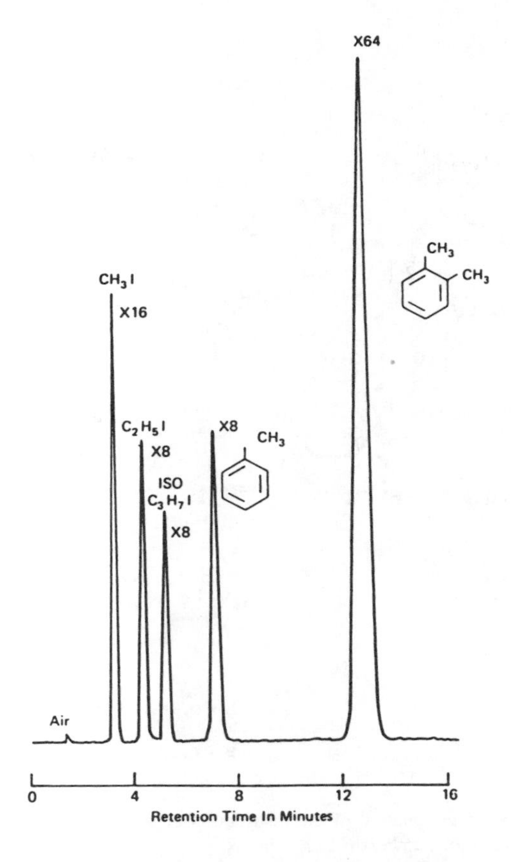

FIGURE 4. Determination of alkoxyl substituents in cellulose ethers using GLC. For experimental details see Reference 31. (Reprinted with permission from Hodges, K. L., Kester, W. E., Wiederrich, D. L., and Grover, J. A., *Anal. Chem.*, 51, 2174, Copyright 1979 American Chemical Society.)

are converted into the methyl esters, which are sufficiently volatile to permit direct analysis after derivatization by GLC. For further information in regard to hydrolysis of carbohydrates see Chapter 3.

C. Derivatization

GLC methods for carbohydrate analysis involve the formation of derivatives of sufficient volatility and adequate thermal stability. TMS ethers and acetates or trifluoroacetate esters are the common derivatives. Another volatile compound of sugar used for GLC purposes is the TMS derivative. Sweeley et al.[36] were the first to report a viable method for analysis of carbohydrates based on the formation of these derivatives.

An alternative approach to acylation is the preliminary conversion of aldoses to the corresponding *O*-methyloximes[37,38] (Figure 5.**vii**), followed by acetylation or TMS of the free hydroxyl groups (Figure 5.**viii**). Since the oximation involves reaction at the C-1 position, the incidence of multiple derivatives from a single sugar is diminished. Although single peaks are obtained for glucose, fructose, mannose, and xylose, some carbohydrates give two peaks, presumably due to formation of *syn*- and *anti*-isomers.

The derivatization of sugars, with trifluoroacetic anhydride to form the corresponding esters, has been described earlier.[39,40] The advantages are enhanced volatility and sensitivity while one of the disadvantages is the difficulty in preparation.

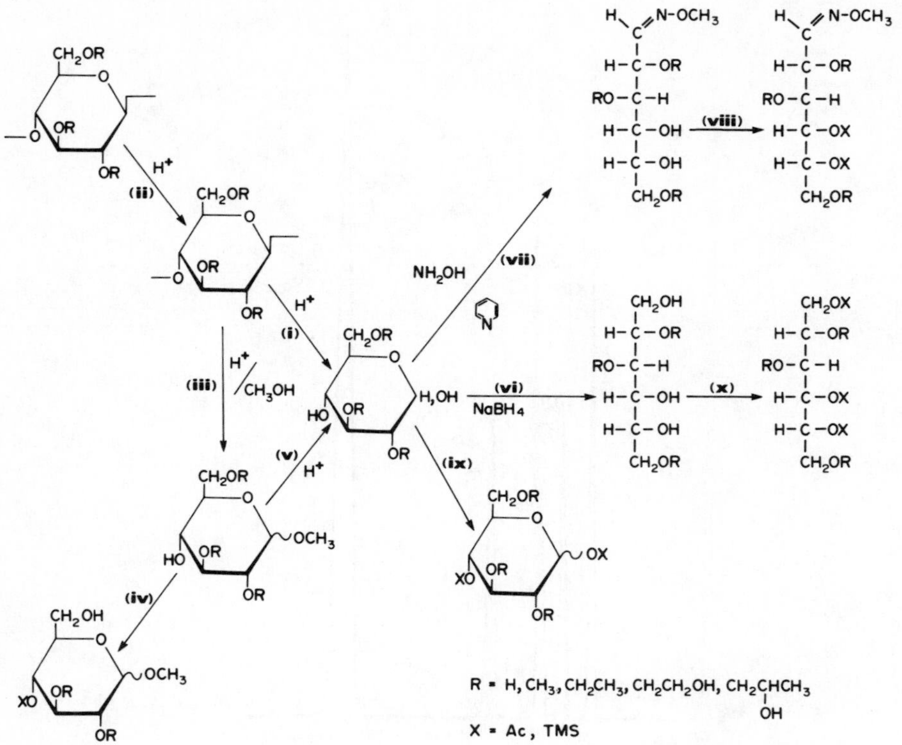

FIGURE 5. Scheme for determination of the distribution of substituents in neutral cellulose ethers using GLC. For explanation of numbers (**i** to **x**) see text.

The reducing sugars, obtained after hydrolysis, can be analyzed as their acetates or TMS derivatives directly (Figure 5.**ix**) or more conveniently converted into acyclic compounds (Figure 5.**vi**) before the hydroxyl groups are derivatized (Figure 5.**x**). This approach eliminates the problem of multiple derivative formation when different ring forms and/or different anomeric forms are generated from reducing sugars or from equilibrium mixtures of methyl glycosides. The acyclic derivatives are characterized by retention times and, if necessary for unusual sugars, by GLC-MS. The two most widely used derivatives are alditol acetates and acetylated aldononitriles.

The partially alkylated monosaccharides obtained on depolymerization of cellulose ethers are preferentially analyzed as alditol acetates by GLC-MS, as shown by Björndal et al.[41] The neutral sugars obtained in acid hydrolysis (Figure 5.**i**) are reduced (Figure 5.**vi**) and acetylated (Figure 5.**ix**) for the analysis.[42] The resulting partially alkylated alditol acetates are stable and can be stored at room temperature for a long period of time. This is in contrast to the TMS derivatives which are not stable and which will decompose rapidly. Identification with the aid of GLC-MS has been described for all common sugars.[43] Separation of the partially alkylated sugars by GLC gives evidence in the methyl-substitution pattern complementary to that obtained from mass spectra. The electron-impact mass spectra do not differentiate between the sugar epimers (such as glucose, galactose, and mannose derivatives) or their alditols, but this information is obtained from the retention times in GLC.

The preparation of the derivatives for GLC described here has been reviewed in an earlier issue in this series.[44] GLC of carbohydrates and their derivatives has also been reviewed earlier.[45] For further information in regard to derivatization of carbohydrates see also Chapters 4 to 9.

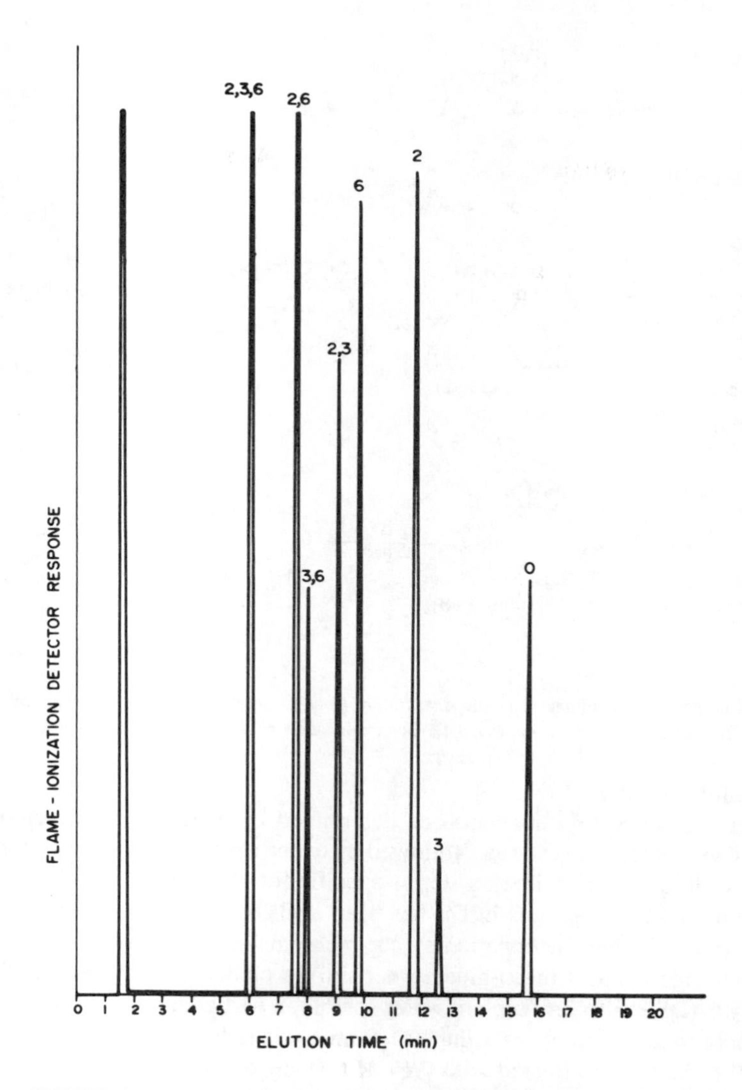

FIGURE 6. Separation of partially methylated alditol acetates obtained by hydrolysis, reduction, and acetylation of methylcellulose (DS = 1.70). The numbers indicate the positions of methyl groups. Fused-silica capillary column with DB-1 phase, 50m × 0.2 mm (Hewlett Packard), 200°C isothermal.

D. Methylcellulose (MC)

The distribution of substituents in MC has been analyzed by several groups by quantitative separation on a carbon column and further fractionation by paper chromatography and paper electrophoresis or by ion exchange resin of the D-glucose derivatives obtained by hydrolysis.[46-48] ^{13}C-NMR has also been used for locating the methyl groups in MC.[49] Recently, MC was analyzed using GLC for separation of the eight partially methylated alditol acetates as well as their TMS derivatives obtained from complete hydrolysis of MC.[34] The GLC chromatogram of this separation is shown in Figure 6. Hydrolysis was performed with 0.67 *M* trifluoroacetic acid at 100°C for 16 hr. In the same study, ^{13}C-NMR spectra of MC were studied. A comparison between the two methods showed that complete hydrolysis and characterization by GLC-MS of the glucose ethers obtained resulted in a more accurate quantitative estimation of the distribution of methyl groups in cellulose compared to NMR.

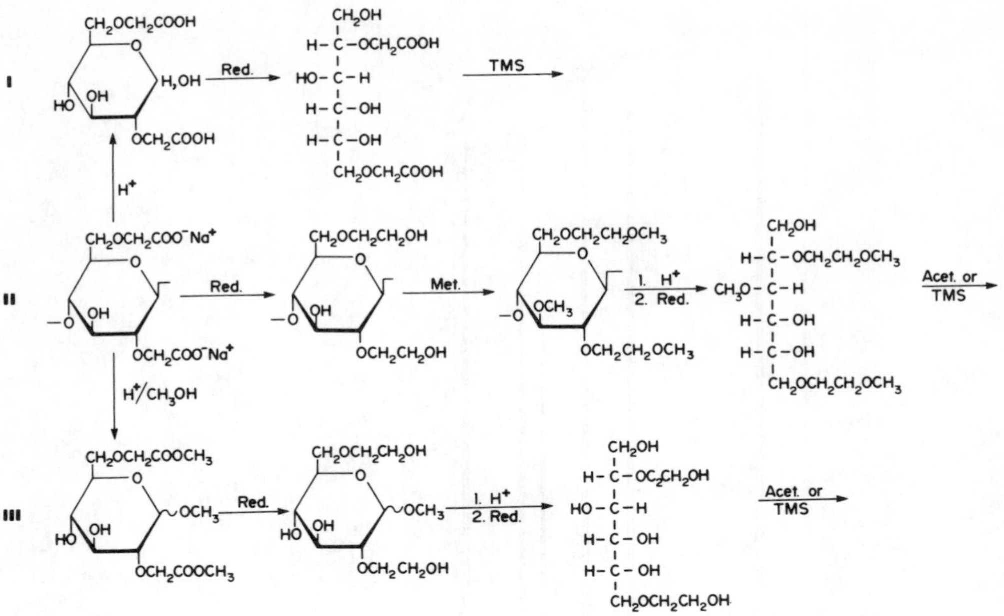

FIGURE 7. Scheme for determination of the distribution of substituents in acidic cellulose ethers using GLC. Carboxymethyl cellulose is shown in this example. For explanation of numbers (**I** to **III**) see text.

E. Ethyl Cellulose (EC)

The amount of ethoxyl groups has been determined by oxidation of the ethoxyl groups to acetic acid with chromium trioxide, followed by quantitative determination of the resulting acetic acid by GLC.[50] This oxidation step is specific for ethoxyl groups.

The location of ethoxyl groups in EC has been analyzed by quantitative separation on a carbon column and further fractionation by paper chromatography and paper electrophoresis or by ion exchange resin of the D-glucose derivatives obtained by hydrolysis.[51] When the commercial EC (DS 2.3 to 2.6) are not water soluble, a partial degradation of the molecular weight was achieved by formolysis using 90% formic acid at 100°C for 2 hr. After evaporation to dryness, the residue was treated with 0.67 *M* trifluoroacetic acid at 100°C for 16 hr. The two treatments resulted in complete hydrolysis. The partially ethylated glucose residues were analyzed as their alditol acetates and TMS ethers on GLC-MS. The eight derivatives were all separated on fused-silica capillary columns.[34] The retention time data of partially ethylated alditol acetates on several packed GLC columns have been published.[52]

F. Carboxymethyl Cellulose (CMC)

The location of substituents in CMC has been analyzed by several groups. Hydrolysis and separation of the mixture of glucose and carboxymethyl glucoses on a carbon-Celite® column has been achieved.[53-55] For determination by GLC of the distribution of the carboxymethyl groups, three different routes are depicted in Figure 7. The distribution of carboxymethyl groups in CMC using GLC (Figure 7.I) has been determined by Buytenhuys and Bonn.[35] To obtain conditions for complete hydrolysis, the hydrolysate was monitored by gel permeation chromatography. The resulting monomers were silylated and separated by GLC. Despite several peaks for each sugar, they were able to determine quantitatively the amounts of the eight glucosidic monomers except for the 2,3 and 3,6 glucosides, which did not separate. The TMS derivatives of the eight alditols obtained by hydrolysis and reduction with sodium borohydride have been shown to separate completely on different fused-silica capillary columns.[34]

Methods have been described for locating the *O*-carboxymethyl groups in *O*-carboxymethyl guar.[56] Guar consists of a backbone of β-(1 → 4)-linked D-mannopyranosyl residues, of which ≈60% are substituted at *O*-6 with a single α-D-galactopyranosyl group. This polymer was depolymerized by methanolysis, the *O*-carboxymethyl groups were reduced, and the mixture of methyl glycosides and *O*-(2-hydroxyethyl)-substituted methyl glycosides was converted into a mixture of per-*O*-acetylated alditols and partially *O*-(2-acetoxyethyl)ated, and partially *O*-acetylated alditols (Figure 7.**III**). Analysis of these alditols by GLC-MS allowed the positions of substitution of the *O*-carboxymethyl groups on the galactosyl and mannosyl residues to be determined. However, this method did not distinguish between *O*-carboxymethyl substitution on 4-linked and 4,6-linked mannosyl residues. This limitation was overcome by carboxy-reduction of the *O*-carboxymethyl groups (Figure 7.**II**) using the carbodiimide method,[57] the product was methylated, the glycosyl residues were hydrolyzed, the sugar was reduced, and the alditols were acetylated to yield a mixture of partially *O*-acetylated, partially *O*-methylated alditols and partially *O*-acetylated, partially *O*-(2-methoxyethyl)ated, partially *O*-methylated alditols. These derivatives, when separated and quantitated by GLC and identified by GLC-MS, gave information on the type of carboxymethyl substitution in guar. This procedure, applied to CMC, would give the distribution of the carboxymethyl groups.

G. Hydroxyethyl Cellulose (HEC)

HEC is prepared by reacting alkali cellulose with ethylene oxide. The determination of the substituents in HEC is more complicated compared to the previously mentioned ethers, because of the presence of glucose ethers with polyethylene oxide substituents. The molar substitution in commercial HEC ranges from 1.5 to 3.0.

The distribution of substituents in HEC has been determined by quantitative separation on a carbon column of the D-glucose derivatives after hydrolysis.[58] A group separation of the compounds, obtained by hydrolysis of HEC, by partition chromatography on an ion-exchange column in its sulfate form, followed by GLC of three main fractions, was used to obtain a quantitative determination of all major glucose derivatives.[59] Among these, 1,2-*O*-ethylidene-D-glucose, which is formed by intramolecular glucosidation of 2-*O*-hydroxyethyl-D-glucose, was detected.

Another approach would be to methylate the HEC and then analyze the products after complete hydrolysis, reduction, and acetylation. The resulting *O*-acetyl, partially *O*-methylated alditols, with partially *O*-acetylated, partially *O*-(methoxyethyl)ated, partially *O*-methylated alditols, were analyzed by GLC-MS and quantitated by GLC. Preliminary results showed that these derivatives were more volatile than the TMS derivatives of the monosaccharide ethers obtained from the native HEC and the separation of all the compounds and accurate determination of the substituents could be achieved in a relatively short time.[34]

H. Hydroxypropyl Cellulose (HPC)

HPC is prepared by reacting alkali cellulose with propylene oxide at elevated temperatures and pressures. The determination of substituents in HPC is also complicated due to the presence of glucose ethers with polypropylene oxide groups.

A method has been described for locating the *O*-(2-hydroxypropyl) groups in *O*-(2-hydroxypropyl) guar.[60] For the structure of guar see Section III.F. Per-*O*-methylation of the *O*-(2-hydroxypropyl) guar followed by hydrolysis afforded a mixture of partially *O*-methylated monosaccharides and partially *O*-(2-methoxypropyl)ated, partially *O*-methylated monosaccharides. These monosaccharide derivatives were reduced, and the alditols acetylated, to give a mixture of partially *O*-acetylated, partially *O*-methylated alditols with partially *O*-acetylated, partially *O*-(2-methoxypropyl)ated, partially *O*-methylated alditols. These alditol derivatives were identified by GLC-MS, and quantitated by GLC. This procedure, applied to HPC, would give the distribution of the hydroxypropyl groups.

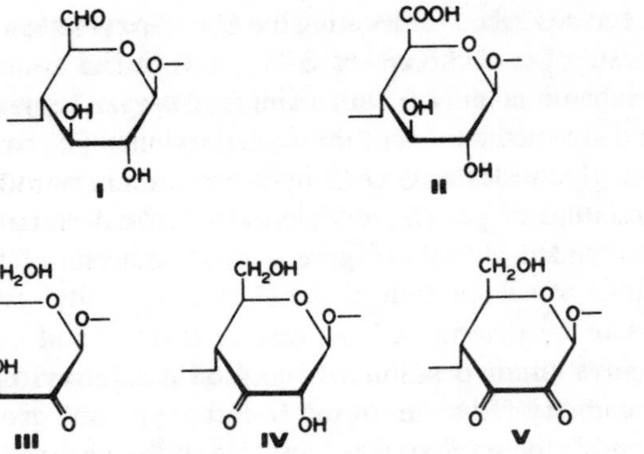

FIGURE 8. Some oxidized cellulose units found in cellulose derivatives. For explanation of numbers (I to V) see text.

I. Mixed Cellulose Ethers

In order to obtain properties from cellulose ethers to meet specific needs of industrial applications, combinations of two or more ethers have been manufactured. Hydroxypropyl (HPMC) and hydroxybutyl methylcelluloses (HBMC) have been prepared by reacting methylcellulose with propylene oxide and butylene oxide, respectively. Ethylation of these polysaccharides followed by hydrolysis, reduction, acetylation, and analysis by GLC-MS, resulted in DS, molar substitution, and the distribution of both the methyl, hydroxypropyl, and hydroxybutyl groups in these cellulose derivatives.[34]

Another commercial derivative is carboxymethyl hydroxyethyl cellulose (CMHEC), which is prepared by partial carboxymethylation of hydroxyethyl cellulose. Methylation of this polysaccharide, followed by hydrolysis, reduction, TMS, and analysis by GLC-MS, determined the position of both the carboxymethyl and hydroxyethyl groups as well as DS and molar substitution.[34]

Ethyl hydroxyethyl cellulose (EHEC) is prepared by partial ethylation of hydroxyethyl cellulose. Methylation of this polysaccharide, followed by hydrolysis, reduction, acetylation, and analysis by GLC-MS, resulted in the distribution of both the ethyl and hydroxyethyl groups.[34]

IV. MISCELLANEOUS DERIVATIVES

A. Oxyderivatives

Cellulose can be oxidized by several methods transforming the hydroxyl groups into aldehydes, ketones, and acids or combinations of these, while the glucosidic linkages and the ring structures remain intact. Recently, these oxidized polysaccharides have provided a convenient approach to the synthesis of Schiff-base or stable amine derivatives. These modifications could, therefore, be of substantial importance in future industrial applications, e.g., immobilization techniques. The oxidation of cellulose and determination of functional groups have been reviewed recently.[61] The degree of oxidation and proportion of these products can also be determined by GLC-MS. Some oxidized anhydroglucose units are depicted in Figure 8.

*1. 6-Aldehydro-Derivatives (Figure 8.***I***)*

The oxidation in the 6 position of D-galactose-containing polysaccharides by galactose oxidase is perhaps one of the most widely used enzymatic modification methods. A method to monitor the oxidation of this position is to transform the 6-aldehydo-aldose, obtained after hydrolysis, to the corresponding methyloxime, followed by acetylation or TMS and analysis by GLC.[62] Two other methods have been used to determine the extent (DS) to which aldehyde groups were incorporated into acetylated 6-aldehydo-cellulose. The 2,4-dinitrophenylhydrazone was prepared and from the nitrogen content, the DS was calculated.[63] A convenient and more reliable method for determining DS based on the carbonyl group was by reduction of the 6-aldehydo-cellulose with sodium borodeuteride, hydrolysis of the resulting 6-deuterio cellulose acetate, and preparation of a derivative suitable for determining, by MS, the position and extent of incorporation of deuterium.[63]

*2. 6-Carboxy-Derivatives (Uronic Acids) (Figure 8.**II**)*

Further oxidation of the 6 position will lead to a carboxyl group resulting in a uronic acid. The uronic acids can be determined qualitatively and quantitatively by GLC. After complete hydrolysis or methanolysis, the resulting monosaccharides, including uronic acids, can be analyzed as their TMS derivatives after conversion into ammonium or sodium salts[64,65] or as their methyl ester methyl glycosides.[64] The problem with multiple peaks can be eliminated by reduction of the uronic acids to aldonic acids, followed by lactonization and TMS of the lactones.[66] Analysis of the aldonic acids themselves as TMS ethers, without prior lactonization, has also been reported.[62]

A more complicated way is to carboxyl reduce the uronic acid to the corresponding aldose. This can be performed in the polymeric state using the carbodiimide method.[57] The reducing agent, sodium borodeuteride, will label the 6 position of the resulting glucose units which carried a carboxyl group originally.

*3. 2-,3-Keto- and 2,3-Diketo-Derivatives (Figures 8.**III**, **IV**, and **V**)*

Selective oxidation of the 2 and 3 position has been accomplished.[67] These ketoglycosyl units are very labile and decompose into colored products at pH values above 7 and they are also sensitive to strong acids.[68] The carbonyl content (degree of oxidation) can be determined by the hydroxylamie method.[69] Determination of the site and degree of oxidation in the partly oxidized homopolysaccharide has been achieved through reduction with sodium borodeuteride, hydrolysis, and study of the resulting hexoses in the form of their TMS *O*-methyl oximes, or alditol acetate derivatives.[67] 2-Keto-D-glucose residues (Figure 8.**III**) will be transformed into D-glucose and D-mannose and 3-keto-D-glucose residues (Figure 8.**IV**) into D-allose and D-glucose.

4. Dialdehyde-, Dicarboxy-, and Tricarboxy-Derivatives

Oxidation of cellulose with sodium metaperiodate affords, quantitatively, dialdehyde cellulose, and further oxidation with bromine and nitrogen tetroxide give di- and tricarboxy-cellulose, respectively.[70] These oxycelluloses constitute an important and versatile compound for potential application in the synthesis of other cellulose derivatives.

The degree of oxidation of these compounds can be determined by GLC. The dialdehyde compound was reduced with sodium borohydride and the resulting derivative was hydrolyzed with 2 *M* trifluoroacetic acid for 2 hr at 120°C.[71] The resulting glycerol and erythritol were analyzed by GLC after converting them into their TMS derivatives. Hydrolysis of the dialdehyde cellulose, without prior reduction, would result in erythrose and ethandial (glyoxal), which could be analyzed directly as their TMS derivatives by GLC.

The dicarboxycellulose after hydrolysis would result in 2,3,4-trihydroxybutyric acid (threonic acid) and oxyacetic acid (glyoxylic acid). They could be analyzed by GLC after converting them into their TMS derivatives. The corresponding compounds, obtained by hydrolysis of tricarboxycellulose, would be 2,3-dihydroxybutanedioc acid (tartaric acid) and glyoxylic acid (oxyacetic acid), respectively. They could be analyzed as their TMS derivatives by GLC. Unoxidized residues of these derivatives would show up as glucose, which also could be quantified as its TMS derivative by GLC.

V. MASS SPECTROMETRY (MS)

A. Introduction

MS has become an important and versatile technique in carbohydrate chemistry for identification of compounds and has been reviewed earlier.[72-74] Combined GLC-MS has become increasingly important for investigation of complex mixtures. The development of fused-silica capillary columns has led to complete separations of similar compounds using very small quantities of material.

Electron impact (EI) MS has, to date, been the technique used mostly in GLC-MS of carbohydrates. Carbohydrate derivatives give weak or no molecular ions on EI-MS. Molecular weight, however, may be determined by chemical ionization (CI), fast atom bombardment (FAB), field desorption (FD) or field ionization (FI) techniques. These four types of MS techniques also give simpler spectra than does EI, but usually give less structural information. The MS that will be discussed in this chapter are from compounds depicted in the two schemes in Figures 5 and 7. They include several derivatives of neutral and acidic alditols and glycosides.

B. Alditols

In order to avoid the formation of anomeric mixtures of derivatives, reducing sugars can be converted into acyclic compounds. This can be accomplished by reduction of the monosaccharides with sodium borohydride. This will simplify separation of mixture of sugars analyzed by GLC-MS. Methyl glycosides can also be transformed into alditols after acid hydrolysis and reduction (see Figure 5). However, caution must be exercised, since two different aldoses may give the same alditol owing to symmetry properties. This can be eliminated by using sodium borodeuteride as the reducing agent, which will label the anomeric carbon C-1 for aldoses and C-2 for ketoses.

Systematic studies on the MS of alditol acetates, alditol methyl ethers and alditol TMs ethers[72-77] have revealed a simple and similar mode of fragmentation. These derivatives do not give a molecular ion and will not differentiate between stereoisomers (glucose, galactose, and mannose). In contrast, direct observation of molecular ions is frequently possible in chemical ionization MS using isobutane or ammonia as reagent gas.[78,79] Primary fragments are formed by α-cleavage of the alditol chain, and secondary fragments are produced by subsequent elimination (s) of acetic acid, methanol, and trimethylsilanol, respectively.

MS of partially methylated alditol acetates has been extensively studied and the mode of fragmentation has been determined by using the deuterium-labeling technique.[80] A valuable collection[43] of MS, together with retention time data, has been published for partially methylated alditol acetates from hexoses, pentoses, 6-deoxyhexoses and some heptoses and dideoxyhexoses.

The major fragment ions and their origins of the EI-MS of 1,4,5-tri-*O*-acetyl-2,3,6-tri-*O*-ethylglucitol, obtained by hydrolysis of ethylcellulose, are illustrated in Figure 9. The intensities of the primary fragments given by the alditol acetates are much lower than the intensities of the corresponding fragments from partially alkylated alditol acetates. The most stable ions (highest intensities) seem to be derived from fission between adjacent carbon atoms carrying alkoxyl groups (Scheme I.1). This could be explained by the stabilizing effect of the alkoxyl group on the generated positive ion. Less stable ions are derived by fission between adjacent carbon atoms carrying alkoxyl and acetoxyl or TMS groups (Scheme I.2) with the former group preferentially carrying the positive charge. Fission between carbon atoms carrying acetoxyl or TMS groups results in low intensity fragments (Scheme I.3).

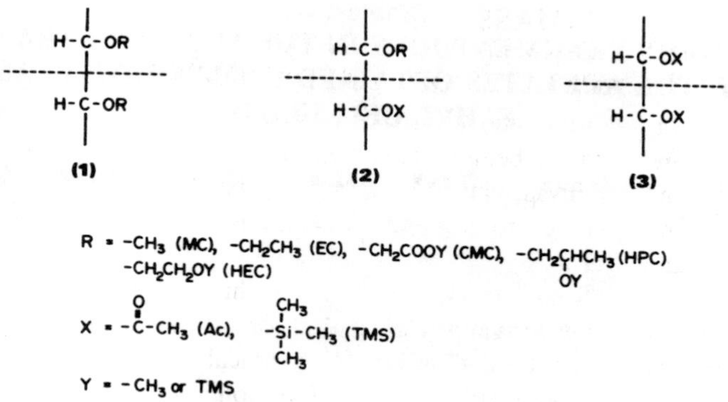

Scheme I

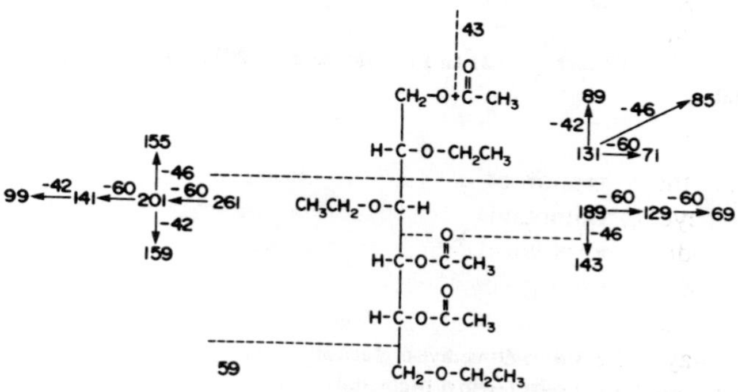

FIGURE 9. Major fragmentation ions from 1,4,5-tri-*O*-acetyl-2,3,6-tri-*O*-ethyl-D-glucitol obtained from ethyl cellulose.

The secondary fragments observed may be derived from primary fragments by single or consecutive eliminations for ethyl cellulose, of acetic acid (60), ketene (42), ethanol (46), or acetaldehyde (44) (see Figure 9). The corresponding compounds for methylcellulose would be methanol (32) and formaldehyde (30). The fragmentation of partially ethylated alditol acetates follows the same pattern as for the methylated and carboxymethyl derivatives.[34,40] The major peaks from partially methylated and ethylated alditol acetates obtained from methylcellulose and ethyl cellulose are shown in Table 1.

The mass fragmentation of partially methylated alditol trimethylsilyl (TMS) ethers follows a similar pattern.[34] The primary fragments formed by α-cleavage of the alditol chain and secondary fragments are produced by subsequent elimination(s) of trimethylsilanol (m/z 90). The number of carbons in the alditol chain can be determined from $(M\text{-}CH_3)^+$ and $(M\text{-}MeSiOH)^+$ ions. The major peak in most of the spectra was the TMS ion (m/z 73).

Aldononitrile acetates and partially methylated aldononitrile acetates have been analyzed by GLC-MS.[81] The MS of aldononitrile acetates contain characteristic peaks and are easy to interpret. Although none of the derivatives yields the molecular ion, all gave ions at m/z M-73 and M-98, corresponding to the loss of CH_2OAc and –CHOAcCN moieties, respectively. The base peak was m/z 43 (CH_3CO^+). The fragmentation of the partially methylated aldononitrile acetates was shown to be analogous to that described for methylated alditol acetates, i.e., the fission takes place mainly between carbon atoms carrying vicinal methoxyl groups.

The derivatives of the types commonly used for neutral alkylated sugars, namely partially alkylated alditol acetates and partially methylated aldononitriles, are insufficiently volatile to be suitable for analysis of uronic acids, aldehydo, or carboxymethyl groups.

Table 1
PRIMARY FRAGMENT IONS IN THE MASS SPECTRA OF ALDITOL ACETATES OBTAINED FROM METHYL- AND ETHYL CELLULOSE

Position of alkyl group	45 m/z	117 m/z	161 m/z	189 m/z	233 m/z	261 m/z
Methylcellulose						
2,3,6[a]	+	+	+		+	
2,3		+				+
2,6	+	+				
3,6	+			+	+	
2		+				
3				+		+
6	+					

	59 m/z	131 m/z	189 m/z	203 m/z	261 m/z	275 m/z
Ethyl cellulose						
2,3,6[b]	+	+	+		+	
2,3		+	+			+
2,6	+	+				
3,6	+			+	+	
2		+				
3				+		+
6	+					

[a] 1,4,5-tri-*O*-acetyl-2,3,6-tri-*O*-methyl-D-glucitol
[b] 1,4,5-tri-*O*-acetyl-2,3,6-tri-*O*-ethyl-D-glucitol

GLC-MS has been shown to offer a convenient method for separation and identification of hydroxy dicarboxylic acids as open chain TMS derivatives.[82,83] The different acids can be readily identified from their characteristic spectra. The most prominent fragmentations involving the rupture of one bond are the loss of a silicon-linked methyl group $(M-15)^+$ and the formation of α-cleavage ions by carbon chain cleavage.

Aldonic acids can also be characterized by GLC-MS of the corresponding per(trimethylsilyl)ated aldonolactones.[84] It was found tht 1,4- and 1,5-lactones give significant differences in their spectra and it was concluded that the diastereoisomers studied could be readily distinguished from each other.

Alditol trifluoroacetates have also found use in GLC-MS.[85,86] The primary fragmentation pattern is similar to that of alditol acetates, but the molecular ion is generally present. Secondary fragments are formed by elimination of trifluoroacetic acid or trifluoroacetic anhydride.

C. Glycosides

The MS of TMS glycosides have been investigated in great detail by using deuterium-labeling techniques and high-resolution MS.[87] Glycosides obtained by peracetylation, permethylation and per(trimethylsilyl)ation follow the same fragmentation pathways and show ions derived from a number of specific degradation pathways.[72-75,88,89] The molecular ion appeared to be weak, but stronger fragments appeared at $(M-CH_3)^+$, $(M-CH_3-Ac)^+$, $(M-CH_3-CH_3)^+$ and $(M-CH_3-Me_3SiOH)^+$. Stereoisomers of carbohydrate derivatives give similar MS, and the small differences in peak intensity sometimes observed do not generally allow an unambiguous assignment of configuration. Thus α- and β-hexopyranoside derivatives give virtually identical MS. Acids will be transformed into methyl esters during conditions used for methanolysis. The mass fragmentation of the resulting methyl ester methyl glycosides of uronic acids has been reported.[90,91]

REFERENCES

1. **Felcht, U.-H.**, Cellulose ethers — synthesis, application and analytical aspects, in *Cellulose and Its Derivatives,* Kennedy, J. F., Philips, G. O., Wedlock, D. J., and Williams, P. A., Eds., John Wiley & Sons, New York, 1985, 273.
2. **Wadsworth, L. C. and Daponte, D.**, Cellulose esters, in *Cellulose Chemistry and Its Application,* Nevell, T. P. and Zeronian, S. H., Eds., John Wiley & Sons, New York, 1985, 344.
3. **Bouveng, H. O., Garegg, P. J., and Lindberg, B.**, Position of the *O*-acetyl groups in birch xylan, *Acta Chem. Scand.*, 14, 742, 1960.
4. **Meier, H.**, Isolation and characterization of an acetylated glucomannan from pine (*Pinus silvestris* L.), *Acta Chem. Scand.*, 15, 1381, 1961.
5. **Bethge, P. O. and Lindstrom, K.**, Determination of *O*-acetyl groups in wood, *Svensk Papperstidn.*, 76, 645, 1973.
6. **Sugihara, J. M.**, Relative reactivities of hydroxyl groups of carbohydrates, *Adv. Carbohydr. Res.*, 8, 1, 1953.
7. **Bonner, W. A.**, C1-C2 Acetyl migration on methylation of the anomeric 1,3,4,6-tetra-*O*-acetyl-D-glucopyranoses, *J. Org. Chem.*, 24, 1388, 1959.
8. **Bouveng, H. O.**, Phenylisocyanate derivatives of carbohydrates. II. Location of *O*-acetyl groups in birch xylan, *Acta Chem. Scand.*, 15, 96, 1961.
9. **Leigh, W. R. D. and Krzeminski, Z. S.**, A method for locating the *O*-acetyl groups in secondary cellulose acetates, *J. Chem. Soc. C:*, 1700, 1966.
10. **de Belder, A. N. and Norrman, B.**, The distribution of substituents in partially acetylated dextran, *Carbohydr. Res.*, 8, 1, 1968.
11. **Hakomori, S.-I.**, A rapid permethylation of glycolipid, and polysaccharide catalyzed by methylsulfinyl carbanion in dimethyl sulfoxide, *J. Biochem.*, 55, 205, 1964.
12. **Björndal, H., Lindberg, B., and Rosell, K.-G.**, Distribution of substituents in partially acetylated cellulose, *J. Polym. Sci. Part C.*, 36, 523, 1971.
13. **Lindberg, B. and Rosell, K.-G.**, Hydrolysis and migration of *O*-acetyl groups during preparation of chlorite holocellulose, *Svensk Papperstidn.*, 77, 286, 1974.
14. **Lindberg, B., Rosell, K.-G., and Svensson, S.**, Positions of *O*-acetyl groups in birch xylan, *Svensk Papperstidn.*, 76, 30, 1973.
15. **Lindberg, B., Rosell, K.-G., and Svensson, S.**, Positions of *O*-acetyl groups in pine glucomannan, *Svensk Papperstidn.*, 76, 383, 1973.
16. **Kenne, L., Rosell, K.-G., and Svensson, S.**, Studies on the distribution of *O*-acetyl groups in pine glucomannan, *Carbohydr. Res.*, 44, 69, 1975.
17. **McComb, E. A. and McCready, R. M.**, Determination of acetyl in pectin and in acetylated carbohydrate polymers, *Anal. Chem.*, 29, 819, 1957.
18. **Goodlett, V. W., Dougherty, J. T., and Patton, H. W.**, Characterization of cellulose acetates by nuclear magnetic resonance, *J. Polym. Sci. Part A*, 9, 155, 1971.
19. **Ennor, K. S., Honeyman, J., and Stening, T. C.**, Sugar nitrates, *Chem. Ind. (London)*, 1308, 1956.
20. **Ennor, K. S. and Honeyman, J.**, Sugar nitrates. Part V. Removal of nitrate groups, *J. Chem. Soc.*, 3586, 1958.
21. **Wu, T. K.**, Carbon-13 and proton nuclear magnetic resonance studies of cellulose nitrates, *Macromolecules*, 13, 74, 1980.
22. **Patterson, P. M., Patterson, D. J., Blackwell, J., Koenig, J. L., Jamieson, A. M., Carignan, Y. P., and Turngren, E. V.**, High resolution solid-state ^{13}C-nmr spectroscopy of cellulose nitrates, *J. Polym. Sci. Polym. Phys. Ed.*, 23, 483, 1985.
23. **Lipkin, D., Philips, B. E., and Abrell, J. W.**, The action of hydrogen fluoride on nucleotides and other esters of phosphorus(V) acids, *J. Org. Chem.*, 34, 1539, 1968.
24. **Ohashi, K., Terada, T., Kohno, T., Hosomi, S., Mizoguchi, T., and Uehara, K.**, Enzymatic isomerization and epimerization of D-erythrose 4-phosphate and its quantitative analysis by gas chromatography/mass spectrometry, *Eur. J. Biochem.*, 142, 347, 1984.
25. **Brade, H.**, Identification and quantification of methyl ester groups in methylated sugar acids and phosphates by g.l.c.-m.s. after alkaline transesterification with sodium ethoxide, *Carbohydr. Res.*, 149, 451, 1986.
26. **Percival, E.**, Desulfation of polysaccharides,, *Methods Carbohydr. Chem.*, 8, 281, 1980.
27. **Schweiger, R. G.**, New cellulose sulfate derivatives and applications, *Carbohydr. Res.*, 70, 185, 1979.
28. **Lahaye, M., Yaphe, W., and Rochas, C.**, ^{13}C-N.m.r. spectral analysis of sulfated and desulfated polysaccharides of agar type, *Carbohydr. Res.*, 143, 240, 1985.
29. **Nicholson, M. D. and Merritt, F. M.**, Cellulose ethers, in *Cellulose Chemistry and Its Applications,* Nevell, T. P. and Zeronian, S. H., Eds., John Wiley & Sons, New York, 1985, 364.
30. **Jullander, I. and Lagerström, O.**, Ethyl ethers and their analytical determination, *Methods Carbohydr. Chem.*, 3, 303, 1963.

31. **Hodges, K. L., Kester, W. E., Wiederrich, D. L., and Grover, J. A.,** Determination of alkoxyl substitution in cellulose ethers by Zeisel-gas chromatography, *Anal. Chem.*, 51, 2172, 1979.
32. **Miller, T. G. and Hronek, R. J.,** Determination of carboxymethyl substitution in cellulose ethers by Zeisel reaction and liquid chromatography, *Anal. Chem.*, 57, 2091, 1985.
33. **Albersheim, P., Nevins, D. J., English, P. D., and Karr, A.,** A method for analysis of sugar in plant cell-wall polysaccharides by gas-liquid chromatography, *Carbohydr. Res.*, 5, 340, 1967.
34. **Rosell, K.-G.,** Unpublished results.
35. **Buytenhuys, F. A. and Bonn, R.,** Distribution of substituents in CMC, *Papier (Darmstadt)*, 31, 525, 1977.
36. **Sweeley, C. C., Bentley, R., Makita, M., and Wells, W. W.,** Gas-liquid chromatography of trimethylsilyl derivatives of sugars and related substances, *J. Am. Chem. Soc.*, 85, 2497, 1963.
37. **Laine, R. A. and Sweeley, C. C.,** Analysis of trimethylsilyl *O*-methyloximes of carbohydrates by combined gas-liquid chromatography-mass spectrometry, *Anal. Biochem.*, 43, 533, 1971.
38. **Laine, R. A. and Sweeley, C. C.,** *O*-Methyl oximes of sugars. Analysis as trimethylsilyl derivatives by gas-liquid chromatography and mass spectrometry, *Carbohydr. Res.*, 27, 199, 1973.
39. **Vilkas, M., Hiu-I-Jan, Boussac, G., and Bonnard, M. C.,** Cromatographie en phase vapeur de sucres a l'etat de trifluotacetates, *Tetrahedron Lett.*, 1441, 1966.
40. **Bourne, E. J., Tatlow, C. E. M., and Tatlow, J. C.,** Studies of trifluoroacetic acid. II. Preparation and properties of some trifluoroacetyl esters, *J. Chem. Soc.*, 1367, 1950.
41. **Björndal, H., Hellerqvist, C. G., Lindberg, B., and Svensson, S.,** Gas-liquid chromatography and mass spectrometry in methylation analysis of polysaccharides, *Angew. Chem. Int. Ed. Engl.*, 9, 610, 1970.
42. **Sawardeker, J. S., Sloneker, J. H., and Jeanes, A.,** Quantitative determination of monosaccharides as their alditol acetates by gas liquid chromatography, *Anal. Chem.*, 37, 1602, 1965.
43. **Jansson, P.-E., Kenne, L., Liedgren, H., Lindberg, B., and Lönngren, J.,** A practical guide to methylation analysis of carbohydrates, *Chem. Commun. Univ. Stockholm*, 8, 1, 1976; *Chem. Abstr.*, 87, 136, 1977.
44. **Churms, S. C.,** Carbohydrates, Volume I, in *Handbook of Chromatography*, Zweig, G. and Sherma, J., Eds., CRC Press, Boca Raton, Fla., 1982.
45. **Dutton, G. G. S.,** Applications of gas-liquid chromatography, *Adv. Carbohydr. Chem. Biochem.*, 28, 11, 1973.
45a. **Dutton, G. G. S.,** Applications of gas-liquid chromatography, *Adv. Carbohydr. Chem. Biochem.*, 30, 9, 1974.
46. **Croon, I. and Lindberg, B.,** The distribution of substituents in partially methylated cellulose. Homogeneous and heterogeneous reaction with methyl sulfate, *Svensk Papperstidn.*, 60, 843, 1957.
47. **Croon, I.,** Distribution of substituents in partially methylated celluloses. IV. Reaction with diazomethane in moist ethereal solution, *Svensk Papperstidn.*, 62, 700, 1960.
48. **Croon, I.,** The distribution of substituents in partially methylated celluloses. III. Heterogeneous reaction with methyl chloride, *Svensk Papperstidn.*, 61, 919, 1959.
49. **Parfondry, A. and Perlin, A. S.,** ^{13}C-N.m.r. spectroscopy of cellulose ethers, *Carbohydr. Res.*, 57, 39, 1977.
50. **Jacin, H. and Slanski, J. M.,** Quantitative determination of ethoxyl in *O*-ethylcellulose and in ethylhydroxyethyl cellulose, *Anal. Chem.*, 42, 801, 1970.
51. **Croon, I. and Flamm, E.,** The distribution of substituents in ethyl cellulose, *Svensk Papperstidn.*, 61, 963, 1958.
52. **Sweet, D. P., Albersheim, P., and Shapiro, R. H.,** Partially ethylated alditol acetates as derivatives for elucidation of the glycosyl linkage-composition of polysaccharides, *Carbohydr. Res.*, 40, 199, 1975.
53. **Timell, T. E.,** A method for estimating the distribution of the substituents in partially substituted carboxymethylcelluloses, *Svensk Papperstidn.*, 55, 649, 1952.
54. **Timell, T. E. and Spurlin, H. M.,** The distribution of the substituents in partially substituted carboxymethylcellulose, *Svensk Papperstidn.*, 55, 700, 1952.
55. **Croon, I. and Purves, C. B.,** The distribution of substituents in partially substituted carboxymethyl cellulose, *Svensk Papperstidn.*, 62, 876, 1959.
56. **McNeil, M., Szalecki, W., and Albersheim, P.,** Quantitative methods for determining the points of *O*-(carboxymethyl) substitution in *O*-(carboxymethyl)-guar, *Carbohydr. Res.*, 131, 139, 1984.
57. **Taylor, R. L. and Conrad, H. E.,** Stoichiometric depolymerization of polyuronides and glycosaminoglycuronans to monsaccharides following reduction of their carbodiimide-activated carboxyl groups, *Biochemistry*, 11, 1383, 1972.
58. **Croon, I. and Lindberg, B.,** An investigation of the distribution of substituents in hydroxyethyl cellulose, *Svensk Papperstidn.*, 59, 794, 1956.
59. **Ramnäs, O. and Samuelson, O.,** Determination of the substituents in hydroxyethyl cellulose, *Svensk Papperstidn.*, 71, 674, 1968.

60. **McNeil, M. and Albersheim, P.,** A quantitative method to determine the points of *O*-(2-hydroxypropyl) substitution in *O*-(2-hydroxypropyl)guar, *Carbohydr. Res.*, 131, 131, 1984.
61. **Nevell, T. P.,** Oxidation of cellulose, in *Cellulose Chemistry and Its Applications,* Nevell, T. P. and Zeronian, S. H., Eds., John Wiley & Sons, New York, 1985, 243.
62. **Petersson, G.,** Gas-chromatographic analysis of sugars and related hydroxy acids as acyclic oxime and ester trimethylsilyl derivatives, *Carbohydr. Res.*, 33, 47, 1974.
63. **Clode, D. M. and Horton, D.,** Synthesis of the 6-aldehydo derivative of cellulose, and a mass-spectrometric method for determining position and degree of substitution by carbonyl groups in oxidized polysaccharides, *Carbohydr. Res.*, 19, 329, 1971.
64. **Raunhardt, O., Schmidt, H. W. H., and Neukom, H.,** Gas chromatography investigation of uronic acids and uronic acid derivatives, *Helv. Chim. Acta,* 50, 1267, 1967.
65. **Kennedy, J. F., Robertson, S. M., and Stacey, M.,** GLC of the *O*-trimethylsilyl derivatives of hexuronic acids, *Carbohydr. Res.*, 49, 243, 1976.
66. **Perry, M. B. and Hulyalkar, R. K.,** The analysis of hexuronic acids in biological materials by gas-liquid partition chromatography, *Can. J. Biochem.*, 43, 573, 1965.
67. **Bosso, C., Defaye, J., Gadelle, A., Wong, C. C., and Pedersen, C.,** Homopolysaccharides interaction with the dimethyl sulphoxide-paraformaldehyde cellulose solvent system. Selective oxidation of amylose and cellulose at secondary alcohol groups, *J. Chem. Soc. Perkin Trans. 1,* 1579, 1982.
68. **Theander, O.,** The oxidation of glycosides. VIII. The degradation of methyl α-D-3-oxo-glucopyranoside, methyl β-D-3-oxo-glucopyranoside, and methyl β-D-2-oxo-glucopyranoside, by lime water, *Acta Chem. Scand.*, 12, 1887, 1958.
69. **Green, J. W.,** Determination of carbonyl groups, *Methods Carbohydr. Chem.*, 3, 49, 1963.
70. **Tajima, K.,** New approaches in oxycellulose chemistry: Preparation and reactions, *J. Appl. Polym. Sci. Appl. Polym. Symp.*, 37, 709, 1983.
71. **Ishii, T.,** Preparation of 2,3-dialdehyde polysaccharides from lignocellulosic materials, *Mokuzai Gakkaishi,* 31, 361, 1985.
72. **Lönngren, J. and Svensson, S.,** Mass spectrometry in structural analysis of natural carbohydrates, *Adv. Carbohydr. Chem. Biochem.*, 29, 41, 1974.
73. **Kochetkov, N. K. and Chizhov, O. S.,** Mass spectrometry of carbohydrate derivatives, *Adv. Carbohydr. Chem.*, 21, 39, 1966.
74. **Kochetkov, N. K. and Chizhov, O. S.,** Mass spectrometry of carbohydrates, *Methods Carbohydr. Chem.*, 6, 540, 1972.
75. **Chizhov, O. S., Molodtsov, N. V., and Kochetkov, N. K.,** Mass spectrometry of trimethylsilyl ethers of carbohydrates, *Carbohydr. Res.*, 4, 273, 1967.
76. **Petersson, G.,** Mass spectrometry of alditols as trimethylsilyl derivatives, *Tetrahedron,* 25, 4437, 1969.
77. **Petersson, G. and Samuelson, O.,** Determination of the number and position of methoxyl groups in methylated aldohexoses by mass spectrometry of their trimethylsilyl derivatives, *Svensk Papperstidn.*, 20, 731, 1968.
78. **McNeil, M. and Albersheim, P.,** Chemical-ionization mass spectrometry of methylated hexitol acetates, *Carbohydr. Res.*, 56, 239, 1977.
79. **Murata, T. and Takahashi, S.,** Characterization of *O*-trimethylsilyl derivatives of D-glucose, D-galactose and D-mannose by gas-liquid chromatography chemical-ionization mass spectrometry with ammonia as reagent gas, *Carbohydr. Res.*, 62, 1, 1978.
80. **Björndal, H., Lindberg, B., and Svensson, S.,** Mass spectrometry of partially methylated alditol acetates, *Carbohydr. Res.*, 5, 433, 1967.
81. **Dmitriev, B. A., Backinowsky, L. V., Chizhov, O. S., Zolotarev, B. M., and Kochetkov, N. K.,** Gas-liquid chromatography and mass spectrometry of aldononitrile acetates and partially methylated aldononitrile acetates, *Carbohydr. Res.*, 19, 432, 1971.
82. **Petersson, G.,** Mass spectrometry of aldonic and deoxyaldonic acids as trimethylsilyl derivatives, *Tetrahedron,* 26, 3413, 1970.
83. **Petersson, G.,** Mass spectrometry of hydroxy dicarboxylic acids as trimethylsilyl derivatives. Rearrangement fragmentations, *Org. Mass Spectrom.*, 6, 565, 1972.
84. **Petersson, G., Samuelson, O., Anjou, K., and von Sydow, E.,** Mass spectrometric identification of aldonolactones as trimethylsilyl ethers, *Acta Chem. Scand.*, 21, 1251, 1967.
85. **Chizhov, O. S., Dmitriev, B. A., Zolotarev, B. M., Chernyak, A. Ya., and Kochetkov, N. K.,** Mass spectra of alditol trifluoroacetates, *Org. Mass Spectrom.*, 2, 947, 1969.
86. **König, W. A.,** Gaschromatographie und Massenspektrometrie von Kohlehydraten, *Z. Naturforsch.*, 29C, 1, 1974.
87. **DeJongh, D. C., Radford, T., Hribar, J. D., Hanessian, S., Bieber, M., Dawson, G., and Sweeley, C. C.,** Analysis of trimethylsilyl derivatives of carbohydrates by gas chromatography and mass spectrometry, *J. Am. Chem. Soc.*, 91, 1728, 1969.

88. **Biemann, K., DeJongh, D. C., and Schnoes, H. K.,** Application of mass spectrometry to structure problems. VIII. Acetates of pentoses and hexoses, *J. Am. Chem. Soc.*, 85, 1763, 1963.
89. **Heyns, K., Sperling, K. R., and Grützmacher, H. F.,** Combination of gas-chromatography and mass spectrometry for analysis of partially methylated sugar derivatives. The mass spectra of partially methylated methyl glycosides, *Carbohydr. Res.*, 9, 79, 1969.
90. **Kováĉik, V., Bauer, S., Rosik, J., and Kováĉ, P.,** Mass spectrometry of uronic acid derivatives. III. The fragmentation of methyl ester methyl glycosides of methylated uronic and aldobiouronic acids, *Carbohydr. Res.*, 8, 282, 1968.
91. **Kováĉik, V., Mihalov, V., and Kováĉ, P.,** Identification of methyl (methyl *O*-acetyl-*O*-methyl(hexopyranosid)uronates, by mass spectrometry, *Carbohydr. Res.*, 54, 23, 1977.

Chapter 12

MISCELLANEOUS GAS CHROMATOGRAPHIC (GC) ANALYSES OF CARBOHYDRATES

Peter Englmaier, Susumu Honda, and Christopher J. Biermann

TABLE OF CONTENTS

I. ANALYSIS OF SUGAR DERIVATIVES

The analysis of sugar phosphates and glycosides is considered in this section. The analysis of sugar phosphates most often involves trimethylsilyl (TMS) derivatives, whereas the glycosides are analyzed as TMS, acetate, or trifluoroacetate derivatives. More information on these derivatives may be obtained from the appropriate chapters in this book.

A. Sugar Phosphates

Sugar phosphates are important in regards to intermediary carbohydrate metabolism. In this capacity, these compounds occur freely; consequently, acid hydrolysis of sugar phosphate polymers is not usually a consideration for their analysis. Related compounds, like the phosphated inositols and nucleotides, are often analyzed by similar methods as the sugar phosphates. Sugar phosphates, along with neutral sugars, have been extracted from nematode tissue using distilled water.[1] The identity of the extracted sugars was confirmed by mass spectrometry (MS) of sample and standard monosaccharides. Most other investigations employed commercial standards in method development studies.

Sugar phosphates are derivatized in one of several general methods. The first method involves permethylation of the phosphate group(s) with diazomethane in methanol followed by silylation of the hydroxyl groups.[2] The second method involves the use of bis(trimethylsilyl)acetamide (BSA) as the silylation reagent,[3,4] though some workers claim that BSA tends to give inconsistent results for simple sugars.[5] Bis(trimethylsilyl)trifluoroacetamide (BSTFA) has also been used in acetonitrile at 80°C for 10 min.[6] The last method, and perhaps the most common, involves direct silylation using hexamethyldisilazane (HMDS) and trimethylchlorosilane (TMCS) in pyridine at elevated temperatures.[1,7-9] The use of dimethylsulfoxide (DMSO) and cyclohexane allowed silylation with HMDS and TMCS to occur at room temperature.[5] In the latter study, sensitivity was 0.007 μmol for fructose 6-phosphate, due to some decomposition of this derivative in the gas chromotagraph (GC). In one study, each of several types of substituted oximes were prepared prior to silylation in order to reduce the number of products.[6] Figure 1 shows the structure of hexa-*O*-trimethylsilyl-D-glucopyranose 6-phosphate as an example of direct silylation of a sugar phosphate.

Some of the more interesting results from various studies are analysis of phosphated sugars is a useful method of identifying the presence of enzymes such as cyclase, and the sensitivity may be increased dramatically with the flame ionization thermionic detector.[4] The aldose-1-phosphates require derivatization by suitable silylation reagents in order to get good results.[6,7] Consecutive injections into the GC of derivatized sugar phosphates should be made until the results are consistent, since sites on the column may irreversibly absorb some of the derivatives.[4]

TMS derivatives of sugar phosphates have been analyzed using nonpolar phases such as SE-30. Elution temperatures of the sugar phosphates are on the order of 170 to 230°C. There is very little literature regarding the use of capillary columns for the separation of sugar phosphates. The MS of sugar phosphates and related compounds has also been studied.[1,6,10]

B. Sugar Glycosides

Methyl glycosides have already been considered in Chapter 3 in the section on methanolysis. A wide variety of other glycosides are, of course, possible. The analysis of phenolic glycosides was used for chemotaxonomical screening of northern willow twigs.[11] The phenolic glycosides (e.g., salicin, fragilin, picein, salidrosid, triandrin, tremuloidin, and populin) were extracted, purified, derivatized by silylation, and analyzed on a capillary column with SE-54 stationary phase. The glycosides were eluted at temperatures of 190 to 295°C.

In another study, hexopyranosides and hexofuranosides were studied as silylated, ace-

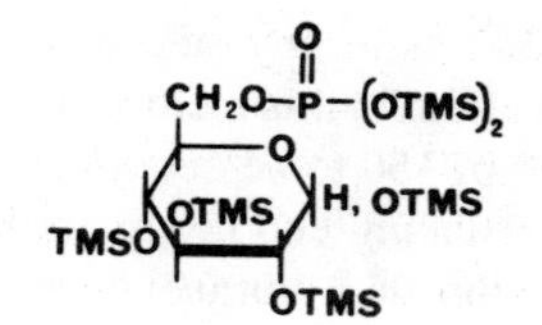

FIGURE 1. The structure of hexa-*O*-trimethylsilyl-D-glucopyranose 6-phosphate.

tylated, and trifluoroacetylated derivatives.[12] The retention times of methyl, ethyl, phenyl, *m*-tolyl, guaiacyl, and 2-naphthyl D-glucosides and D-galactosides were investigated, as well as the methyl glycosides of D-mannose and D-glucosiduronic acid.

II. BUTANE BORONIC ESTER DERIVATIVES

Cyclic butaneboronic esters have been used for GC analysis of substances with diol groups. Because $BuB(OH)_2$ (butaneboronic acid) is difunctional, steric requirements often favor the formation of one anomer.[13,14] Consequently, chromatograms are simplified compared to direct silylation or analysis of monosaccharides as other glycosides. L-Fucose, L-arabinose, D-xylose, D-glucose, and D-fructose gave single peaks; D-galactose and D-mannose show minor peaks in addition to a single major peak. Hexoses require silylation after treatment with butaneboronic acid to form suitably volatile derivatives. Silylation of the pentitols after reaction with butaneboronic acid simplified their analysis as well.[14] In related work, the butaneboronate-TMS derivatives of polyhydroxyalkylpyrazines have been investigated.[15] Fairly specific detection of boron is possible using the alkali flame ionization detector (AFID). The conditions for optimal response with this detector were investigated.[16] More information on these derivatives is available in a review article,[17] and MS of these derivatives has been investigated.[14,15]

III. DETERMINATION OF THE DEGREE OF POLYMERIZATION/END-GROUP ANALYSIS

The analysis of oligo- and polysaccharides for molecular weight determination involves a method of end-group analysis. Thus, one usually obtains the identity of the reducing end group at the same time the average degree of polymerization is determined. Generally, the reducing end group is modified so that it can be distinguished from the other monosaccharides. One simple way is to reduce the end group to give the corresponding alditol.[18] The polymer is then hydrolyzed. The alditols are then separated from the aldoses by use of a strongly basic ion exchange resin which binds the aldoses.[18] The aldoses are removed from the resin, reduced to the corresponding alditols, and analyzed separately from the other alditols by gas-liquid chromatography. The relative amount, as well as the structure of the original reducing end group, and, therefore, the degree of polymerization, can be calculated.

For polymers containing amino sugars, this procedure may be modified by nitrous acid deamination prior to acid hydrolysis, with subsequent analysis of the alditols, anhydroaldoses, and aldoses as peracetylated alditols, anhydroaldoximes, and aldononitriles, respectively, at the same time.[19] This simplifies the analysis somewhat and eliminates problems such as selective absorption on the ion exchange resin.[18] Other derivatives, which do not require deamination, may also be used.

In the case of polymers without amino sugars, the analysis simply consists of end-group reduction followed by analysis of the aldoses as aldononitrile acetates and the alditols as alditol acetates.[20] Since aldoses yield a different derivative from the alditols, the aldoses do not require separation from the alditols at any step of the analysis. This approach is also useful for silylation of aldoses and alditols; this method allowed the degree of polymerization (DP) of polysaccharides up to a DP of 150 to be determined with a fair degree of accuracy.[21] More elaborate derivatives of the reducing end group are possible, such as the formation of the methoxime followed by reduction with borane, to give the corresponding amine prior to acid hydrolysis.[22]

IV. SEPARATION OF D- AND L-ENANTIOMERS

The separation of any chiral pair by gas, or any, chromatography depends on the interaction of one chiral center, or molecule, relative to another. This is accomplished by introducing a chiral center in the molecule to be analyzed or by introducing a chiral center onto the stationary phase of the column. These methods are considered separately.

A. Use of Chiral Stationary Phases

Chiral stationary phases used in the analysis of monosaccharides include *N*-propionyl-L-valine-*tert*-butylamide polysiloxane for the analysis of alditol acetates[23] and L-valine-*tert*-(*S*)-α-phenylethylamide and its diastereoisomer bonded to XE-60 for the analysis of alditol trifluoroacetates,[24] as well as aldoses directly trifluoroacetylated or trifluoroacetylated after the formation of methyl glycosides.[25] It should be noted that when alditols are formed from some aldoses (ribose and xylose for example) chirality is lost since a symmetrical polyol is formed.[24]

B. Use of Chiral Derivatives

The introduction of chiral centers into sugars by suitable derivatization techniques allows the separation of sugar diastereoisomers without the use of chiral stationary phases. Many types of derivatives for this purpose have been investigated. Diastereoisomers of aldoses may be separated by oxidizing the aldose to the corresponding aldaric acid and forming an ester with a suitable chiral alcohol, such as one form of 2-butanol, 3-methyl-2-butanol, or 3,3-dimethyl-2-butanol, and derivatizing the remaining hydroxyl groups with acetic anhydride.[26] The more hindered the alcohol, the better the separation of diastereoisomers. These derivatives were separated on a capillary column with Carbowax® 20M. The derivatization procedure is somewhat involved, and hexoses are not sufficiently volatile for analysis on this column. Also, (−)-2-butanol has also been used to form glycosides of monosaccharides which were then analyzed as TMS derivatives on SE-30 capillary columns.[27] This method was applied to neutral sugars, uronic acids, and 2-acetamido-2-deoxy sugars, but is complicated by the presence of both anomers for each sugar. Additional considerations were the expense and lack of optical purity of R- and S-2-butanol, though these are no longer considerations.

Little[28] formed the dithioacetals of sugars by their reaction with (+)-1-phenylethanethiol which are acyclic derivatives. The method was applied to neutral aldoses which were subsequently acetylated or silylated and separated by capillary GC. The derivatives were separated on a SE-30 or SE-54 phase. Unfortunately, the acetylated derivatives were not investigated on more polar phases which may have given better resolution of the products, although good results were obtained with the SE-54 stationary phase.

Schweer prepared (−)-menthyloxime pertrifluoroacetates of sugars.[29] Typical of oximes in general, each sugar formed a *syn* and *anti* derivative. The derivatives were separated on a capillary column with OV-225. In most cases the *syn* and *anti* forms were not completely resolved, though this does not cause any difficulty, it merely simplifies the chromatogram somewhat. In an analogous study, Schweer prepared similar derivatives with (−)-bornyloximes separated on the same column with good results.[30] In all cases studied, the D- and

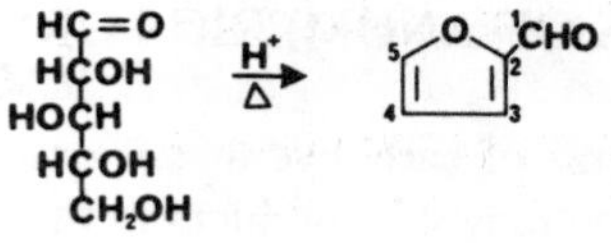

FIGURE 2. The conversion of xylose to 2-furaldehyde under acidic conditions and elevated temperature.

L-monosaccharides could be distinguished. Nitrogen was used as the carrier gas; had helium or hydrogen been used, perhaps some of the overlap could have been eliminated or at least the analysis time reduced. The TMS ethers of α-methylbenzylaminoalditols of 24 monosaccharides, including ketoses and *N*-acetyl amino sugars, were analyzed on a capillary Carbowax® 20M column.[31] Wtih 20 of the 24 monosaccharides, the diastereoisomers were resolved.

V. THERMAL/PYROLYSIS METHODS

A well-known reaction of pentoses is their conversion to 2-furaldehyde (furfural) under acidic conditions and elevated temperatures. The reaction is near quantitative if the furfural is distilled from the reaction mixture as it is produced. Aldohexoses give 5-(hydroxymethyl)furfural (HMF), although the yield is much lower since HMF is difficult to distill from the reaction mixture and undergoes secondary reactions. This reaction is the basis of an ingenious method to measure pentoses.[32] Figure 2 shows the reaction of xylose to furfural.

In this method, an aqueous fermentation sample was centrifuged, acidified to 0.25 *M* sulfuric acid, and then injected into a glass lined metal GC column packed with Chromosorb® 101 and operated at 200°C. Arabinose and xylose were measured as furfural, while rhamnose and fucose were measured as 5-methyl-2-furaldehyde. The sensitivity was on the order of 0.1 g/ℓ; the range was linear up to 3 g/ℓ of the pentoses. The hexoses glucose, galactose, and fructose did not interfere, although valeric acid did interfere. Schultz[33] attempted a similar method with hexoses (measuring HMF), but could not get reproducible results due to the complications of low product volatility and the occurrence of secondary reactions.

Pyrolysis gas chromatography (PGC) involves heating a sample at elevated temperatures in the absence of oxygen, followed by GC analysis of the liberated fragments. In practice this often involves the use of a specialized probe in the injector port of the GC. By studying the fragments, much can be learned about the original molecule. This method allows otherwise nonvolatile compounds to be studied by GC, if the pattern of the fragments gives useful information. Gardner et al., successfully studied the complicated structure of lignin by this method.[34]

The subject of analytical pyrolysis in molecules of biological importance[35] and in carbohydrates[36] has been reviewed. The technique is useful for such purposes as quantitative analysis of cellulose esters by analysis of the free acid which is liberated at 700°C, determination of the degree of substitution in hydroxyethyl starch by the amount of acetaldehyde liberated during pyrolysis, and determination of the amount of methylation in pectins by assay of the liberated methanol during pyrolysis.[35]

Morgan and Jacques applied this method to the study of mono-, di-, and trisaccharides.[37] This method allow distinction of (1 → 4) glycosidic linkages from (1 → 6) glycosidic linkages in disaccharides. They also took advantage of the reactivity of boric acid with carbohydrates to get modified fragments during pyrolysis. Related substances such as nucleotides and nucleosides may also be studied by PGC in milligram quantities.[35,38] Structural features are deduced from the proportion of various products.

VI. ANALYSIS OF ANOMERIC FORMS OF SUGARS

The analysis of anomeric forms of carbohydrates is easily accomplished by GC. In fact, with some derivatives, it is something of a problem in that each monosaccharide may yield two, three, or even more forms due to the various anomeric forms present. A monosaccharide present in two ring sizes (i.e., pyranose and furanose forms) with two anomers for each ring size and an acyclic form would give five peaks if silylated directly.

Bentley and Botlock[39] followed the activity of mutarotase by preparing penta-*O*-TMS derivatives of glucose solutions. The key to success is to prevent mutarotation during derivatization. This is accomplished by diluting a small volume of aqueous solution with dimethylformamide (DMF), freezing at liquid air temperature, and then derivatizing. This method has advantages over polarimetry in that both species are measured directly and turbid solutions can be analyzed.

VII. DIRECT DETERMINATION OF SUGAR ALCOHOLS

Following the hydrogenolysis of saccharides, some of the more volatile components could be analyzed by GC with 15% diglycerol on Chromosorb® W.[40] In this way, ethylene glycol, propane-1,2-diol, and butane-2,3-diol could be separated by programming from 70 to 135°C. This method may have use in analysis of fermentation broths where diols are the desired products.

In another study,[41] sugar alcohols in biological media were determined using a 0.5 *m* column packed with Polypak® I, 120 to 200 mesh (Hewlett-Packard) up to 250°C, with an injector temperature of 300°C, and nitrogen carrier gas. Sugar alcohols as large as mannitol were claimed to be separated by this method.

VIII. STRUCTURAL ANALYSIS BY PERIODATE OXIDATION

A. Introduction

Structural analysis by periodate oxidation has been discussed by Sharon.[42] Periodate oxidation, and its modified form known as Smith-degradation, involves periodate oxidation of sugars, followed by reduction with borohydride, and mild acid hydrolysis of the oligo- or polysaccharide. Free vicinal hydroxyl groups are subject to oxidation at the carbon-carbon linkage to give dialdehydes. An aldehyde next to a hydroxyl group is also subject to oxidation at the carbon-carbon linkage to give the corresponding carboxylic acid and aldehyde. Thus, sugars which are 3-*O*-substituted or 2,4-di-*O*-substituted do not contain free vicinal hydroxyl groups and are not subject to periodate oxidation. Based on the composition of the degradation products of oxidized sugar units and the amount of periodate consumed, much may be learned about the structure of complex carbohydrates.

The important periodate oxidation and Smith-degradation products from carbohydrates are glyceraldehyde, glycerol, ethylene glycol, erythritol, and others. Dutton et al.[43] lists the expected degradation products from arabinoxylan, glucomannan, galactoglucomannan, and arabinoxylan and analyzed these products as TMS derivatives.

Baird et al. analyzed some of these products as acetylated aldononitriles or alditols.[44] They discussed estimation of chain length based on the composition of the products. More recently, Turner and Cherniak studied these products in a similar fashion.[45]

B. Products of Periodate Oxidation

The structures of oligo- and polysaccharides, as well as the carbohydrate moieties of glycoconjugates, are complex because there are variations of monosaccharide composition, sequence, and position of attachment of the glycosidic linkage resulting in large numbers methyl cellobioside → (IO_4^-) → dialdehyde compound I (erythrose, glyceraldehyde, glyoxal, glyoxal)

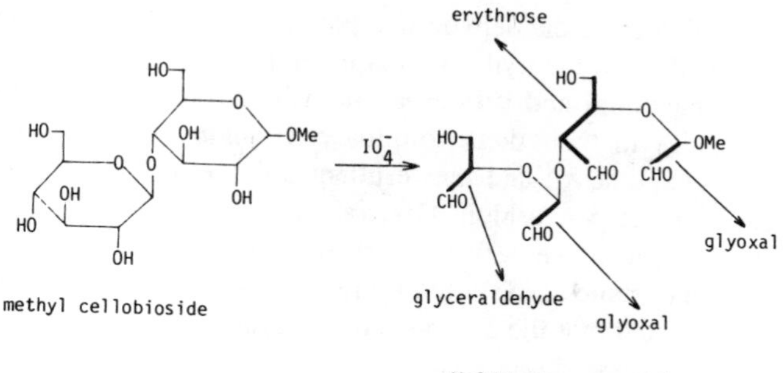

Scheme 1. Periodate oxidation of methyl cellobioside. Note that compound I is composed of one molecule each of glyceraldehyde and erythrose, and two molecules of glyoxal.

Table 1
THEORETICAL AMOUNTS OF CONJUGATED ALDEHYDES IN THE PRODUCTS OF PERIODATE OXIDATION OF METHYL ALDOHEXOBIOSIDES

Linkage type	Amount of aldehyde (mol/mol of aldohexobioside)				
	Glyoxal	Glyceraldehyde	Hydroxymalonaldehyde	Tetrose	Methyl hexobioside
1 → 2	1	2	1	0	0
1 → 3	1	1	0	0	1
1 → 4	2	1	0	1	0
1 → 6	2	2	0	0	0

of isomers. For example, the number of possible isomers of an aldohexobiose, based on the difference of monosaccharide composition is 256 (16^2), because each monosaccharide unit may be one of the possible 16 aldohexoses. Each biose built up from two different aldohexoses has two isomers based on sequence, hence the total number of isomers of an aldohexobiose, based on monosaccharide composition and sequence, amounts to 496 (= [2 × 256] − 16). Since there are six types of glycosidic linkages, i.e., 1 → 1, 1 → 2, 1 → 3, 1 → 4, 1 → 5, and 1 → 6, each of these isomers has six positional isomers. In addition, each type of linkage, except the 1 → 1 linkage, has two anomeric configurations. The 1 → 1 linkage may have three different modes, α,α,β,β;αβ and β,α if the two monosaccharides are different. The number of possible isomers of higher oligo- and polysaccharides are tremendous. Of course, all of these isomers do not occur in nature, but the number of isomers to be considered in practical analysis is still quite large.

Periodate oxidation has been one of the powerful tools to elucidate complex carbohydrate structures. The early studies of periodate oxidation analysis were, however, focused on the estimation of periodate consumption, because it was easily determined by simple techniques such as redox titration and spectrophotometry. Recent advancement of GC has made it possible to also determine the resultant products. The oxidation products are conjugated aldehydes, as exemplified by methyl cellobioside shown in Scheme 1.

The dialdehyde Compound I is composed of one molecule each of glyceraldehyde and erythrose and two molecules of glyoxal, linked through hemiacetal bonds. The aldehyde composition of the dialdehyde compound, produced by periodate oxidation of an aldohexobiose, is varied, dependent on the type of interglycosidic linkage, as indicated in Table 1. Therefore determination of the aldehyde composition can lead to elucidation of the linkage type.

The component aldehydes in dialdehyde compounds are partially released with aqueous solutions of mineral acids, but the hydrolysis is incomplete even at high temperatures. The use of a mixture of ethanethiol and trifluoroacetic acid, however, facilitates conversion of the component aldehydes to their diethyldithioacetals under very mild conditions.[46] The conversion is almost quantitative, and the resultant dithioacetals can easily be derivatized to their TMS ethers in a one-pot fashion. Glyoxal survives trimethylsilylation and is found as the bis(diethyldithioacetal). The derivatized products can be analyzed simultaneously by GLC using nonpolar phases such as OV-1. Figure 3 shows the results obtained from methyl glucobiosides. It is indicated that the component aldehydes were beautifully separated from each other in all cases of these disaccharides.

Table 2 gives some experimental data obtained from various glycosides of mono- and oligosaccharides, together with a few glycoproteins.[48]

All the molar ratios of component aldehydes in the products of periodate oxidation of the glycosides, obtained by this method, were in good agreement with the theoretical values given in parentheses, except for a few data involving hydroxymalonaldehyde.

The diethyldithioacetal method has another advantage that is also applicable to analysis of monosaccharide composition of intact and permethylated samples of carbohydrates. Figure 4 explains the multilateral use of this method for structure elucidation of tomatine, a model oligosaccharide glycoside.[47]

When this method was applied to the products of periodate oxidation, glycolaldehyde, glyceraldehyde, and glyoxal were found in molar proportions of 1.0:1:2.8, together with a slow-eluting compound, presumably assigned to the derivative of glucosylthreose. On the other hand, the native sample was hydrolyzed with 2 *M* trifluoroacetic acid, and the hydrolyzate derivatized to the TMS diethyldithioacetal. Xylose, glucose, and galactose were found in a 1:1.8:1.1 *M* ratio. Further, the sample was permethylated with the dimethylsulfinium carbanion in dimethylsulfoxide, hydrolyzed, and derivatized to the TMS dithioacetal in a similar manner. GC analysis indicated the presence of 2,3,4-tri-*O*-methylxylose, 2,3,4,6-tetra-*O*-methylglucose, 2,3,6-tri-*O*-methylgalactose and 4,6-di-*O*-methylglucose in a 0.8:1:1.0:0.9 *M* ratio. From these multiple experiments the structure of the oligosaccharide moiety of this glycoside was unequivocally determined as:

```
xylopyranose-(1→3)\
                   >glucopyranose-(1→4)-galactopyranose
glucopyranose-(1→2)/
```

Anomeric configurations of the interglycosidic linkages are unknown from these experiments. They should be determined by other methods such as nuclear magnetic resonance and examinations of the specificities to glycosidases.

Periodate oxidation analysis of reducing oligosaccharides suffers from a problem of over oxidation, which proceeds in a non-Malapradian fashion, leading to incorrect determination of their structures. The authors solved this problem by prior conversion of the reducing oligosaccharides to 1,5-anhydroalditol derivatives via acetobromo sugars.[49] It can readily be achieved by treatment of oligosaccharide samples with hydrogen bromide in acetic acid, followed by reduction of the resultant acetobromo sugars with lithium aluminum hydride. Periodate oxidation of the 1,5-anhydroalditol derivatives gave dialdehyde compounds containing a moiety characteristic of the position of the substitution in the anhydroalditol residue, as indicated in Figure 5.

Dithioacetalation, followed by trimethylsilylation of the periodate oxidation products resulting from the 1 → 2 (sophorose), 1 → 3 (laminarabiose), 1 → 4 (cellobiose), and 1 → 6 (gentiobiose) linked glucobioses, gave the derivatives of dialdehyde 1, 1,5-anhydroglucitol, dialdehyde 2, and dialdehyde 3, respectively, as shown in Figure 6.[49]

In addition to these characteristic fragments derived from the reducing ends of oligosac-

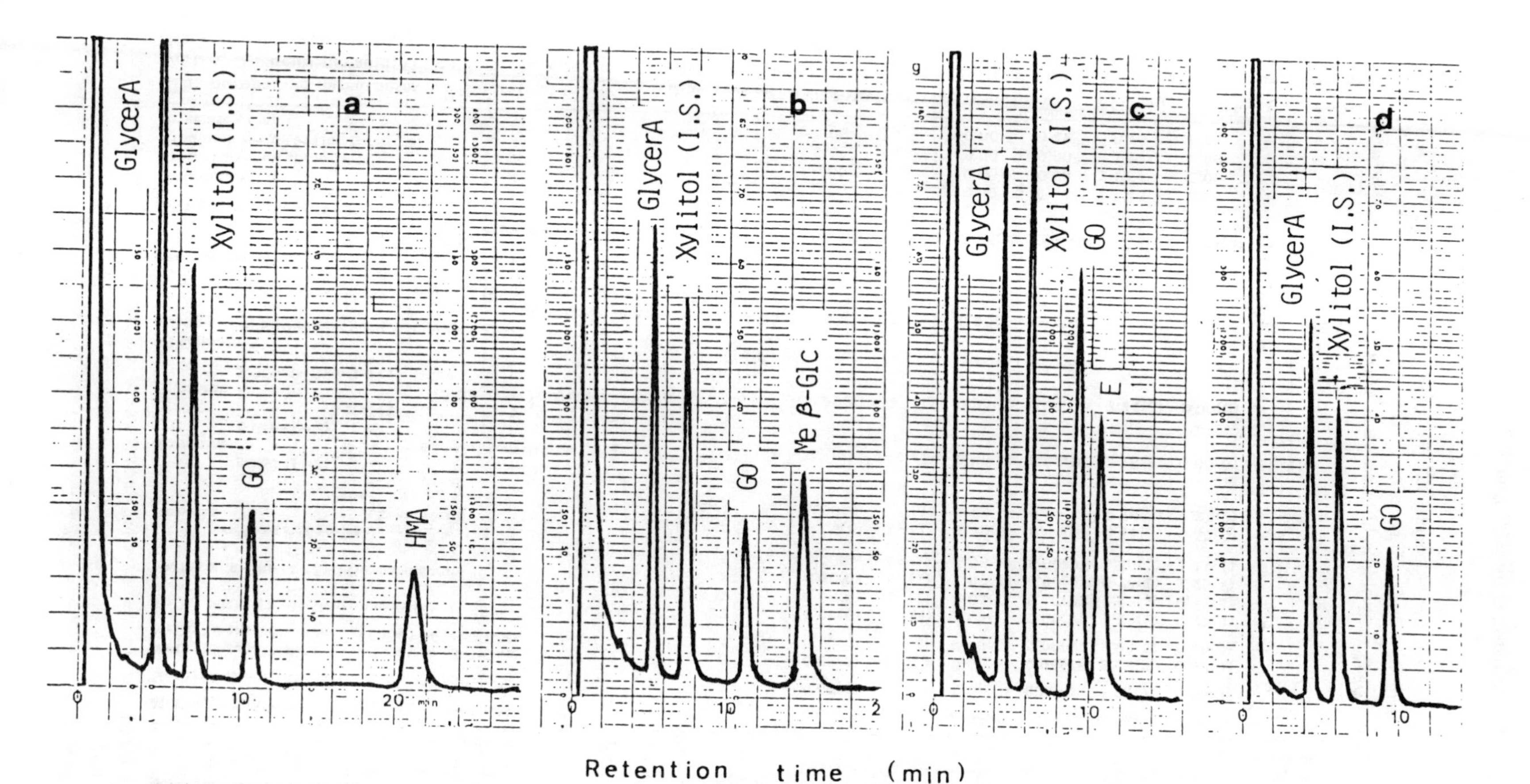

FIGURE 3. Analysis of the components of the periodate oxidation products, obtained from methyl glycosides of sophorose (a, 1→2 linkage), laminarabiose (b, 1→3 linkage), cellobiose (c, 1→4 linkage), and gentiobiose (d, 1→6 linkage), by the dithioacetal method. GO, derivativie of glyoxal; GlycerA, derivative of glyceraldehyde; HMA, derivative of hydroxymalonaldehyde; E, derivative of erythrose. Column, 3% silicone OV-1 on Chromosorb® W (2 m, 3 mm I.D.); a column temperature, 180°C; carrier, nitrogen (50 mℓ/min); detection, flame ionization.

Table 2
DETERMINATION OF THE CONJUGATED ALDEHYDES IN THE OXIDATION PRODUCTS OF VARIOUS OLIGOGLYCOSIDES[a]

Glycoside and its structure	Reaction time (h)	Molar proportion of aldehydes: GlycolA	GlycerA	LactA	E	GO	HMA	Other aldehyde
Methyl β-sophoroside β-Glc-(1→2)-β-Glc-Me	3	0.00(0)	2 (2)	0.00(0)	0.00(0)	1.06(1)	0.88(1)	—
Methyl β-laminaribioside β-Glc-(1→3)-β-Glc-Me	3	0.00(0)	1 (1)	0.00(0)	0.00(0)	1.04(1)	0.00(0)	Me-Glc 0.97(1)
Methyl β-cellobioside β-Glc-(1→4)-β-Glc-Me	3	0.00(0)	1 (1)	0.00(0)	0.95(1)	2.09(2)	0.00(0)	—
Methyl β-gentiobioside β-Glc-(1→6)-β-Glc-Me	3	0.00(0)	2 (2)	0.00(0)	0.00(0)	2.06(2)	0.00(0)	—
Rutin α-Rha-(1→6)-β-Glc-quercetin	3	0.00(0)	1 (1)	1.06(1)	0.00(0)	1.83(2)	0.28(0)	—
	5	0.00(0)	1 (1)	1.02(1)	0.00(0)	1.92(2)	0.12(0)	—
Naringin α-Rha-(1→2)-β-Glc-naringenin	3	0.00(0)	1 (1)	1.00(1)	0.00(0)	0.75(1)	1.25(1)	—
	5	0.00(0)	1 (1)	1.05(1)	0.00(0)	0.90(1)	1.11(1)	—
Robinin α-Rha-(1→6)-β-Gal- / α-Rha- kaempferol	3	0.00(0)	1 (1)	1.76(2)	0.00(0)	2.45(3)	0.57(0)	—
	5	0.00(0)	1 (1)	1.86(2)	0.00(0)	2.76(3)	0.26(0)	—
Tomatine β-Xyl-(1→3)\ β-Glc-(1→2)/ β-Glc-(1→4)-β-Gal-tomatidine	3	0.96(1)	1 (1)	0.00(0)	0.00(0)	2.78(3)	0.00(0)	Glc-Thr
Digitonin β-Xyl-(1→3)\ β-Glc-(1→3)-β-Gal-(1→2)/ β-Glc-(1→4)-β-Gal-digitogenin	3	0.95(1)	1 (1)	0.00(0)	0.00(0)	2.84(3)	0.00(0)	—
Mi-saponin A α-Rha-(1→3)-β-Xyl-(1→4)-α-Rha-(1→2)-α-Ara- / β-Glc- protobassic acid	3	0.95(1)	1 (1)	0.90(1)	0.00(0)	2.81(3)	1.00(1)	Xyl-DHBA
Sakuraso-saponin β-Glc-(1→2)\ α-Rha-(1→2)-α-Rha-(1→2)-β-Gal-(1→4)/ β-GlcUA-protoprimulagenin A	3	0.00(0)	2 (2)	1.96(2)	0.00(0)	1.88(2)	1.90(2)	—
Ovalbumin (egg)	3	0.00	1	0.00	0.00	1.23	0.00	—
Bromelain (pineapple stem)	3	0.00	1	0.00	0.38	1.64	0.00	—
Ribonuclease B (bovine pancreas)	3	0.00	1	0.00	0.00	1.27	0.00	—
Invertase (yeast)	3	0.00	1	0.00	0.48	0.49	0.54	—

[a] The numbers in parentheses are theoretical values. Ara, L-arabinose; Xyl, D-xylose; Gal, D-galactose; Glc, D-glucose; Rha, L-rhamnose; GlcUA, D-glucuronic acid; GlycolA, glycolaldehyde; GlycerA, D-glyceraldehyde; LactA, L-lactaldehyde; E, D-erythrose; Thr, D-threose; GO, glyoxal; HMA, hydroxymalonaldehyde; DHBA, dihydroxybutyraldehyde.

From Honda, S., Takai, Y., and Kakehi, K., *Anal. Chim. Acta*, 105, 153, 1979. With permission.

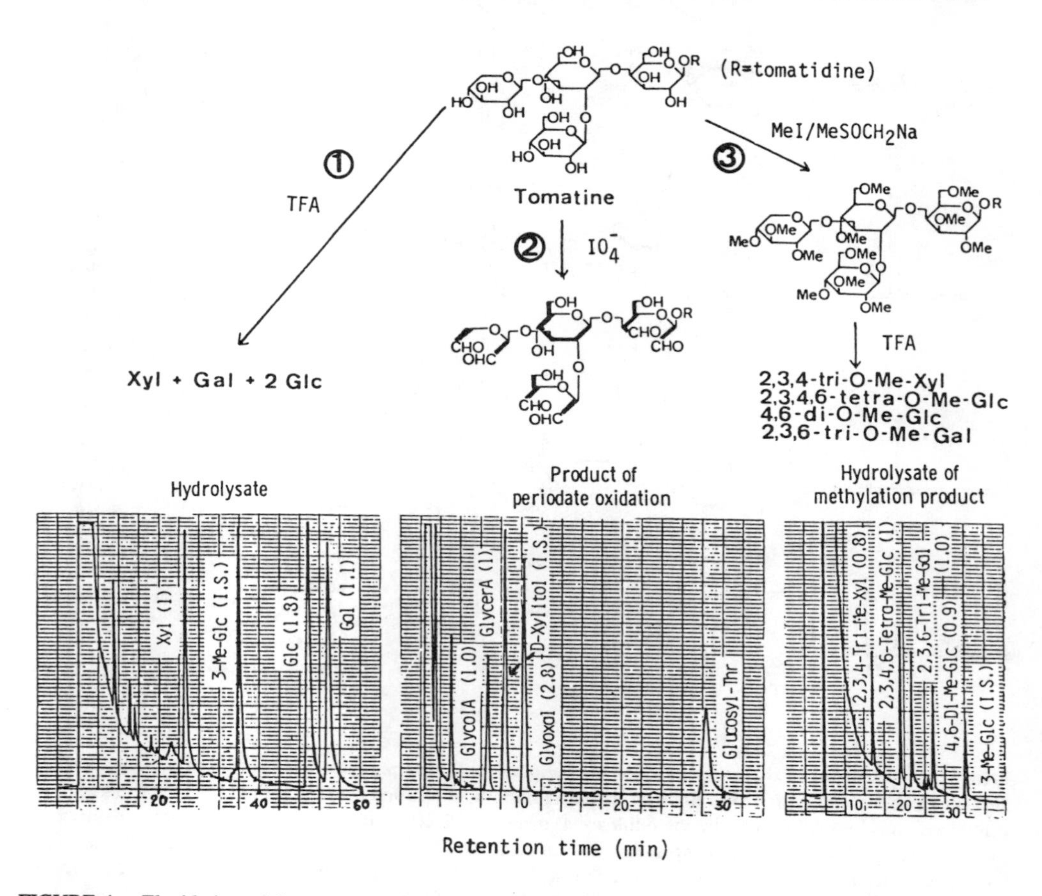

FIGURE 4. Elucidation of the structure of tomatine by simultaneous determination of the component monosaccharides in the acid hydrolysate, the component aldehydes of the dialdehyde compound produced by periodate oxidation, and the partially methylated monosaccharides formed by acid hydrolysis of permethylated derivative. The GC conditions for the determination of the component aldehydes of the dialdehyde compound were the same as those described in Figure 3. The conditions for the determination of the component monosaccharides and the partially methylated monosaccharides derived from the permethylate were as follows. Column, a glass-capillary column coated with silicon SF-96 (50 m, 0.28 mm I.D.); column temperature, 225°C; carrier, nitrogen (1.5 mℓ/min); detection, flame ionization.

charides, ordinary small aldehydes, such as glycolaldehyde, glyoxal, glyceraldehyde, hydroxymalonaldehyde, and tetroses were also formed at the nonreducing ends and the middle part of the oligosaccharides, in conjugated forms. These aldehydes could be determined in the same manner as that of the periodate oxidation products of oligosaccharide glycosides mentioned above. In the cases of glucobioses mentioned above (Figure 5), all bioses gave the peaks of the derivatives of glyceraldehyde (Peak α) and glyoxal (Peak β) in approximately equimolar proportions.

One difficulty arises in periodate oxidation of certain polysaccharides with branching such as starches, glycogens, and guaran. The problem is that in 1 → 4 polysaccharides with branching at the C-6 position of some residues, the dialdehyde which is formed tends to form 6-member hemiacetal rings with hydroxyl groups of adjacent chains (if they are not oxidized).[50] This interferes with the periodate oxidation, and a certain number of "resistant" (to periodate oxidation) glucose units are formed (about 1/3 the number of branch points).

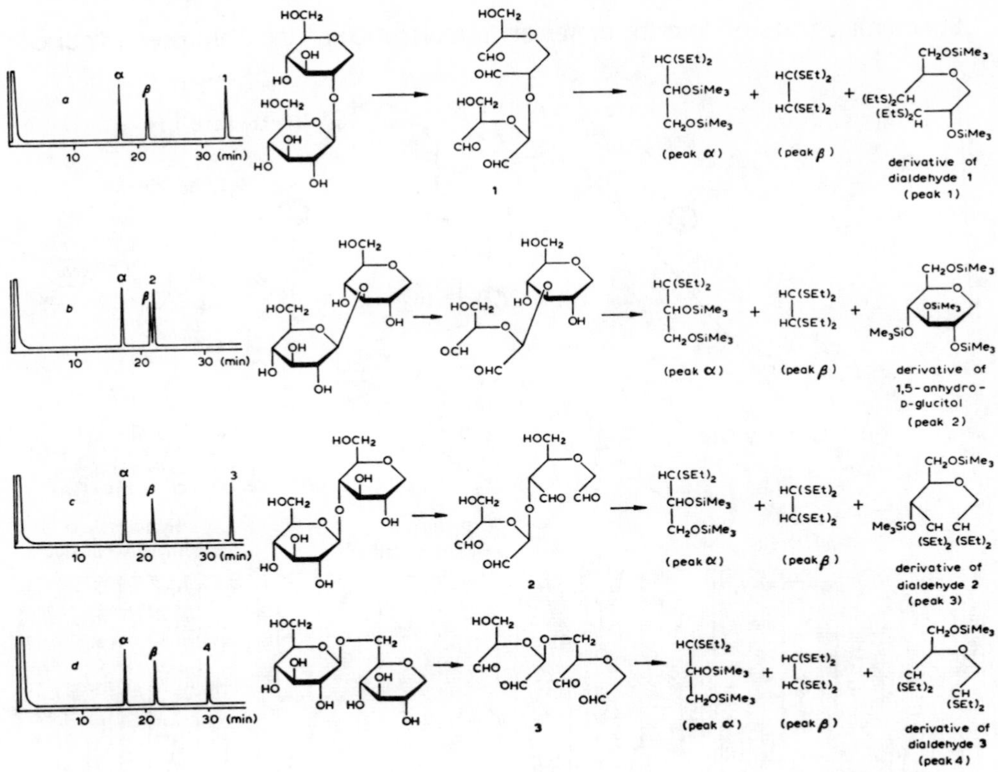

FIGURE 5. Analysis of the components of the periodate oxidation products, obtained from 1,5-anhydroalditol derivatives of sophorose (a, 1 → 2 linkage), laminarabiose (b, 1 → 3 linkage), cellobiose (c, 1 → 4 linkage), and gentiobiose (d, 1 → 6 linkage), by the dithioacetal method. GC conditions were the same as those described in Figure 3, except that the column temperature was raised from 100°C to 250°C with a gradient of 5°C/min. After 30 min temperature was kept constant. (From Honda, S., Kakehi, K., and Kubono, Y., *Carbohydr. Res.*, 75, 61, 1979. With permission.)

This problem may be circumvented by sodium borohydride reduction after periodate oxidation, followed by further periodate oxidation. For a more complete understanding of this phenomenon, consult the work of Ishak and Painter[50-52] and references cited therein.

IX. CARBOHYDRATE DITHIOACETALS

The synthesis of dithioacetals is one of the various methods used to obtain simple chromatograms of aldoses. As they lose their asymmetry at C_1 on conversion to dithioacetals, each aldose appears as one single peak as is analogous to alditol acetates,[53] and the formation of *syn* and *anti* forms as obtained with oximation is avoidable. Volatile derivatives may be achieved by TMS or acylation.[28]

The reaction requires acidic catalysis, preferably with trifluoroacetic acid.[46,53] Stronger acids cause the formation of by-products such as monothioacetals, while weak acids do not work effectively. Suitable thiols are ethanethiol[46,53] and propanethiol, their TMS dithioacetals are as volatile as TMS-oxime aldoses. Furthermore, these derivatives offer the possibility of chiral separation as reported by Little.[28] A suitable chiral thiol such as (+)-1-phenylethanethiol is easily coupled with aldoses, and the diastereomeric derivatives separate well on common, low-polarity columns. Compared with the separation of trifluoroacetates on chiral partition media as proposed by König[25,54] the columns need not be self-made and

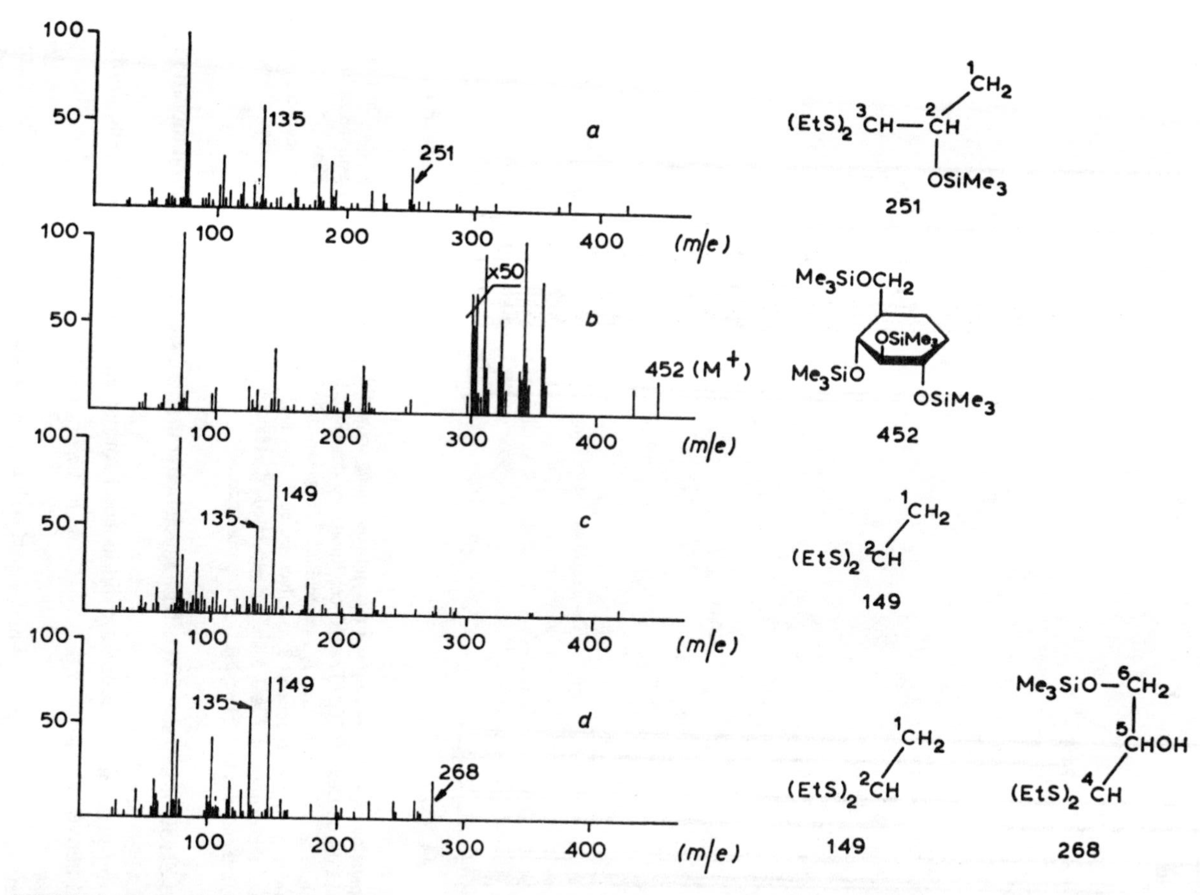

FIGURE 6. MS of the compounds of peaks 1 (a), 2 (b), 3 (c), and 4 (d) in Figure 3. Ionization potential, 70 eV; temperature of ion source, 270°C. (From Honda, S., Kakehi, K., and Kubono, Y., *Carbohydr. Res.*, 75, 61, 1979. With permission.)

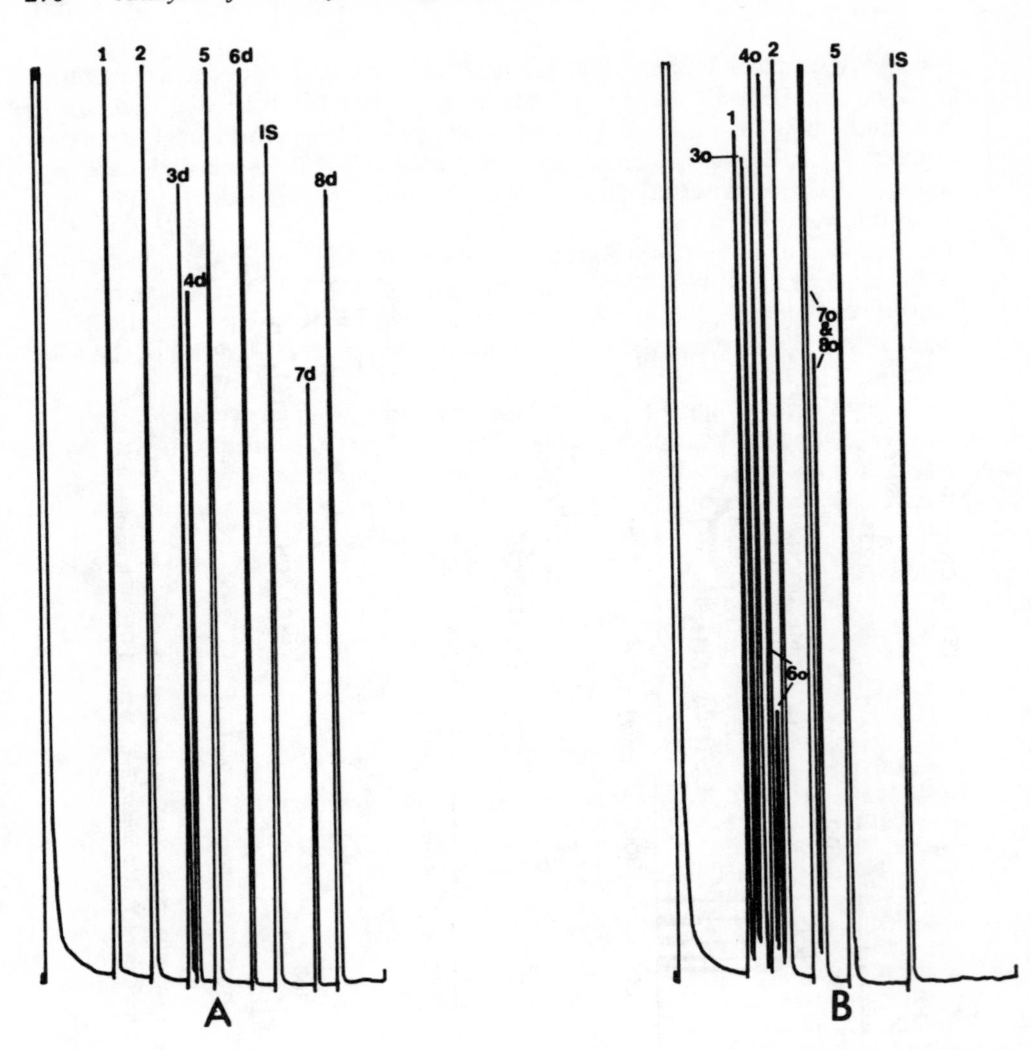

FIGURE 7. Separation of a carbohydrate test mixture containing some aldoses as dipropyl-dithioacetal-TMS (a) and as oxime-TMS (b) on Dexsil® 300. The GC conditions are described in the text. Detector: flame ionization. Run time: 14 min. The components are: 1, D-arabinitol; 2, L-quebrachitol; 3, D-arabinose; 4, D-ribose; 5, myo-inositol; 6, L-rhamnose; 7, D-glucose; and 8, D-galactose. 6, 7, and 8 appear as two peaks with oximation (syn and antiform), and the oximes of 7 and 8 are not separated. d indicates aldose-dithioacetals, o means aldose-oximes. IS is the internal standard, phenyl-β-D-glucopyranoside.

phases such as SE 30 or Dexsil® 300 offer a much better thermal stability than chiral modified silicones.

The following general procedure for dithioacetal formation is suitable with all the thiols recommended above:

1. Add a mixture of 4 parts thiol and 1 part trifluoroacetic acid and shake well. Incubate for 30 min at room temperature. 50 μℓ of the mixture is suited for samples up to 0.5 mg.
2. Add 50 μℓ of pyridine to terminate the reaction. Internal standard (use a solution containing 0.1 mg per sample of phenyl-β-D-glucopyranoside in pyridine) is substituted, if it is desired.

3. Usually the sample is now silylated. Since trifluoroacetic acid is present in the sample, silylation with HMDS[55] works well. Add 60 $\mu\ell$ of HMDS, shake well, and heat to 75°C for 10 min. The clear solution is now ready for injection. After complete removal of the derivatization agents, an acylation is also suitable.[28] Acid anhydrides are generally used; for trifluoroacetylation *N*-methyl-bis (trifluoroacetamide) (MBTFA) will also work well.
4. The final step involves chromatography. The partition medium depends on the type of volatile derivatives. With TMS nonpolar phases[28] such as SE 30 (capillary or packed column) and Dexsil® 300 (packed column) are suitable. Use either bonded phase capillaries or 3% loadings on fully inert diatomite supports with a flame ionization detector.

Figure 7 shows a separation of some aldoses as their dipropyl dithioacetal TMS on a SE 30 packed column, programmed from 160 to 300°C at 10°/min. The method is well suited to polysaccharide hydrolyzates with the special feature of enantiomeric separation using a chiral thiol as the derivatizing agent. A selective monitoring of aldoses in mixtures with cyclitols etc., analogous to an oximation, is possible.

REFERENCES

1. **Womersley, C.,** A micromethod for the extraction and quantitative analysis of "free" carbohydrates in nematode tissue, *Anal. Biochem.*, 112, 183, 1981.
2. **Wells, W. W., Katagi, T., Bentley, R., and Sweeley, C. C.,** Gas chromatography of sugar phosphates, *Biochim. Biophys. Acta*, 82(2), 408, 1964.
3. **Horning, M. G., Boucher, E. A., and Moss, A. M.,** *J. Gas Chromatogr.*, 5, 297, 1967.
4. **Sherman, W. R., Goodwin, S. L., and Zindo, M.,** The GC of some completely trimethylsilylated inositol and other sugar phosphates, *J. Chromatogr. Sci.*, 9, 363, 1971.
5. **Leblanc, D. J. and Ball, A. J. S.,** A fast one-step method for the silylation of sugars and sugar phosphates, *Anal. Biochem.*, 84, 574, 1978.
6. **Harvey, D. J. and Horning, M. G.,** Characterization of the trimethylsilyl derivatives of sugar phosphates and related compounds gas chromatography-mass spectrometry, *J. Chromatogr.*, 76, 51, 1973.
7. **Eisenberg, F., Jr. and Bolden, A. H.,** Gas chromatography of sugar phosphates and sugar nucleotides, *Anal. Biochem.*, 29, 284, 1969.
8. **Hashizume, T. and Sasaki, Y.,** Gas chromatography of sugar phosphates, *Anal. Biochem.*, 15, 346, 1966.
9. **Hashizume, T. and Sasaki, Y.,** Gas chromatographic separation of ribonucleotides by means of trimethylsilyl (TMS) derivatives, *Anal. Biochem.*, 15, 199, 1966.
10. **Zinbo, M. and Sherman, W. R.,** Gas chromatography and mass spectrometry of trimethylsilyl sugar phosphates, *J. Am. Chem. Soc.*, 92(7), 2105, 1970.
11. **Julkunen-Tiitto, R.,** Chemotaxonomical screening of phenolic glycosides in northern willow twigs by capillary gas chromatography, *J. Chromatogr.*, 324, 129, 1985.
12. **Yoshida, K., Honda, N., Iino, N., and Kato, K.,** Gas-liquid chromatography of hexopyranosides and hexofuranosides, *Carbohydr. Res.*, 10, 333, 1969.
13. **Eisenberg, F., Jr.,** Cyclic butaneboronic acid esters: novel derivatives for the rapid separation of carbohydrates by gas-liquid chromatography, *Carbohydr. Res.*, 42, 1, 1975.
14. **Wood, P. J., Siddiqui, I. R., and Weisz, J.,** The use of butaneboronic esters in the gas-liquid chromatography of some carbohydrates, *Carbohydr. Res.*, 42, 1, 1975.
15. **Tsuchida, H., Kitamura, K., Komoto, M., and Akimori, N.,** Gas-liquid chromatography and mass spectrometry of trimethylsilyl ethers and butaneboronate-trimethylsilyl derivatives of polyhydroxyalkylpyrazines, *Carbohydr. Res.*, 67, 549, 1978.
16. **Greenhalgh, R. and Wood, P. J.,** The detection of boron and the response of some boronate derivatives of carbohydrates with an alkali flame ionization detector, *Carbohydr. Res.*, 82, 410, 1973.
17. **Eisenberg, F., Jr.,** Gas chromatography of carbohydrates as butaneboronic acid esters, *Methods Enzymol.*, 28, 168, 1972.
18. **Tanka, M.,** Binding of alditols to the hydroxyl form of Dowex-1 × 8, a strongly basic ion-exchange resin: an improved method for estimation of the degree of polymerization of neutral oligosaccharides and polysaccharides, *Carbohydr. Res.*, 88, 1, 1981.

19. **Varma, R. and Varma, R. S.,** A simple procedure for the combined gas chromatographic analysis of neutral sugars, hexosamines, and alditols: determination of the degree of polyermization of oligo- and polysaccharides and chain weights of glycosaminoglycans, *J. Chromatogr.*, 139, 303, 1977.
20. **Morrison, I. M.,** Determination of the degree of polymerization of oligo- and polysaccharides by gas-liquid chromatography, *J. Chromatogr.*, 108, 361, 1975.
21. **Dutton, G. G. S., Reid, P. E., Rowe, J. J. M., and Rowe, K. L.,** Determination of the degree of polymerization of oligo- and polysaccharides by gas-liquid chromatography, *J. Chromatogr.*, 47, 195, 1970.
22. **Das Neves, H. J. C., Bayer, E., Blos, G., and Frank, H.,** A gas-chromatographic method for identification of the reducing units of disaccharides via reduction of the methoximes with borane, *Carbohydr. Res.*, 99, 70, 1982.
23. **Green, C., Doctor, V. M., Holzer, G., and Orô, J.,** Separation of neutral and amino sugars by capillary gas chromatography, *J. Chromatogr.*, 207, 268, 1981.
24. **König, W. A. and Benecke, I.,** Enantiomer separation of polyols and amines by enantioselective gas chromatography, *J. Chromatogr.*, 269, 19, 1983.
25. **König, W. A., Benecke, I., and Sievers, S.,** New results in the gas chromatographic separation of enantiomers of hydroxy acids and carbohydrates, *J. Chromatogr.*, 217, 71, 1981.
26. **Pollock, G. E. and Jermany, D. A.,** The resolution of racemic carbohydrate diastereomers by gas chromatography, *J. Gas Chromatogr.*, 6, 412, 1968.
27. **Gerwig, G. J., Kamerling, J. P., and Viliegenthart, J. F. G.,** Determination of the absolute configuration of monosaccharides in complex carbohydrates by capillary G.L.C., *Carbohydr. Res.*, 77, 1, 1979.
28. **Little, M. R.,** Separation, by G.L.C., of enantiomeric sugars as diastereoisomeric dithioacetals, *Carbohydr. Res.*, 105, 1, 1982.
29. **Schweer, H.,** Gas chromatographic separation of carbohydrate enantiomers as (−)-menthyloxime pertrifluoroacetates on silicone OV-225, *J. Chromatogr.*, 243, 149, 1982.
30. **Schweer, H.,** Gas chromatographic separation of enantiomeric sugars as diastereomeric trifluoroacetylated (−)-bornyloximes, *J. Chromatogr.*, 259, 164, 1983.
31. **Oshima, R., Kumanotani, J., and Watanabe, C.,** Gas-liquid chromatographic resolution of sugar enantiomers as diastereoisomeric methylbenzylaminoalditols, *J. Chromatogr.*, 259, 159, 1983.
32. **Playne, M. J. and Wallis, A. F. A.,** Determination of pentose sugars in fermentation solutions by a simple gas-chromatographic procedure, *Biotechnol. Lett.*, 4(10), 679, 1982.
33. **Schultz, T. P.,** Mississippi State Forest Products Utilization Laboratory, Unpublished data, 1982.
34. **Gardner, D. J., Schultz, T. P., and McGinnis, G. D.,** The pyrolytic behavior of selected lignin preparations, *J. Wood Chem. Tech.*, 5(1), 85, 1985.
35. **Irwin, W. J. and Slack, J. A.,** Analytical pyrolysis in biomedical studies, *Analyst,* 103(1228), 673, 1978.
36. **Bayer, F. L. and Morgan, S. L.,** Biopolymers in analytical pyrolysis gas-chromatography, in *Pyrolysis and GC in Polymer Analysis,* Liebman, S. A. and Levy, E. J., Eds., Marcel Dekker, New York, 1985, 298.
37. **Morgan, S. L. and Jacques, C. A.,** Characterization of simple carbohydrate structure by glass capillary pyrolysis gas chromatography and cluster analysis, *Anal. Chem.*, 54, 741, 1982.
38. **Turner, L. P.,** Characterization of nucleotides and nucleosides by pyrolysis-gas chromatography, *Anal. Biochem.*, 28, 288, 1969.
39. **Bentley, R. and Botlock, N.,** A gas chromatographic method for analysis of anomeric carbohydrates and for determination of mutarotation coefficients, *Anal. Biochem.*, 20, 312, 1967.
40. **van Ling, G., Ruijterman, C., and Vlugter, J. C.,** Catalytic hydrogenolysis of saccharides. I. Qualitative and quantitative methods for the identification and determination of the reaction products, *Carbohydr. Res.*, 4, 380, 1967.
41. **Dooms, L., Declerck, D., and Verachtert, H.,** Direct gas chromatographic determination of polyalcohols in biological media, *J. Chromatogr.*, 42, 349, 1969.
42. **Sharon, N.,** *Complex Carbohydrates: their Chemistry, Biosynthesis, and Functions,* Addison-Wesley, Reading, Mass., 1975, 92.
43. **Dutton, G. G. S., Gibney, K. B., Jensen, G. D., and Reid, P. E.,** The simultaneous estimation of polyhydric alcohols and sugars by gas-liquid chromatography: applications to periodate oxidized polysaccharides, *J. Chromatogr.*, 36, 152, 1968.
44. **Baird, J. K., Holroyde, M. J., and Ellwood, D. C.,** Analysis of the products of Smith degradation of polysaccharides by g.l.c. of the acetylated, derived aldononitriles and alditols, *Carbohydr. Res.*, 27, 464, 1973.
45. **Turner, S. H. and Cherniak, R.,** Total characterization of polysaccharides by gas-liquid chromatography, *Carbohydr. Res.*, 95, 137, 1981.
46. **Honda, S., Fukuhara, Y., and Kakehi, K.,** Gas chromatographic determination of conjugated aldehydes in products of periodate oxidation of pyranose sugars as diethyl dithioacetals, *Anal. Chem.*, 50, 55, 1978.
47. **Honda, S., et al.,** unpublished results.

48. **Honda, S., Takai, Y., and Kakehi, K.,** Periodate oxidation analysis of carbohydrates. XII. Rapid determination of aldehydes in the oxidation products of oligosaccharides by the dithioacetal method, *Anal. Chim. Acta,* 105, 153, 1979.
49. **Honda, S., Kakehi, K., and Kubono, Y.,** Prevention of overoxidation in periodate oxidation of reducing carbohydrates by their conversion into 1,5-anhydroalditol derivatives, *Carbohydr. Res.*, 75, 61, 1979.
50. **Ishak, M. F. and Painter, T.,** The origin of periodate-resistant D-glucose residues in starches and glycogens, *Carbohydr. Res.*, 32, 227, 1974.
51. **Ishak, M. F. and Painter, T. B.,** Anomalous periodate oxidation limit of guaran, *Acta Chem. Scand.*, 27(4), 1268, 1973.
52. **Ishak, M. F. and Painter, T. B.,** Interresidue lactones formed by treatment of periodate-oxidized polysaccharides by aqueous bromine, *Acta Chem. Scand.*, 27(4), 1268, 1973.
53. **Honda, S., Yamauchi, N., and Kakehi, K.,** Rapid gas chromatographic analysis of aldoses as their diethyl dithioacetal trimethylsilylates, *J. Chromatogr.*, 169, 287, 1979.
54. **König, W. A., Benecke, I., and Bretting, H.,** Gaschromatographische Trennung enantiomerer Kohlenhydrate an einer neuen chiralen stationären Phase, *Angew. Chem.*, 93, 688, 1981.
55. **Brobst, K. M. and Lott, C. E., Jr.,** Determination of some components in corn syrup by gas-liquid chromatography of the trimethylsily derivatives, *Cereal Chem.*, 43, 35, 1966.

INDEX

A

D

E

F

H

I

N

O

P

Q

R

S

U

V

W

X

Z